Smart Learning with Educational Robotics

Linda Daniela
Editor

Smart Learning with Educational Robotics

Using Robots to Scaffold Learning Outcomes

Springer

Editor
Linda Daniela
Faculty of Education, Psychology, & Art
University of Latvia
Rīga, Latvia

ISBN 978-3-030-19912-8 ISBN 978-3-030-19913-5 (eBook)
https://doi.org/10.1007/978-3-030-19913-5

This Springer imprint is published by the registered company Springer Nature Switzerland AG.
The registered company address is: Gewerbestrasse 11, 6330 Cham, Switzerland

Foreword

The prosperity of today's societies depends on the capacity of educators to deal with the changes and increasing complexity of the educational needs of the twenty-first-century citizens. Our society is dependent on the technological development. Therefore, it would be logical for educational institutions to promote digital literacy that goes beyond learning technology as users or consumers. They should teach basic skills that will allow new generations to become technology producers. This monograph explains that infusing robotics into the curriculum would help to reach this target. Educational robotics facilitate smart learning because technology is used to empower learners to develop innovative talents that involve computational thinking, programming skills, and collaboration in the construction of robots.

Thus, it is not surprising that instruction through robotics has received increasing attention from educators all over the world, especially in recent years, regardless of the educational level. Many education professionals have begun to accept the challenge of incorporating robotics into life in educational institutions due to their educational benefits. However, educational robotics is an area that is still in an initial phase of development. Today, not all educators are prepared to implement robotics in the classroom. Therefore, it is advantageous to organise, synthesise, and communicate updated knowledge about educational robotics, in order to make it easier for novice educators in educational robotics to understand teaching supported by robots in the classroom and provide experts with other perspectives and avant-garde lines of work.

Professor Linda Daniela is correct in identifying the need to elaborate a monograph on educational robotics. The monograph entitled *Intelligent Learning with Educational Robotics* has a different focus from other manuals on smart learning. It brings together experts in educational robotics from different parts of the world with the purpose of explaining the value of educational robotics in addressing the challenges of learning and teaching in the twenty-first century. This monograph offers a theoretical and updated review that will allow the reader to understand what is meant by educational robotics, its history, types, and educational benefits. In addition, the work offers a broad and diverse set of experiences and ideas at different educational levels, providing insight into the efficient implementation of educational robotics. There is no doubt that reading this book will contribute to the satisfaction of education professionals who want to know about the current advances in educational robotics in order to better prepare future generations.

Gutiérrez Braojos Calixto
Department of Research Methods in Education
University of Granada, Granada, Spain

Foreword

The last decades have been exciting with regard to innovations and technology advancement for education. However, it is also a challenging time for learning designers, teachers, and educational researchers, ensuring that students are ready for an ever-changing world and fully capable of becoming tomorrow's progressive leaders, productive workers, and responsible citizens.

The dawn of the first educational robot can be traced back to the late 1960s. Not only the robot technology has been advanced over the last 50 years, but also the pedagogical approaches and methodologies have been further developed.

Recently, robots have become increasingly popular as an educational tool for various age groups ranging from preschool to primary school over K–12 classrooms to graduate university education. Also, the targeted learning outcomes utilising educational robots are broad, including general interest in science and technology, supporting and enhancing STEAM learning activities, as well as fostering specialised applications such as software engineering or control theory. The pedagogical orchestration of educational robots includes teacher-led demonstrations, guided workshops, or discovery and problem-solving scenarios. The learning activities are often multifaceted including design, construction, and programming for solving a specific problem.

Empirical research focussing on educational robots have documented a greater engagement of students in STEAM learning activities. Other empirical studies show support for critical thinking and complex problem-solving and increased comprehension of complex concepts and procedures. In addition, as artificial intelligence for robots is further developed, data analytics, adapted behaviour to specific learning needs, and enhanced social interaction, including educational robots, are currently a focal point of empirical research.

In this edited volume, *Smart Learning with Educational Robotics*, Professor Linda Daniela brings together international experts on educational robotics showcasing their latest concepts, methodologies, and empirical findings. The contributions focus on students from early childhood to higher education. The chapter authors use empirical research methodologies, including existing, experimental, and emerging conceptual frameworks, from various fields, in order to tackle phenomena for understanding learners' cognitive functions, optimal learning design for educational robotics classrooms, or increasing acceptance and adoption of educational robotics among teachers. Further examples include advancing learning beyond the classroom walls, the design of competitive environments for learning, or building confidence and interest among students through educational robotics.

The synthesis of the latest innovations and fresh perspectives on pedagogical constructs makes *Smart Learning with Educational Robotics* a cutting-edge reading for the researchers and educators in educational robotics, STEAM education, and beyond. Despite the potential and applications of educational robotics being showcased in this edited volume, it is imperative to note that a meaningful integration of educational robotics in pedagogical scenarios shall have a supporting purpose for learning processes, knowledge construction, and learning outcomes.

Dirk Ifenthaler (http://www.ifenthaler.info) is Professor and Chair of Learning, Design, and Technology at the University of Mannheim, Germany, and UNESCO Deputy Chair of Data Science in Higher Education Learning and Teaching at Curtin University, Australia. His previous roles include Professor and Director, Centre for Research in Digital Learning, at Deakin University, Australia; Manager of Applied Research and Learning Analytics at Open Universities, Australia; and Professor for Applied Teaching and Learning Research at the University of Potsdam, Germany. He was a 2012 Fulbright Scholar in Residence at the Jeannine Rainbolt College of Education, University of Oklahoma, USA. His research focuses on the intersection of cognitive psychology, educational technology, data analytics, and organisational learning. His research outcomes include numerous coauthored books, book series, book chapters, journal articles, and international conference papers, as well as successful grant funding in Australia, Germany, and USA. He is the Editor in Chief of the Springer journal *Technology, Knowledge and Learning* (www.springer.com/10758) and Editorial Board Member of several international journals.

Dirk Ifenthaler
University of Mannheim, Mannheim, Germany
Curtin University, Perth, Australia

Preface

Progress in technology development creates new opportunities as well as new challenges for the educational environment, as it has to be able to transform the learning process so as to prepare the future generations for life and work with technology, both by using the opportunities created by technology and by creating new technological solutions. Despite the frequent slogans that technology will deprive people of their jobs, it is clear that technological progress is driven by people's innovative thinking and creative solutions, but people need to be ready for collaboration with technology. It is essential for the educational environment to accept this progress and offer innovative learning methods so that students can develop computational thinking, creative thinking, and digital competence. Educational researchers have come to the conclusion that it is not only technology and technological solutions that are important, but also the students' readiness to learn. Smart learning was defined by Spector (2014) as the scenario where technological possibilities are added to the learning environment and are able to support students' learning. Daniela (2019) has defined SMART pedagogical principles, where, when working with a technology-enhanced learning environment, it is important to take into account not only technological advances but also the developmental peculiarities and learning taxonomies and use innovative pedagogical methods to facilitate the learning process in a technology-enhanced environment where the knowledge construction process is provided.

One of the possibilities to make the learning process creative, innovative, and, at the same time, an asset that promotes the acquisition of current skills and competences is by using robotics, where new solutions are created that promote learning. These ideas were initiated by Papert in developing constructionism ideas and using the Logo language in the learning process (Papert 1980), thus attempting to prove that the education process can be successfully enhanced by active learning and offer students a variety of hands-on activities with computers and other topical technologies.

Nowadays, the ideas for incorporating robotics activities into the learning process are no longer a novelty, but there is still the question of how to use them to promote the development of certain competencies and which pedagogical principles should be taken into account in order to improve the students' motivation to look for new innovative solutions that need to be considered to ensure inclusive education in reality, rather than to try to involve everyone in innovative activities, sometimes without thinking about the specific needs of the students (Daniela and Lytras 2018).

There are different approaches in the classification of robots, one of which is to define the three groups: (i) industrial robots, (ii) assistive robots, and (iii) educational robots. It is not always easy to classify the robots into one of the three groups, but the important issue is that there are huge variety of them and we have to face the reality that students have to learn about them, about the possibilities that one can achieve by programming, about what ethical and legal aspects we have to consider, and about how we can ensure sustainable development and how we can support progress.

The present book, *Smart Learning with Educational Robotics: Using a Robot to Scaffold Learning Outcomes*, provides ideas on how educational robotics (ER) can be used both for working with students in compulsory education and for analyzing the use of educational robotics in higher education. The authors of the chapters have also tried to analyze the students' cognitive development during activities with ER, thus emphasizing the possibilities of using robotics to promote learning and to help develop competences that are important today, where it is necessary to understand how different technologies work, how they can be used to make people's lives easier, and how technology and technological solutions can be used to help students to construct their knowledge. The authors discuss the possibilities of how ER can be classified, and the book concludes with a proposed ER taxonomy.

This book, with its ideas discussed by various authors, offers some insight into the topicalities of ER, but it is also clear that it is still an area that is in a continuous developmental phase and it is necessary to continue researching outcomes from the perspective of knowledge and competence development, as well as from the perspective of the risks that we face. The academics should continue the development of evaluation tools to provide proofs on outcomes and propose solutions to mitigate the possible risks which are stressed in the European Civil Law Rules in Robotics (Nevenjans 2017).

There are 15 chapters included in the book. A brief description of each of the chapters follows.

Dave Catlin, in his chapter "Beyond Coding: Back to the Future with Education Robots," explores the history of education robots – specifically the ideas of Seymour Papert – and suggests that as technology develops, the need for coders will (in the long term) dwindle, but the power of robots to help educate children for the future will increase.

In the next chapter, "Educational Robotics for Reducing Early School Leaving from the Perspective of Sustainable Education," Linda Daniela and Raimond Strods analyze the possibilities of ER from the perspective of sustainable education. It is concluded that the use of ER enhances the motivation to learn in students at high risk of early school leaving and encourages them to construct knowledge actively and independently, thus reducing their risk of early school leaving and, in the long term, ensuring the achievement of the 4th Sustainable Development Goal, particularly with regard to sustainable education.

David Scaradozzi, Laura Screpanti, and Lorenzo Cesaretti, in their chapter "Towards a Definition of Educational Robotics: A Classification of Tools, Experiences and Assessments," analyze the scientific literature reporting experiences in the field of ER. They provide a broad classification of experiences on the use of robots for education, a classification of the available robots used in the ER context, and a classification of existing evaluation methods to carry out the assessment of ER activities. Starting from the distinction between robotics in education (RiE) and ER, this chapter contributes to the discussion of what ER means and consists of.

The chapter "Introducing Maker Movement in Educational Robotics: Beyond Pre-Fabricated Robots and 'Black Boxes'" by Dimitris Alimisis, Rene Alimisi, Dimitrios Loukatos, and Emmanouil Zoulias introduces the ideas developed during the eCraft2Learn project, where they researched, designed, piloted, and evaluated an ecosystem intended to introduce digital fabrication and maker movement in formal and informal education, in order to make robots transparent for children and finally help them make their own robotic artifacts.

In their chapter "Modbloq: Design of a Modular Robot made with 3D Printing for Educational Purposes," Pedro de Oro and Silvia Nuere propose the design and explain how to make a modular prototype educational robot. They conclude that the physical characteristics of the design, as well as the programming language chosen, make it interesting for educational purposes, both for primary and secondary education.

Francisco Bellas, María Salgado, Teresa F. Blanco, and Richard J. Duro, in their chapter "Robotics in Primary School: A Realistic Mathematics Approach," present a methodology, based on realistic mathematics, for the integration of ER in primary schools. This methodology was tested during one semester at the Sigüeiro Primary School (Spain) in the subject of mathematics with students of different ages, ranging from 7 up to 11 years old. Two different educational robots, with different features, were used to highlight that the methodology is independent of the robotic platform used.

The chapter "Crab Robot: A Comparative Study Regarding the Usage of Robotics in STEM Education" by Icleia Santos, Elaine Cristina Grebogy, and Luciano Frontino de Medeiros explains how to use the crab robot design as it is of very low complexity, is really inexpensive, has quick application, and allows the students to engage in a project that also reuses materials, making an appropriate connection between the discipline of science with environmental education.

The authors of the chapter "Innovative Tools for Teaching Marine Robotics, IoT and Control Strategies Since the Primary School" present a project developed to teach robotics, STEM, and the Internet of Things (IoT). Moreover, by directly involving people in themes about the marine environment, they raise awareness and provide knowledge about roboethics, blue careers, and ocean literacy.

In the chapter "Robot Programming to Empower Higher Cognitive Functions in Early Childhood," the authors describe a new approach in ER aimed at empowering higher cognitive functions in school. They conclude that the available evidence has suggested that robot programming could be a powerful tool for improving EFs.

However, in order to be effective, it must be used by embedding EF exercises within an area of major cognitive development for a certain age group or within a domain that is dysfunctional for a certain disorder.

In the chapter "Activities With Educational Robotics: Research Model And Tools For Evaluation Of Progress," Linda Daniela, Raimonds Strods, and Ilze France provide a research model and five research tools (structured observation protocol, evaluation of the possible risks of early school leaving to be filled in by the teachers before and after activities, students' questionnaires to be filled in before and after activities) for evaluating the outcomes of organized after-school robotics activities. The research model and tools were tested and approbated by students who are at risk of early school leaving and students who participated in robotics activities to develop computational thinking.

The chapter "The Use of Robotics for STEM Education in Primary Schools: Teachers' Perceptions" by Ahmad Khanlari aims to better understand elementary teachers' perspectives on the use of robotics for STEM education. The results obtained during the research indicated that the participants' perceptions changed as a result of participating in the workshop, learning about robotics, and being involved in hands-on robotics activities.

Mounir Ben Ghalia, in his chapter "Using Robots to Introduce First-Year College Students to the Field of Electrical Engineering," describes the curriculum of an introductory course for first-year students in an electrical engineering program. He concludes that although most of the first-year students did not have a background in robot programming, it is possible to guide them to write programs that solved complex robot navigation challenges.

The following chapter, "Designing a Competition Robot as a Capstone Project for Electrical and Computer Engineering Students," provides an insight into robotics-based capstone design projects and comes to the conclusion that designing and building robots for capstone design projects support a number of student learning outcomes. These include the following: (i) the ability to apply engineering design to produce solutions that meet specified needs; (ii) the ability to function effectively in a team whose members together provide leadership, create a collaborative and inclusive environment, establish goals, plan tasks, and meet objectives; (iii) the ability to communicate effectively with a range of audiences; and (iv) the ability to create and use software both as an analysis and design tool and as part of systems containing hardware and software.

Ivana Đurđević Babić, in her chapter, focuses on students who are trained for teaching educational programming languages but who did not have a lot of or any opportunity to work with educational robots. Since ER is gaining more attention at all levels of education, it is almost certain that they will have the desire, but also the need, to use ER in some segment of their future work.

The concluding chapter is prepared by experts of ER who propose an "EduRobot Taxonomy" and briefly explain its ideas and arguments before using it to classify some of the robots cited in this book.

Rīga, Latvia Linda Daniela

References

Daniela, L. (2019). Smart pedagogy for technology enhanced learning. In L. Daniela (Ed.), *Didactics of smart pedagogy: Smart pedagogy for technology enhanced learning* (pp. 3–22). Cham: Springer.

Daniela, L., & Lytras, M. D. (2018). Educational robotics for inclusive education. *Technology, knowledge and learning* (pp. 1–7). https://doi.org/10.1007/s10758-018-9397-5

Nevenjans, N. (2017). European civil law rules in robotics. *European commission*. Available at: http://www.europarl.europa.eu/committees/fr/supporting-analyses-search.html. Accessed 25th January 2019.

Papert, S. (1980). *Mindstorms: Children, computers and powerful Ideas*. New York: Basic Books.

Spector, J. M. (2014). Conceptualizing the emerging field of smart learning environments. *Smart Learning Environments, 1*(1), 1–10.

Contents

About the Authors

Joshua Acosta received his bachelor's degree of Science in Electrical Engineering from the University of Texas Rio Grande Valley, USA, in May 2018. He was a member of the 2017–2018 robotics team who competed in the 2018 IEEE Region 5 Robotics Contest held in Austin, Texas. After graduation, he joined Texas Instruments, a global semiconductor design and manufacturing company.

Rene Alimis holds a degree in Computer Engineering (from the University of Thessaly, GR) and an MA in ICT in Education (from the Institute of Education, UCL, London) and has more than 8 years of experience in the field of educational innovation within EU-funded projects (H2020, Erasmus+, Lifelong Learning). She has collaborated with well-known Greek research centers (CERTH, IRETETH) and higher education departments in Europe in the context of EU-funded projects. She is also a certified ICT trainer for adults and holds a Postgraduate Certificate in Education; since 2010, she works closely with teachers and adults of all ages and various cultural and professional backgrounds in numerous Greek public centers for lifelong learning and private companies. She has also carried out research studies in primary and secondary schools in Greece and the UK, exploring how ICT interventions can support students in developing computational thinking skills in a creative and playful way.

Dimitris Alimisi after his resignation in 2013 from a full professorship position in educational technology at the Higher School of Pedagogical & Technological Education (Patras, Greece), is currently the scientific leader at the EDUMOTIVA (European Lab for Educational Technology). He has also served as adjunct professor at the Hellenic Open University in Science Education, at the University of Athens in ICT in Education, and at the University of Patras in STEM Education and as teacher trainer in several courses in Greece and other European countries. His research interests include learning technologies with a special focus on STEAM education, educational robotics, and maker movement in education. He has served for years as coordinator or partner in several EU research and educational projects (FP7, HORIZON 2020, ERASMUS+, and more). Moreover, he is the coeditor of proceedings books and special issues in educational robotics, the co-organizer and co-chair of the International Conference of Educational Robotics (EDUROBOTICS), and a member of program committees in many national and international conferences.

Julian Mauricio Angel-Fernandez was born in Bogotá, Colombia, and holds two bachelor's degrees of 5 years each from the Universidad de los Andes, Bogota, Colombia, in Electronic Engineering and Computer and Systems Engineering. He also holds a master's degree in Computer and Systems Engineering from the same institution. He obtained his European PhD with merit at Politecnico di Milano, Milan, Italy, in Computer and System Engineering with a focus on human-robot interaction. He worked as a lecturer at the Pontificia Universidad Javeriana, Bogota, Colombia, and also worked in several projects, including ER4STEM, which was founded by the European Commission under the program Horizon 2020, and focus on the critical use of robotics on education. His interests are multi-robot system, human-robot interaction, and educational robotics.

Francisco Bellas received his MS degree in Physics from the University of Santiago de Compostela, Spain, in 1999, and his PhD in Computer Science from the University of A Coruña, where he is currently an *associate professor* in the Department of Computer Science and member of the Integrated Group for Engineering Research, in 2003. His research interests include cognitive robotics, evolutionary robotics, and collective intelligence. During the last few years, he has been involved in educational projects related with robotics, mainly in The Robobo Project, a new concept of educational robot that aims to bring intelligent robotics to secondary schools. As the coordinator of this project, he was awarded with the Best Paper Award in the Robotics in Education Conference (Rie 2017), with the first prize in the Vodafone Connecting for Good Galicia 2017 initiative, and with a European project in the 2018 ROS industrial call.

Mounir Ben Ghalia is a professor in the Department of Electrical Engineering at the University of Texas Rio Grande Valley, USA. He served as the principal investigator and the director of the National Science Foundation-funded program titled Research Experiences for Teachers in Emerging and Novel Engineering Technologies in the Rio Grande Valley. His funded program supported the professional development of STEM teachers to introduce engineering education in pre-college classrooms. His research focuses on control systems, robotics, and engineering education; he has published in these areas and has mentored several engineering students in the area of robotics. He is the recipient of the prestigious University of Texas System Regents' Outstanding Teaching Award.

Teresa Fernández Blanco is professor of Mathematics Didactics at the University of Santiago de Compostela. She holds a degree in Mathematical Sciences from the University of Santiago de Compostela in "Numerical methods for equations in partial derivatives: applications in engineering" and a PhD in Mathematical Sciences in the Experimental Sciences and Mathematics program. She received an Extraordinary Doctorate Award.

Her main research interests focus on visualization and spatial reasoning and modeling with special attention to new technologies. She is collaborator in the project ERASMUS + KA2-KO-187-1/2015 (Kids Inspiring Kids for STEAM) and in the project H2020SEAC-2015-1-710577 (STEM4YOUTH: promotion of STEM education through key scientific challenges and their impact on our lives and career prospects).

She is currently supervisor of several PhD theses that integrate STEAM and is the main researcher of the project EDU2017-84979-R entitled "Teaching and learning mathematics in adolescents at risk of exclusion. Teaching intervention through STEAM."

John-John Cabibihan received his PhD degree in Bioengineering, with specialization in biorobotics, from the Scuola Superiore Sant'Anna, Pisa, Italy, in 2007. Concurrent with his PhD studies, he received another international scholarship grant in 2004 from the École Normale Supérieure Paris-Saclay, France. Therein, he spent 1 year with the Laboratoire de Mécanique et Technologie. From 2008 to 2013, he was an assistant professor at the Electrical and Computer Engineering Department, National University of Singapore. He served as the deputy director of the Social Robotics Laboratory and serves as the graduate program coordinator of the Mechanical Engineering Program. In addition, he is currently an associate professor at the Mechanical and Industrial Engineering Department at Qatar University, is member of the Faculty Senate, is lead/co-lead principal investigator of several projects under the National Priorities Research Program of Qatar National Research Fund, and is associate editor/editorial board member of several Scopus indexed journals. His research has attracted more than US$4 million in funding from various funding agencies. He was the general chair of the International Conference on Social Robotics (ICSR) 2018 in Qingdao, China; program chair/co-chair of Ro-Man 2019 in New Delhi, India; ICSR 2017 in Tsukuba, Japan; ICSR 2016 in Kansas City, USA; and ICSR 2012 and ICSR 2010 in Singapore. Over the years, his work has been focused toward assistive and social robotics for the therapy of children with autism, lifelike prosthetics, bio-inspired tactile sensing, and human-robotic touch and gestures. His works were featured in the *British Broadcasting Corp. (BBC) News*, *MIT Technology Review*, *Phys.org*, *Popular Science*, *New Scientist*, and *Discovery News*, among others.

Emanuela Castro received her master's degree in Psychology at the University of Pisa in July 2014 and her PhD degree at Scuola Superiore Sant'Anna in BioRobotics in December 2018, working on a thesis about human-robot interaction and educational robotics. Currently, she is collaborating with IRCCS Fondazione Stella Maris, working on neuropsychology, cognitive rehabilitation, and developmental psychology.

Dave Catlin completed an apprenticeship before going to Bradford University where he graduated in mechanical engineering in 1971. He worked as a contract design engineer until, inspired by Seymour Papert, he founded Valiant Technology with Professor Tom Stonier and designed the Valiant Turtle in 1983, the Classic Roamer in 1989, and the latest Roamer in 2012. His company has shipped several hundred thousand robots to schools in 32 different countries. He has worked with schools on five different continents while researching and developing new ways to use robots to help children learn.

Francesca Cecchi is a senior researcher at the BioRobotics Institute of the Scuola Superiore Sant'Anna, Italy. She received her master's degree in Biomedical Engineering from the University of Pisa in 2007 and her PhD degree in Biorobotics from the Scuola Superiore Sant'Anna (SSSA) in 2011. Her research fields are neurodevelopmental engineering and educational robotics. She is an author and coauthor of more than 20 publications about mechatronic devices for rehabilitation in infancy and educational robots. She has been involved in national and international research project about tele-rehabilitation of children and educational robotics both on the students' level for measuring the efficacy of educational robotics and on the teachers' side by studying training courses for teachers.

Lorenzo Cesaretti is enrolled in the PhD School of Information Engineering at the Dipartimento di Ingegneria dell'Informazione (DII) at the Università Politecnica delle Marche. He is co-founder and CTO at TALENT srl, an Italian innovative start-up working in the field of educational technologies. He has expertise in robotics, educational robotics, and educational technology. He is involved in more than 40 projects and trainings mainly regarding educational robotics, coding, and creative use of technology in education.

Giovanni Cioni is full professor of Child Neurology and Psychiatry at the University of Pisa, where he specialized in child neurology and psychiatry and in rehabilitation medicine, and scientific director of the National Biomedical Research Hospital (IRCCS), Fondazione Stella Maris. He graduated cum laude in medicine at the University of Florence. He has been research fellow in 1984–1986 at the University of Groningen, Department of Developmental Neurology and attended for stages many research and clinical centers in France and the UK. He is an experienced researcher mainly around the following topics: models, neurophysiological mechanisms, and molecular basis of plasticity of the nervous system in children; new approaches to neurological assessment of the newborn and young child; follow-up of the infant at risk for neurological impairment; maturation of perceptual systems and their disorders; early diagnosis and rehabilitation of disorders of motor, cognitive, and language development; and neuroimaging and electrophysiological techniques of exploration applied to neurological disorders in children. He has been recipient as principal investigator or coordinator of local units of many peer-reviewed projects around child neurology and neurorehabilitation, supported by the national and international organizations (European Union FP7, Italian Ministry of Health, Italian Telethon, National Banks, Cure Autism Now Foundation, Mariani

Foundation, other). His publications indexes are the following (Google Scholar January 2019): total number of publications, 362 (last 5 years 93); total H-index, 56 (last 5 years 14); and total number of citation, 10384 (last 5 years 674). For executive functions on the national and international scientific organizations, he served in the European Academy of Childhood Disability (chair); IMAGO7 for ultrahigh field MRI (member of the scientific committee); European Society of Child Neurology; Italian Society of Child and Adolescence Neuropsychiatry (vice president); and The GM Trust. He has been recipient of many scientific awards: 1995, Annual Award of Pediatrics University of Turku (Finland); 1998, Heinz Prechtl Award for Developmental Neurology, University of Graz (Austria); 2003, Danish Society for Cerebral Palsy Award; 2011, Castang Award for Developmental Disabilities; 2016, University of Pisa Cherubino Award for scientific merits; and 2018, Karel and Berta Bobath International Award (2018).

Daniele Costa holds a master's degree in Aerospace Engineering and in Mechanical Engineering and a PhD in Underwater Robotics from the Università Politecnica delle Marche. His research interests are in biomimetic autonomous marine vehicles. He has experience in multibody systems, hydrodynamics, finite element analysis, compliant structures, seal design, and composite materials.

Andrew Csizmadia is senior lecturer in Computer Education at Newman University, Birmingham, in the United Kingdom and is seconded part time to the British Computer Society (BCS) as academic lead for the BCS Certificate in Computer Science Teaching certification.

Linda Daniela is professor and senior researcher, chair of the Council for PhD Defence in Pedagogy, and head of the Scientific Institute of Pedagogy at the University of Latvia. She also serves as expert in Education at the Council of Sciences of the Republic of Latvia.

Her expertise spans smart pedagogy, virtual education, smart education, digital learning materials, educational technologies for learning, educational robotics, educational processes and solutions for reducing social exclusion from the educational processes, and behavioral problems.

Professor Daniela is an author and coauthor of more than 70 publications about processes in all dimensions of education. She has been involved in more than 30 research projects. At the moment, she is leading the research projects as follows: "Human, Technologies and Educational Quality," "Motivating secondary school students towards STEM careers through robotic artifact making (RoboScientists)," "Developing Teaching Materials for Preschool Teaching Undergraduates on Computational Thinking and Introduction to Coding (EarlyCode)," "eMedia: MEdia literacy and Digital citizenship for all," "Audio learning materials for preschool education," "The gap between political development documents and real practice of digitalization of higher education," and "MyHUB – a one-stop-shop on inclusion practices, tools, resources and methods for the pedagogical staff at formal and nonformal educational institutions."

Paolo Dario is professor of Biomedical Robotics and coordinator of the PhD Program in BioRobotics of the Scuola Superiore Sant'Anna, Pisa, Italy. He received his doctor engineering degree in Mechanical Engineering from the University of Pisa and Honorary Doctor Engineering Degree in Biomedical Engineering from the Campus Bio-Medico University in Rome and has been and is visiting researcher, professor, and fellow at various universities and scientific institutions in Europe, the USA, the Middle East, and Asia. His current research interests are in the field of biorobotics and bionics and include surgical robotics, micro-/nanodevices for endoscopy, bio-inspired devices and systems, and assistive and companion robots. He is the author of 400+ journal publications (Scopus), coauthor of 50+ international patents, and co-founder of 5 start-up companies. His H-Index is 59 (Scopus). Moreover, he has been the coordinator of many large national and European projects; served as editor in chief, associate editor, as member of the editorial board of many international journals in biomedical engineering and in robotics, and as president of the IEEE Robotics and Automation Society; and received many prizes and awards, including the 1996 Joseph Engelberger Award for Medical Robotics, the 2014 IEEE RAS George Saridis Leadership Award, and the 2017 IEEE RAS Pioneer Award for Biorobotics. He is also founding editorial board member of the journal *Science Robotics*, associate editor of the IEEE Transactions on Biomedical Engineering, IEEE fellow, and fellow of the European Society of Medical and Biological Engineering. In 2018, he has been nominated editor in chief of the new IEEE Transactions on Medical Robotics and Bionics.

Maria Chiara Di Lieto PhD in Basic and Developmental Neuroscience at University of Pisa (Italy), is a developmental psychologist and cognitive behavioral psychotherapist. She works as an independent contractor and external consultant at IRCCS Fondazione Stella Maris in Calambrone (Pisa, Italy). Her expertise spans cognitive and neuropsychological assessment and rehabilitation of children and adolescents with neurodevelopmental disorders, such as cerebral palsy, specific learning disorders, and language impairment; new rehabilitative technologies, such as educational robotics, to promote executive functions and cognitive tele-rehabilitation and to train neuropsychological processes and learning skills; and compensatory informatic programs to support academic learning. She is an author and coauthor of some scientific publications and research projects about educational robotics, tele-rehabilitation, and developmental neuropsychology. Furthermore, she has been involved in design and trial of some tele-rehabilitative programs in collaboration with Cooperativa Anastasis and has been trainer in some national training courses about executive functions and new technologies for clinical rehabilitation and educational practice.

Ivana Đurđević Babić is an associate professor of Information and Communication Sciences. She teaches courses at integrated undergraduate and graduate 5-year university class teacher studies. Her scientific and research interests include ICT in education, data mining methods, and machine learning techniques.

She actively participated at scientific and professional conferences and was a reviewer of professional and scientific articles.

Currently, she holds the position of chair president of the chair of Mathematics and Informatics as part of the Department of Natural Sciences at the Faculty of Education and is the head of the Commission for Quality Enhancement and Assurance at the Faculty of Education, University of Osijek. She is a member of the Croatian Operational Research Society (CRORS).

Richard J. Duro received his MS and PhD degrees in Physics from the University of Santiago de Compostela, Spain, in 1989 and 1992, respectively. He is currently a *full professor* in the Department of Computer Science and head of the Integrated Group for Engineering Research at the University of A Coruña. His research interests include higher-order neural network structures, signal processing, and autonomous and evolutionary robotics. During the last few years, he has been involved in educational projects related with robotics, including the "Adopt a robot" initiative within the DREAM FET-OPEN European Project, and in the development of accessible but powerful educational robots capable of running AI-related applications.

Heinrich Foltz is professor of Electrical Engineering at the University of Texas Rio Grande Valley, USA, where he has been faculty since 1994. He teaches electronics, electromagnetics, RF/microwave, and design laboratories.

Ilze France is senior researcher of the Interdisciplinary Center for Educational Innovation at the University of Latvia. She also serves as expert in Education at the Council of Sciences of the Republic of Latvia.

Her expertise spans curriculum development, inquiry-based learning, education through science, development of methodological and teacher support materials, digital learning materials, educational robotics, and lesson-based professional development.

Senior Researcher France is an author and coauthor of more than 26 publications about processes in different dimensions of education and of 37 methodological materials and textbooks. She has been involved in six research projects. At the moment, she is an expert at the National Centre for Education of Latvia EU Project "Competency Approach to Learning Content."

Salvador Garza received his bachelor's degree of Science in Electrical Engineering from the University of Texas Rio Grande Valley, USA, in May 2018. He was a member of the 2017–2018 robotics team who competed in the 2018 IEEE Region 5 Robotics Contest held in Austin, Texas. After obtaining his undergraduate degree, he enrolled in the university graduate program.

Elaine Cristina Grebogy is PhD student in Computer Science at the Universidade Federal do Paraná (UFPR) and specialist in Management of the School Environment at the Universidade Castelo Branco (2009). She finished his MSc in Education and New Technologies at the Centro Universitário Internacional (UNINTER) in 2017 and graduated in Pedagogy at the Universidade Castelo Branco in 2007. She has experience in education, with emphasis on educational management and literacy. She is currently the deputy principal of a municipal school in the city of São José dos Pinhais. In addition, she is also participant in Computing Simulations and Educational Robotics research group at the UNINTER, working in educational robotics projects and teacher training courses.

Stephanie (Stephie) Kaye Holmquist is president and CEO of Holmquist Educational Consultants, Inc., a national career and technology education consulting firm in business since 1998. She is also the director of the Collaboratory, associate director of the David C. Anchin Center, and adjunct instructor at the University of South Florida, College of Education. She received her PhD from the University of South Florida in Career and Workforce Education with an emphasis in Curriculum and Instruction, with a focus on Educational Robotics. She has coauthored STEM curriculum, consults nationally in the area of career and technical education and academic integration, and specializes in STEM/STEAM education and professional development. Her programs have won numerous state and national awards. She began her career as the director of the Professional Children's Theatre in Tampa, FL, where she then became the outreach coordinator and director of K.I.D.S. at the Museum of Science and Industry, and then director of Education and Marketing at Great Explorations, The Hands-On Museum in St. Petersburg, Florida. Moreover, she was an adjunct instructor in Aviation Technology as well as Career and Technical Education at St. Petersburg College.

Emanuela Inguaggiato is a child neurologist and psychiatrist. She completed her PhD at Scuola Superiore Sant'Anna (Pisa, Italy) in 2017 in Innovative Strategies in Biomedical Research. She works as medical doctor and researcher at the Stella Maris Foundation of IRCCS in Calambrone (Pisa, Italy), in which her work has focused mainly on early diagnosis and intervention in children with neurodevelopmental disorders (such as ADHD, learning and motor disorders). One of her main fields of research involved the executive functions and the new technologies applicable in the intervention, for example, tools and platforms, that can also be used in a tele-rehabilitation approach. In addition, she is an author and coauthor of some scientific publications and research projects about early intervention in children at risk for neurodevelopmental disorders, educational robotics, and tele-rehabilitation.

Martin Kandlhofer received his MSc in Software Engineering in 2010 and his PhD in Computer Science in 2017 from Graz University of Technology, where he is currently member of the research group Autonomous Intelligent Systems.

His area of research is the preparation and evaluation of educational robotics and AI topics for different age groups on different educational levels (K–12, undergraduate).

Ahmad Khanlari is a PhD candidate in the Department of Curriculum, Teaching, and Learning (CTL) and the Collaborative Program in Engineering Education at the University of Toronto. He received his bachelor's degree and his first master's degree in Electrical and Communication Engineering and his second master's degree in Educational Technology from Memorial University of Newfoundland. Prior to starting his MEd program at Memorial University, he cofounded an educational technology company in 2003 and worked there as the education manager to develop curriculums for teaching robotics and other STEM-related subjects. He also worked as a consultant engineer in the R&D department to design educational kits, such as educational robots, to facilitate teaching and learning of STEM-related topics. His research areas lie in educational robotics, STEM education, engineering education, educational data mining, learning analytics, educational technology, online learning, knowledge building communities, wireless communications, and satellite communication.

Dimitrios Loukatos received his diploma in Electrical and Computer Engineering and his PhD degree in Telecommunications and Computing, both from the National Technical University of Athens (NTUA), Greece. He is currently member of the laboratory teaching staff of the Agricultural Engineering Laboratory of the Agricultural University of Athens. He has worked as a research associate with the NTUA and the National Centre for Scientific Research "Demokritos." His research interests include hardware/software platforms evaluation and optimization, management and applications of wired and/or wireless (sensor) networks, and robotics and applications. His most recent work is focused on the area of IoT, physical computing, and artificial intelligence. In parallel, he has more than 20 years of teaching experience, in secondary, vocational, adult, and higher education, in the areas of computer architecture and programming, embedded systems (Atmel, Arduino, Raspberry Pi, etc.), networking, robotics, human-machine interaction, and STEAM education. He has been involved in many national or European projects.

Luciano Frontino de Medeiros is professor and senior research at the Centro Universitário Internacional (UNINTER), Curitiba, Brazil, and assistant coordinator of the Program of Master in Education and New Technologies. He completed his postdoctoral studies in Artificial Intelligence (2013) at the Universidade Politécnica de Madrid, his PhD in Knowledge Engineering and Management (2010) at the Universidade Federal de Santa Catarina (UFSC), and his MSc degree in Informatics at the Univeridade Federal do Paraná (UFPR) (2001).

His expertise include artificial intelligence, new technologies in education, education at distance, computing simulations, educational robotics, learning objects, systems analysis and development, databases modeling, knowledge engineering, and managing. In addition, he is a DIY enthusiast.

He is also professor, consultant, author, and coauthor of approximately 100 publications involving interdisciplinary projects in education, artificial intelligence, neural networks, knowledge management, and ontologies. He is the head of the

Computing Simulations and Educational Robotics research group at the UNINTER, conducting educational projects joining robotics with recyclable materials and easy-to-use open-source electronic platforms, programming in schools and computational thinking, 3D printing at schools, games and gamification, and quantitative research methodologies for education.

Silvia Nuere is a professor in the Department of Mechanical, Chemical and Industrial Design of the Technical School of Engineering in Industrial Design at the Universidad Politécnica de Madrid and, since 2011, is the creator and director of the scientific journal *ArDIn* (*Arte, Diseño e Ingeniería*) [*Art, Design and Engineering*]. She teaches artistic drawing and design from a point of view of humanism. Her research interests include teaching and learning methodologies based on project-oriented learning and on the necessity of mixing different fields of knowledge as art, design, and engineering. She has promoted this approach to education through more than 30 innovation education projects

She is an author and coauthor of more than 50 publications about artistic learning methodologies and humanistic approach to education. She has also, as an artist, participated in more than 20 collective exhibitions and made several illustrations for the scientific journal *Investigación y Ciencia*, the Spanish edition of the Scientific American Journal.

Pedro de Oro Martín industrial electronic engineer at Carlos III University of Madrid with an Industrial Design Engineering Master at Technical University of Madrid, is a designer and product developer focused on educational robotics and 3D printing with almost a decade of experience. He has been product owner of multiple international robotics products such as the educational robot Zowi or PrintBot Evolution. On his spare time, he develops his enthusiasm for open source, art and design, and their links with robotics.

Chiara Pecini is associate professor at the Department of Educational Sciences, Linguistic and Developmental Psychology, University of Florence (Italy). She has been working as clinical neuropsychologist and researcher in Developmental Neuropsychology at the IRCCS Fondazione Stella Maris Institute from 1998 to 2018.

Her studies span from typical development to developmental neuropsychology and are focused on describing the trajectories of the main cognitive processes for learning, defining the cognitive profiles of children with neurodevelopmental disorders, and developing new intervention tools to empower and rehabilitate basic cognitive processes. Within the last field, in 2013, she started to develop and use new tele-rehabilitation techniques for the intervention on reading and executive functioning. Furthermore, she is an author and coauthor of more than 30 publications about developmental neuropsychology and learning in typical development and has been involved in several research projects.

Samuel Roberts received his bachelor of Science in Electrical Engineering degree from the University of Texas Rio Grande Valley, USA, in May 2018. He was the captain of the 2017–2018 robotics team who competed in the 2018 IEEE Region 5 Robotics Contest held in Austin, Texas. After graduation, Mr. Roberts joined Raytheon Company, an international aerospace and defense company.

Icléia Santos is PhD student in Computer Science at the Universidade Federal do Paraná-UFPR and specialist in Management of Education at the FACEL, Curitiba, and in Psychopedagogy at the Faculdades Bagozzi (2011). She finished her MSc in Education and New Technologies at the Centro Universitário Internacional (UNINTER) in 2017. She holds a degree in Pedagogy from the Faculdade de Administração, Ciências, Educação e Letras (2005). She acted as pedagogical supervisor at the Colégio Técnico Industrial de Araucária, from 2004 to 2006, focusing on vocational education. Currently, she is principal of Emilio de Menezes Municipal School, in São José dos Pinhais, since 2006, working in elementary school. She has active participation in the computing simulations and educational robotics research group at the UNINTER, working mainly in educational robotics projects.

David Scaradozzi is assistant professor and mobility delegate of the Engineering Faculty of the UNIVPM. He is chercheur associé at LSIS – UMR CNRS 6168 – Marseille, France, and member of ISME (Italian Interuniversity Center of Integrated Systems for the Marine Environment, www.isme.unige.it) where he cooperates with the Italian Navy and NATO. His research activities are control and optimization of dynamical systems; robotics and automation (motion and interaction control problems in distributed agents, rapid prototyping and mechatronics); underwater robotics and marine technologies, focusing on tools for 3D scientific documentation of sea operative surveys for marine protected areas and archaeological sites study using divers, AUVs, ROVs, and other technological devices; educational robotics, with special interests devoted to all the aspects regarding the study and development of new robotic tools; and lesson plans for teaching e-STrEM (environmental Science Technology robotics Engineering Maths) subjects, in formal and nonformal education, to foster innovation at school, raising awareness on environmental sustainability and training the twenty-first-century skills. He has worked within the national and international projects and has of about 80 publications in refereed international journals, books, and conferences.

Laura Screpanti is a PhD student at the School of Information Engineering of the Dipartimento di Ingegneria dell'Informazione (DII), Università Politecnica delle Marche. She has expertise in bioengineering and educational robotics.

Giuseppina Sgandurra has a first honor degree cum laude in Medicine and Surgery in 2002 at the University of Catania and a specialization in Child Neurology and Psychiatry in 2008 at the University of Pisa; moreover, she received her PhD in New Strategies in Biomedical Research from Scuola Superiore Sant'Anna in 2012.

Currently, she is junior researcher full time (RTD-A) at the Department of Clinical Neuroscience, IRCCS Fondazione Stella Maris, and Department of Clinical and Experimental Medicine, University of Pisa. She works in the Laboratory Neuroengineering – mechatronics and robotics in rehabilitation. Her main research topics are new approaches for diagnosis and treatment of motor and cognitive development disorders, neuroimaging techniques applied to the child neurology, and movement disorders. Her current research focus builds upon the new perspectives of the following: (i) assessment of the upper limb asymmetries in infants and children with hemiplegia, (ii) rehabilitation as action observation training for the rehabilitation of the upper limb in children with UCP, and (iii) educational robotics approaches for rehabilitation in developmental disorders. She is experienced in leading design of new technological tools for quantitative assessment and rehabilitation and personalization of exercises for the rehabilitation of developmental disabilities, the clinical assessment with new clinical and technological tools, and the design of RCTs. She has received a grant (278k) from the Italian Ministry of Health for the proposal "Action-Observation Therapy (AOT) and Information and Communications Technologies (ICT) for Home Rehabilitation of Children with Hemiplegia." She has been also involved in the EU project CareToy and in the Tuscany regional project Mechtoy with a lead role in the management of clinical specification and validation. Furthermore, she is author and coauthor of scientific papers also in extenso published on international indexed journals.

Mariantonietta Valzano is a primary school teacher at I.C. Largo Cocconi. She has large expertise in educational robotics. Since 2010, she has been carrying out the experimental project "Robotica a scuola," teaching robotics as a school subject to students from 6 to 11 years old and involving other teachers in the replication of the same project. She is currently working also at the development of the robotics curriculum with other researchers at the Università Politecnica delle Marche.

Silvia Zingaretti holds a PhD in Information Engineering. She has knowledge about computer vision algorithms, structure from motion, 3D reconstruction, and camera calibration techniques.

Emmanouil Zoulias is task manager and scientific assistant in Edumotiva. He holds a BSc and PhD in Biomedical and Educational Informatics and Technologies (NTUA, Greece) as a specialist on data mining. His main active research focuses on biomedical and educational informatics and technologies. He is a developer, administrator, and integrator of various web2-based platforms. He has served for almost 4 years as IT expert in the Public Health System of Greece, for 10 years as IT expert and member in the National Centre of Public Administration and Local Governance, and for 2 years out of them as CIO. His teaching experience includes, among others, educational technology courses in the field of web development and ICT skills. He has been involved in various European projects in the field of health, e-health, educational technologies, telemedicine, and tele-education in collaboration with several research organizations. He is currently scientific partner of the School of Nursing in the University of Athens.

Chapter 1
Beyond Coding: Back to the Future with Education Robots

Dave Catlin

Abstract Jeannette Wing's 2013 call for education to make coding a key skill coincided with a boom in new education robots. Not surprisingly most of these new robots focus on developing student's computational thinking abilities and programming know-how. Is that all robots can offer? To find the answer I'll explore the history of education robots: specifically the ideas of Seymour Papert. What we'll find is something with far more potential than providing learners with a way of developing their coding skills. And against accepted wisdom, I'll suggest that as technology develops the need for coders will (in the long term) dwindle but the power of robots to help educate children for the future will increase.

Keywords Logo · Seymour Papert · Jeannette Wing · Education Robots · STEM · Computational thinking · Coding · EduRobot taxonomy · Machine learning · Artificial intelligence

Introduction

The Constructionism 2018 Conference hosted a panel session entitled 'Inside the Trojan Horse – A discussion Among the Next Generation of Constructionist'. The room was full of Seymour Papert 'groupies', perhaps more respectfully, academics who Papert inspired, worked with and contributed much to his ideas of constructionism, Logo and robots. Yet, some were like one panellist, who admitted until recently he'd never heard of Papert. When I mention Papert or Logo to young teachers I'm often greeted with blank stares. In this chapter, I want to take you back to Papert's ideas and together explore how much they offer future education. To old hands, I apologise for running through some historic stories and ideas which they probably know. But I keep coming across articles about this history and Papert's

D. Catlin (✉)
Valiant Technology Ltd., London, UK
e-mail: dave@valiant-technology.com

L. Daniela (ed.), *Smart Learning with Educational Robotics*,
https://doi.org/10.1007/978-3-030-19913-5_1

approach to education so severely distorted by Chinese whispers that they lose their potency. I want to take the chance to correct these errors. I must also confess that while writing this chapter I'm left to marvel again at Papert's vision and how much I'd forgotten, not known or not fully grasped the first time. While a large part of the chapter devotes itself to this exercise, I believe Papert's ideas will resonate with our future use of education robots.

I understand the controversial statement, 'need for coders will dwindle' needs justification. Again, we can explain this by looking back to the work of another of my mentors, futurist and professor of science and society, the late Tom Stonier. Just to add a little more weight to his views I'll include predictions made by Alan Turing 70 years ago. Comments from both men prove prophetic.

All this preparatory work is about building a platform, from where we can gaze at the future of education robots. On the way, we'll ask what an education robot is. Fortunately, the EduRobot taxonomy defines this in some detail (Catlin et al. 2018a, b). That brings us to the crux of the chapter: how can robots play a major role in future education? What problems do they need to overcome? What dangers lurk ready to thwart our ambition? Again we'll find the answers in earlier work. The 'educational robotic application (ERA) principles' summarise, into ten axioms, five decades of practical classroom experience of using robots. ERA explains the value of robots and we'll explore how they work together to support effective education programmes.

We can't take this journey without bumping into issues of education policy, much of which justifies Papert's well-known dislike of conventional schooling. We'll find ourselves at a crossroad. Powerful, machine learning robots are already 'attending teacher training college'. Papert's vision offers a more appealing alternative to using this technology. But before we do any of this, we need to review what's going on today.

Jeannette Wing's Revolution

In 2006, Jeannette Wing, then President's Professor of Computer Science at Carnegie Mellon University, delivered a seminal paper to 'The Association of Computer Machinery'. She stated that thinking methods and disciplines used by computer scientists would benefit students of all subjects. She reasoned the widespread use of computers in all disciplines made coding literacy an important skill. But, more importantly, the way computer scientists worked involved computational thinking, a problem-solving template that all subjects could use. 'It [Computational Thinking] represents a universally applicable attitude and skill set everyone, not just computer scientists, would be eager to learn and use'. She continues, 'To reading, writing, and arithmetic, we should add computational thinking to every child's analytical ability' (Wing 2006).

Wing's paper inspired computer scientists, educators and politicians and led to growing interest around the world to promote coding in schools. In August 2010, the UK government tasked the Royal Society to review the status of computing in British schools. Their report found 'The current delivery of computing education in many UK schools is highly unsatisfactory. Although existing curricula for Information and Communication Technology (ICT) are broad and allow scope for teachers to inspire pupils and help them develop interests in Computing, many pupils are not inspired by what they are taught and gain nothing beyond basic digital literacy skills such as how to use a word-processor or a database' (Furber 2012).

The report concluded, 'Every child should have the opportunity to learn Computing at school, including exposure to Computer Science as a rigorous academic discipline'.

By September 2013 the UK government launched its new computing curriculum (Computing Programmes of Study 2013).[1] This document stated a set of clear aims:

'The national curriculum for computing aims to ensure that all pupils:

- Can understand and apply the fundamental principles and concepts of computer science, including abstraction, logic, algorithms and data representation.
- Can analyse problems in computational terms, and have repeated practical experience of writing computer programmes in order to solve such problems.
- Can evaluate and apply information technology, including new or unfamiliar technologies, analytically to solve problems.
- Are responsible, competent, confident and creative users of information and communication technology'.

While the UK was among the first adopters of Wing's revolution, it wasn't alone. In 2017 the European Union (EU) Code Week, held between the 7th and 22nd of October, saw 1.2 million people in more than 50 countries and 4 continents taking part in coding events. This included Australia, USA, Argentina, Brazil, Malaysia and many other non-European countries. An extra 1.3 million young people engaged in Africa Code Week, a spin-off initiative, and in 2018, the EU event had expanded to include 72 countries (CodeWeek EU 2018).

The Economic Need for More Coders

Commentators often cite the future economic prosperity of a country as a reason for striving for computer literacy. For example, in Costa Rica, Carol Angulo and her colleagues at the Omar Dengo Foundation stated 'labour demand in informatics experiences constant growth' (Angulo et al. 2018). They cite economic reports claiming the country lacked 8000 computer scientists in 2017 and between June and August 2016 saw the creation of 5000 new technology jobs.

[1] Strictly speaking this applies not just to England, but Scotland, Wales and Northern Ireland also followed similar polices.

Computational Thinking

Table 1.1 shows one version of what this means.

Wing's idea claims students would benefit from thinking like computer scientists. Several authors, while not dismissing the claim, raised questions and concerns. David Hemmendinger urges his fellow computer scientists to tone down their more zealous rhetoric (Hemmendinger 2010). He points out that all subjects claim meta-cognitive thinking skills belonging to their profession. And some, like mathematicians, have a prior claim over most of the elements claimed by computer science.

Catlin and Woollard (2014) raised a few other concerns:

- Design technology[2] promoted the way designers think as a method all subjects could use – including computer science (for a designer, writing a programme is simply a design problem). This effort failed because art teachers wanted their students to think like artists, science teachers wanted learners to think like scientists, and so on.
- Computational thinking is an emergent behaviour: it grows from writing programmes to solving problems. Bransford and his colleagues confirm this when they comment on the difference between novices and experts who rely on 'internalised mental structures' (Bransford et al. 2000).

A study by Lave and Wenger records how practitioners gain the mental skills of their profession from exposure to its relevant experiences (Lave and Wenger 1991). Computational thinking is such a set of skills which you can't simply 'bolt on' to a novice.

George Polya's million-copy selling mathematics book, 'How to Solve It', presented a set of heuristics representing the way mathematicians think and solve problems (Polya 1990). In his foreword to the latest edition, Professor Ian Stewart gives a compelling critique which I believe applies to all such thinking methods. He starts

Table 1.1 Computational thinking skills based on Southampton University (2013)

Skill	Competencies
Abstraction	Dealing with complexity by stripping away unnecessary detail
Algorithm	Identifying the processes and sequence of events
Decomposition	Breaking complex artefacts, processes or systems into their basic parts
Generalisation	Identifying the patterns and shared by artefacts, processes or systems
Logical analysis	Applying and interpreting Boolean logic
Evaluation	Systematically (through criteria and heuristics) making proven value judgements

This table represents a fuller version of the computational thinking than shown in much of the literature. Many other sources used in K-12 education miss out or subsume logical analysis and evaluation into the other skills. They also refer to generalisation as pattern recognition

[2] Now more popularly known as the maker movement

with, 'There are fashions in the teaching of mathematics. Problem solving came into vogue in the 1980s, …The 1980 yearbook of the National Council of Teachers of Mathematics (NCTM) in the USA reads as if it had been marinated in Polya sauce'. But did this approach work? Stewart claims the anecdotal evidence from practising teachers says yes, but the results from International Mathematics Olympiads says no. Yet, he emphatically supports Polya's ideas. He explains Polya's methods give the mathematician a toolkit for solving a problem, but that is not something you can use dogmatically. How and when you use the heuristics is the internalised art of an experienced mathematician (Stewart 1990, pp. xi–xxx).

As an experienced designer, I believe this story applies to the design process. When education rigidly applied design thinking to maker projects, it often led to sterile work. We make this worse with high-stakes testing which demands the students to show the different parts of such thinking skills irrespective of whether they're all relevant to a specific project. We contrive to replace creativity and design flair with box ticking exercises showing how well we know a process. I suspect the issues apply to computational thinking.

Coding in the Future

It's clear now that there's a shortage in computer programmers. Anyone, leaving school in the next few years would find well-paid work if they had coding skills. Will it be the same when children entering primary school today search for their first job? There are reasons to doubt this.

Alan Turing was the first to raise a relevant issue. In a 1947 lecture to the London Mathematical Society, Turing stated, 'Roughly speaking those who work in connection with the ACE [a computer] will be divided into its masters [programmers] and its servants [users]. Its masters will plan out instruction tables [programs] for it, thinking up deeper and deeper ways of using it. Its servants [users] will feed it with cards [data] as it calls for them'. He continues, 'The masters are liable to get replaced because as soon as any technique becomes at all stereotyped it becomes possible to devise a system of instruction tables [programs] which will enable the electronic computer to do it for itself'. Turing speculates the programmers may refuse to allow computers to steal their jobs. '… [they will] surround the whole of their work with mystery and make excuses, couched in well-chosen gibberish, whenever any dangerous suggestions were made' (Turing 1947).

John Cribben records: Britain's National Physics Laboratory recruited Turing after the Second World War and before he moved to Manchester to work on ACE. While there he wrote a report called the 'Proposed Electronic Calculator'. 'He was interested in developing an adaptable machine that could through its programming, carry out many different tasks; he suggested one program could modify another…' (Cribben 2013).

Commenting in the mid-eighties, David Parnas agreed, '…Of course, automatic programming is feasible. We have known for years that we can implement higher-level programming languages' (Parnas 1985). He equated automatic coding with creating these languages because they're closer to natural language than machine and assembler code.

Forward to 2014, the Pliny Project '... aims to develop a family of systems for automatically detecting and fixing errors in programs, and synthesizing reliable code from high-level specifications' (Rice University 2018). Many programmes need to solve the same coding problems so Pliny reviews code from thousands of existing products and turns them into reusable scripts.

Tomasz Korzeniowski, CEO of CODEBEAT, claims, 'A fundamental shift in software development is underway'. He focuses on using artificial intelligence (AI) and machine learning to '... develop a set of practical tools which combine static code analysis with machine learning. ...Technological development was historically all about the mechanization of manual tasks. Today it's expected to impact cognitive tasks as well. And everyone should make the most out of this innovation, including software developers. Let's not forget that automation isn't new to the industry. In fact, software is already performing many software engineering tasks that used to be done by humans' (Korzonek 2016).

It seems the mood among programmers is starting to support Turing's prediction. An Evans Data Corporation survey of 500 programmers found 29.1% fear AI and machine learning threatens their jobs (Garvin 2016).

We've discussed the possibilities of software creating software, but what about radical changes to computing? Will, for example, quantum computing change the nature of programming? Some people say it will, others disagree and some think we'll see a mixture of the quantum and classic computer programmes (Hayes 2014). Whether this is true or not, it's a brave person who'd bet the coding skills we start to teach today's 5-year-olds will help them find a job 20 years from now.

The Wealth of Information

In 1983, Professor Tom Stonier wrote *The Wealth of Information: A Profile of the Post-industrial Economy*. He claimed advanced Western economies were undergoing a structural (permanent) shift to a wealth-creating information economy (Stonier 1983). He isolated a few patterns which would dictate the transformations affecting employment. In a case study, he examined the fate of a number of industries in my hometown of Bradford, West Yorkshire. He included the wool industry – which provided me with my first job making machines to turn fleece into yarn. Wool manufacturing in Britain dated from the Bronze Age and became a huge contributor to the country's economy in medieval times and a major beneficiary of the industrial revolution. Recognised as the wool capital of the world, Bradford employed 70,000 people in 1976; this fell to 35,000 by 1980 and is now a few thousand. More modern businesses also suffered the same dramatic changes. Stonier cites Thorne Consumer Electronics who in 1974 boasted the largest television factory in Europe. It employed 4700 workers making 10,000 television sets each week. It stopped trading in 1978.

Stonier identifies causes driving permanent changes to the economy in advanced countries:

1. Passing on of knowledge and skills from advanced high-cost countries to countries with lower costs
2. A decline in demand for outdated products
3. Advances in technology opening new possibilities

Over recent years many advanced countries focused their education efforts on science, technology, engineering and maths (STEM) – they believe this will give them a competitive edge. It's difficult to see how this can happen when every country follows the same policy. And Stonier's case studies show people with high STEM skills still lost their jobs and struggled to find work when their industry moved forever. This doesn't mean we shouldn't focus on STEM teaching – far from it. What it does mean is we shouldn't demote the arts, history and other non-technical subjects. For example, the creative industries accounted for $340 billion (14.2%) to the British GVA (gross value added)[3] in 2017. It grew twice as much as any other part of the economy (Bradly 2017).

Stonier predicted this area of economic growth and forecast movement of labour from manufacturing to the service industry. The following decades saw major job increases in retail and call centre employment. However, we now see online shopping and offshore call centres reducing the number of people employed by retail and moving call centre jobs to lower-cost economies. All this made possible by technology not available in 1983. This technology also supports a vibrant offshore coding industry moving programming jobs to low-cost economies all over the world.

What Education Do We Need?

The answer to this critical question will define a role for robots in education. How do we educate students who live in a society undergoing frequent economic upheavals where we can't guarantee full employment, even for those with STEM skills? Stonier answers this by making a distinction between education and training. K-12 education should focus on helping students develop as people finding their interests and talents and turning them into lifelong learners. Such education creates skilled knowledgeable citizens able to adapt to whatever happens in the future. This isn't a new debate. We can go back to the Ancient Greeks: Plato supported training people for their role in life (Cooper 2001). Aristotle thought education should help a student maximise their intellectual and moral 'virtues' (Hutchinson 1995; Hobson 2001). He believed people who'd maximised their talents would find solutions to whatever challenges their society faced.

[3] GVA – it is a measure of total output and income in the economy.

Should We Teach Coding and Computational Thinking?

I've presented an analysis which questions many of the reasons used to justify the coding effort in schools. This isn't an argument to stop what we're doing: it's about changing perspective and clarifying how best robots can play a role in future education.

When programmers say to me, 'It's intuitive…', I'm left wondering intuitive to whom – my grandmother? In the early 1930s, data collected by Alexander Luria studying peasants in the steppes of Central Asia confirmed Vygotsky's ideas on thinking's dependence on social, historical and cultural settings (Luria 1976). All Luria's subjects had little or no education. Their thinking grew from their practical work. He noted shifts in their thinking as they engaged in different experiences forced on them by Stalin's economic reforms. The changing thought patterns also started to change their cultural life.

The intuition programmers talked about refers to members of their cultural community. By giving young students the chance to explore, coding places them at what Lave and Wenger called the 'periphery of such a community'. Lave and Wenger noted as the individual journeyed from community fringes to its heart, both the student and the nature of the community changed. Introducing students to such a community will enable them to grow with future changes (Lave and Wenger 1991).

Seymour Papert

South African born Seymour Papert invented education robots. 'The son of an itinerant South African entomologist researching the tsetse fly Seymour Papert spent his early childhood camping along the East African coast in the 1930s. The Papert family's way of life was straight out of a Hemmingway story. Travelling along the bush trails they hunted their food and fixed their trucks when they broke down' (Crevier 1993). Papert recalls how before he was 2 years old he'd got to know about these trucks. 'I was particularly proud of knowing the parts of the transmission system, the gearbox and most especially the differential' (Papert 1980, p. vi). When he got to understand how gears worked, he loved playing with them and different forms of rotating objects.

Eventually, his family moved to Johannesburg and he started attending school. 'I remember working with differentials did more for my mathematical development than anything I was taught in elementary school. Gears, serving as models, carried many abstract ideas into my head. I clearly remember two examples from school math. I saw multiplication tables as gears, and my first brush with equations with two variables (example $3x + 4y = 10$) immediately evoked the differential' (Papert 1980, p. vi). He believed his experience with gears provided him with useful mental models for learning. He generalised the idea as a rule, 'Anything is easy if you can assimilate it into your models' (Papert 1980, p. vii).

Papert's move from living in the bush with black Africans to Johannesburg's white South African society gave him a cultural shock. He struggled to understand apartheid. How did his neighbours, who in many ways seemed reasonable people, have these bizarre racial ideas? His curiosity led him to generalise the question: how do we get our ideas? It inspired him to study logic; he attended a university course on the subject while still in high school. There he settled a debate with fellow students about whether you could formalise logic: he built a machine that did just that. He didn't realise that such machines existed – but it showed his practical interest in computers and engineering (McCorduck 2004).

He first studied philosophy at the University of the Witwatersrand before transferring to mathematics where he earned a PhD. While working for a second mathematical doctorate, this time at Cambridge University, he met Jean Piaget. Piaget persuaded him to join him in Switzerland to try to figure out how children learn mathematics. After 4 years the allure of computers enticed him to the National Physics Laboratory in London. While working there he met Marvin Minsky who convinced him to move to the Massachusetts Institute of Technology (MIT). It's there he'd weave his formative interests and experiences into his major contribution to education.

Logo

The US Navy commissioned Bolt, Beranek and Newman (BBN)[4] to study how they might use computers to train their sailors. Wally Feurzeig ran the project and Cynthia Solomon was on the team. Solomon, who'd worked for Minsky at MIT, got to know Papert and persuaded Feurzeig to invite him to join the project as a consultant.[5] The group had realised the potential of the computer to help schoolchildren to learn. Solomon recalls, 'Papert made a summer visit to the parents of his first wife in Cyprus. When he came back we all met in Danny Bobrow's apartment and Seymour explained the idea of Logo'. Under Papert's directions, Bobrow wrote the first Logo as a variation on LISP (Papert 1969, 1993, p. 168).

In those days people created high-level computer languages for specific purposes – for example, FORTRAN for science and engineering and COBOL for business. AI researchers favoured LISP. Papert believed 'Logo was designed for learning and it's unique in this respect. No other language was designed for this purpose. BASIC and its

[4]Leo Beranek and Richard Bolt, professors at MIT, with Bolt's former student Robert Newman. People who worked there in the 1960s and 70s told me it was difficult to know who worked for BBN and who for MIT.

[5]I constructed this history from interviews and discussions with Solomon, Feurzeig and Paul Wexelblat and email correspondence with Marvin Minsky, Danny Bobrow and Mike Paterson. Some say Papert co-invented Logo and the Turtle – this wasn't the view of those I interviewed. The Children's Machine reference is Papert's explanation of his Eureka moment in Cyprus – the moment he invented Logo.

variants such as PILOT which are most commonly used were made for totally different purposes and were handed on like cast-off-clothing to be used by the world of education. I think this is a scandal' (Papert 1983, pp. Part 1 – 02:48–03:10).

Papert continues explaining, 'We tried to achieve a number of goals. First of all that it should be easily accessible. There should be corners of the language that you can get into; like baby-talk getting into English that are easy for the youngest beginner. But, it shouldn't be a toy language. It's not that Logo is easy, it's easy to get into, but once you're in there you can progress to the most sophisticated ideas in the world of programming' (Papert 1983, pp. Part 1 – 05:30–05:55).

This idea became summarised in a metaphor, 'Low floors, high ceilings and wide walls' (Resnick 2016). To low floors (easy access) and high ceiling (sophisticated programming), Papert added wide walls – meaning they're many routes between floor and ceiling. This begs a question: why should we force computational thinking on to children? This is an emergent thinking pattern which we should gently cultivate by involving children in worthwhile programming challenges. 'Rather than pushing children to think like adults, we might do better to remember that they are great learners and to try harder to be more like them' (Papert 1993, p. 155).

Papert's invention of Logo took place a decade before personal computers arrived on anyone's desk. BBN housed the computer, a Programmed Data Processor (PDP), in their offices and schools connected to it through a teleprinter. Papert asked deep questions about the role of computers in schools. Alan Kay pithily paraphrased Papert's answer, 'Should the computer program the kid or should the kid program the computer?' (Brand 1972)

Papert and Learning

Papert wanted to find better ways to learn – not teach. 'Why is there no word in English for the art of learning? Webster says pedagogy means the art of teaching. What is missing is the parallel word for learning' (Papert 1993, p. 81). This doesn't mean teaching and instruction don't have their place, but he felt what he called instructionism wasn't necessarily the best way to improve education. 'The word instructionism is meant to mean something rather different from pedagogy or the art of teaching… [it's] the belief that the route to better learning must be the improvement of instruction…' (Papert 1993, pp. 138–139). He added, '…constructionism as one of a family of educational philosophies that denies this "obvious truth"…'.

None of this means you don't 'tell' children anything. It took our cave-dwelling ancestors about 70,000 years[6] to invent the wheel; children don't have time to invent

[6] I base this on the findings in the Blombos Caves in Papert's native South Africa.

and discover everything. Papert thwarts the many arguments posed by his critics and those opposed to constructionist approaches to education (Price n.d.). '[His method] does not call into question the value of instruction. That would be silly: Even the statement… every act of teaching deprives the child of an opportunity for discovery is not a categorical imperative against teaching, but a paradoxically expressed reminder to keep it in check' (Papert 1993, p. 139).

Papert challenged the way schools work. For example, I once went into a school with a couple of researchers from London's Institute of Education. We wanted to set up a small project. We explained our aims to the teacher who responded by saying, 'Yes, we're happy to do this – the examinations are over so we can do the interesting stuff'. This is a symptom of the problem: I don't blame teachers – they're caught in the black hole of accountability and high-stakes testing.

Papert believed, '…we know if we become involved with an area of knowledge, we learn it – with or without school, and in any case without the paraphernalia of curriculum and tests. We also know if we do not become involved with the area of knowledge, we'll have trouble learning it with or without school methods' (Papert 1993, p. 141). He described his efforts as, '...expanding beyond Piaget's cognitive emphasis to include a concern with the effective. It develops a new perspective for education research focused on creating the conditions under which intellectual models will take root' (Papert 1980, pp. vii–viii). Once again we see a basic theme: he recognised how the love affair he had with gears acted as a springboard for his development. But he knew such an idiosyncratic experience wouldn't have universal appeal. He believed the computer could become a more versatile machine tool allowing students to explore ideas and build a wider range of mental models.

The Child Scientist

My daughter once saw the police performing a stop and search on a car; she rationalised the man must have lost his Barbie Doll and the police were helping him find it. Parents everywhere will have similar cute stories of their young offspring trying to make sense of their world. 'The child does not wait with a virginally empty mind until we are ready to stuff it with a statistically validated curriculum. He [/she] is constantly engaged in inventing theories about everything, including himself, schools and teachers' (Papert 2005).

'All of us learn by constructing, exploring, or theory building, but most of the theory building on which we cut our teeth resulted in theories we would have to give up later' (Papert 1980, p. 132). Papert encourages children to hypothesise based on their experience and to test and reformulate their thinking based on new experiences. He sees this as a lifelong process.

Children and Thinking

In the Children's Machine Papert reproduces the work of another of his ex-wives, Sherry Turkle. She reports on two students Jeff and Kevin creating space exploration scenes on a computer. Jeff follows a computational thinking approach: Kevin more of an artistic strategy. Kevin doesn't have a plan: he tries something and if he doesn't like it he changes it. Despite the differences Turkle reports, 'Kevin not only succeeded in creating a space scene, but like Jeff, he learnt a great deal about computer programming and mathematics' (Turkle 2005). Papert picks up the story, 'Kevin is lucky to be in an environment where he is allowed to work in his own style. In many schools he would be under pressure to do things "properly"...' (Papert 1993, p. 148). Both Papert and Turkle claim bullying children into thinking in a particular way will adversely affect their intellectual growth.

Constructivism and Constructionism

'It is easy enough to formulate simple catchy versions of the idea of constructionism; for example, thinking of it as "learning-by-making"' (Papert and Harel 1991). The authors continue, '...[how] constructionism is much richer and more multifaceted, and very much deeper in its implications, than could be conveyed by any such formula'.

Papert believed Piaget's views on constructivism. '... knowledge simply cannot be transmitted or conveyed ready made to another person' (Papert 1993, p. 142). He continues by claiming even if you tell people some information, they reconstruct a personal version and understanding. Learning is something we do – no one can do it for us.

Edith Ackermann describes Piaget's constructivism, '... how children become progressively detached from the world of concrete objects and local contingencies, gradually becoming able to mentally manipulate symbolic objects within a realm of hypothetical worlds' (Akermann n.d.). Papert discusses the transition from concrete to abstract thought. He believes Piaget's stage of concrete operations doesn't limit itself to young children: '...the sophisticates do not resort to "concrete thinking" only in their preliminary gropings toward solving a problem or when they are operating as novices outside their areas of expertise... ...what Levi-Strauss and Piaget identify as "concrete" are present at the core of important and sophisticated intellectual enterprises' (Papert 1993, p. 151).

Papert declares, 'This praise for the concrete is not to be confused with a strategy of using it as a stepping-stone to the abstract. That would leave the abstract ensconced as the ultimate form of knowing. I want to say something more controversial and more subtle in helping to demote abstract thinking from being seen as "the real stuff" of the working of the mind. More often, if not always in the last analysis concrete thinking is more deserving of this description…' (Papert 1993, p. 146). He hastens to add that this isn't a rejection of logic, formal and abstract thinking, but we shouldn't think of them as our ultimate goal. We need a deeper respect for concrete thinking.

Constructionism focuses on concrete thinking: it's about children building new knowledge using what's available in their experience. This may involve building physical artefacts, or creating new ideas and thoughts. We don't think twice about teaching students the scientific method, but that's only one part of scientific effort. Albert Einstein's thought experiments and Schrodinger's famous cat serve as examples of stories created by scientists to help them and us understand: a core idea in constructionism is to let the children create their own stories.

Papert's Paradigm

At the Construit 2017 Conference I engaged another delegate in conversation about Logo. He proclaimed, 'Papert's ideas don't work' . This isn't the first time I've come across this. Branson and colleagues discuss conflicting research (Bransford et al. 2000, pp. 53–55). They cite research which proclaims it doesn't (Vanderbilt 1996). The Vanderbilt team's research stated Papert proposed using Logo would help children transfer knowledge from one problem to another and their tests showed it didn't. However, other studies found the opposite (Klahr and Carver 1988). Branson credits the difference to how well the Vanderbilt students knew the Logo language.

You'll find similar claims and counterclaims in the literature. To cope with this Mike Blamires and I introduced the term Papert's Paradigm when we reviewed the use of robots in special needs education (Catlin and Blamires 2018). We use the term 'paradigm' referring to Kuhn's paradigm shift philosophy of science (Kuhn 1996). In Kuhn's theory, positive or negative research isn't enough to prove or disprove a theory: it's about accumulated evidence and probability. In reviewing the claims made by Pea and Kurland (1984) and the counterclaims made by (Noss 1995) and (Johnstone 2003, pp. 123–133), Blamires and I feel Papert is the clear 'winner'. Yet, it's worth looking at some other aspects of this dispute.

The scope of Papert's intellectual effort is vast, and sometimes it appears contradictory so it's not surprising that it's easily misunderstood. Papert's work is like a diary of his learning journey which changes with time. For example, he disliked teachers as a child and mistakenly continued this early in his career. '[he demonised] teachers by identifying them with the roles that School forced on them. I disliked School's coercive methods, and it was the teachers who applied the coercion. I disapproved of judgment by grading, and it was the teacher who gave the grades'. The response to Mindstorms dramatically changed his attitude, '…my identification of "teacher" with "School slowly dissolved…"' (Papert 1993, p. 59). To understand his message you need to follow how his opinion evolved.

When reviewing the Children's Machine, Daniel Dennett echoes this idea when mentioning Logo. He comments, 'But one can learn even more from 'mistakes' than from a string of successes – that is a central tenet of Papert's vision of learning, and he practices what he preaches' (Dennett 1993). Dennett testifies his experience with undergraduates who were, 'experts at piling in the facts, drilling for the big test – and they were pathologically uncomfortable in any setting where they had to think'. He ran a course using Logo which he thought was spectacularly successful: 'The students forgot their phobias and inhibitions and took flight, creating a trove of idiosyncratic projects, effortlessly learning the fundamentals of programming, and building a robust base on which we could then help them construct a more 'adult' set of edifices'.

We find a different challenge in Richard Mayer's provocative titled paper, 'Should there be a Three-Strike Rule against Pure Discovery Learning' (Mayer 2004). He criticises the ideas of Piaget and Papert while making the case for guided participation. While his statements aren't wrong his criticisms are strawman arguments. First, Piaget's ideas aren't a teaching theory: they describe a cognitive process, whether it does or doesn't involve guidance. Papert had pre-empted Mayer's criticism with, 'But "teaching without curriculum" does not mean spontaneous, free-form classrooms or simply "leaving the child alone". It means supporting children as they build their own intellectual structures with materials drawn from the surrounding culture. In this model, educational intervention means changing the culture, planting new constructive elements in it and eliminating noxious ones' (Papert 1980, pp. 31–32). Papert never prescribed a strategy which said teachers shouldn't help students.

In my experience, UK teachers in the eighties were ahead of the researchers. When they practised constructionist teaching, not involving Logo, they already used guided participation. Interestingly, Roy Pea stated that when he saw Logo working in the way claimed by Papert it was because, 'Someone has provided guidance, support, ideas, for how the language could be used' (Pea 1984, pp. 55–66). This shows Pea knew Logo worked despite his widely publicised criticism. Table 1.2 shows types of interventions used in a Logo research project.

Sylvia Weir noted the early Logo community 'de-emphasised' the role of the teacher – but it did not eliminate it. She said, 'At its best, the interactive computer experiences should drive itself, for at least some of the time, under the steam of the user's intentions. The user should be the initiator, setting the goals and taking responsibility for tracking down the errors in her program (debugging)' (Weir 1987). In that quote, she gives a flavour of what we mean when we say the 'student takes charge of their learning'.

Some modern commentators, like David Ng, look back on Papert and build meanings which don't resonate with my understandings (Ng 2017). Ng says, '... putting children in charge of their own learning is not the core premise of *Mindstorms*. If it was, why would Papert use learning French in France as his analogy for Mathland? How would putting American students in charge of their own learning in French help? Would they learn as naturally and easily as children in France? Agency is necessary, but completely insufficient if the raw materials for learning aren't even available'.

Table 1.2 Classification interventions in a Logo Maths project adapted from (Hoyles and Sutherland 1989)

Category	Description
Motivational	
Reinforcement	'That's good'
Encouragement	'Try it'
Reflection	
Process (forward)	Encourage students to reflect and predict what they had to do
Goal (forward)	Encourage students to keep in mind their eventual goal
Process (backward)	Encourage students to reflect on what they'd done
Goal (backward)	Encourage students to reflect on their goal
Directional	
Nudge	'Are you sure you want to do that'
Method	'Do you remember how you solved that type of problem before?'
Building	Encouraging students to use a specific method they already know
Factual: new	Providing students with new piece of information
Factual: recall	Reminding students of information they already know
New idea	Introducing a new idea – like the repeat programming command or a mathematical method

Courtesy of Hoyles and Sutherland

Weir's statement reflects Mindstorms and makes it clear – the child 'taking charge of their learning' is at the heart of any constructionist enterprise. This doesn't need to involve Logo or computers. Papert only thought of Logo and Turtle robots, '…as a valuable educational object, but its principal role here is to serve as a model for other objects, yet to be invented' (Papert 1980, p. 11). I urge readers to check out Fleet Circus: a project I had the privilege to witness and record (Catlin and Thomson 1998). It involved a class of 10- and 11-year-old students in a Maker Space project designing a circus full of automatons. They controlled most of the designs by programming a control box. Apart from the spectacular videos showing their work, the project shows what it means for children to take charge of their own learning. It is the perfect demonstration of constructionism, the role of the teacher (Trevor Thomson) and solid improvement in test scores.

The circus students didn't limit their effort to the classroom or project time. Of their own choosing they took work home, they asked questions and sought answers, and instead of taking breaks in the schoolyard, you'd find them working away on their project. They'd taken charge of their own learning. So to answer Ng's question about learning French: if you fire up student's passion to learn French they'll learn it.

What Fleet Circus also shows is something else Mr. Ng seems to misunderstand, 'It is about an end to the culture that makes science and technology alien to the vast majority of people... Most branches of the most sophisticated modern culture of Europe and the United States are so deeply "mathophobic" that many privileged children are as effectively (if more gently) kept from appropriating science as their own' (Papert 1980, p. 4). Papert continues prophetically, 'In my vision, space-age objects, in the form of small computers, will cross these cultural barriers to enter the private worlds of children everywhere. …computers can be carriers of powerful ideas and of the seeds of cultural change, how they can help people form new relationships with knowledge that cut across the traditional lines separating humanities from sciences...'.

Papert's Paradigm is about forming constructionist cultures which encourage students to think, create, explore and love learning. It's where students become motivated to find out, become excited and have the confidence to succeed or fail with equanimity. Failed efforts become stepping stones to solutions that work. Debugging is a basic trait of the Logo culture and it's something students do: it's radically different from looking up an answer. They want to know.

Difference between Papert and Wing

Wing's 2006 'revolution' enticed numerous new stakeholders into the education space. But I find many of these lack a deep educational vision. Despite his interest in computers, Papert's main focus is education. If he believed coding had nothing to offer learning, I believe he would've abandoned it. The 'Fleet Circus project' reflects this perspective. The children wrote essays with pen and pencil; they used tools to build automatons and painted and sculptured artwork. They did programme control boxes to animate their models, but programming wasn't the focus of their effort. They used the computer as a tool in the same matter-of-fact way they used pencils or hacksaws. When they didn't know how to do something, they found out. All of this epitomises Papert's Paradigm.

Can Papert's Ideas Work?

'*Seymour Papert dreamed of a learning revolution—why hasn't it happened?*' asked Junaid Mubeen (Mubeen 2017). I have to disagree with the sentiment. It hasn't happened in America or Europe, but it did happen in Costa Rica. Their 'Computers in Education Program' started in 1988 (Fonesca 1999) and still runs today (Angulo et al. 2018). Writing in 2003 Johnstone reports from a population of 3.8 million their programme has affected over 1.5 million children, teachers and adults (Johnstone 2003, pp. 131–133). Why is the Costa Rican programme so successful? I believe it starts with clear political vision supported by sensible administration. When Logo became available in the early eighties, teachers in the UK, still influenced by the Plowden Report which put Piaget at the heart of teaching practice, took to it with enthusiasm (Plowden 1967). Despite heavy investment in computers and teacher training, every election campaign sees candidates vowing to fix education – implying previous efforts failed. Every time the government appoints a new

Secretary of State for Education, they tear down previous efforts and start again. In Costa Rica, they started with a clear vision and steadily improved on their accomplishments for nearly 30 years.

'Nothing could be more absurd than an experiment in which computers are placed in a classroom where nothing else is changed' (Papert 1993, p. 149). Costa Rica made changes and implemented Papert's ideas. When I see cases where individual schools or teachers follow this advice, something special follows.

Education Robots

You might wonder, why I've talked about computers and programming and so far nothing about robots. I believe the principles and practice described by Papert's Paradigm provide the education foundation for robots past, present and future. So let's now look at education robots.

Papert the Father of Education Robots

Papert invented the world's first education robot. Students wrote Logo programmes to solve problems like NIM (Papert calls it Twenty-One) – a maths game played between two people. He recalls, 'I was doodling at the computer as I often do by writing little programs with no particular importance or difficulty in themselves. You could call it just playing. I don't know what such activity does for the mind, but I assume it's the same as what happens when one draws patterns or pictures with pencil on paper while thinking or listening to a lecture. What happened this time came from thinking that writing programs can be like drawing in many ways. In a way the Twenty-one program is a representation – might one say a kind of picture of the form of a mental process, just as a pencil and paper drawing can be a representation of a physical shape' (Papert 1993, pp. 174–178). Once again we witness Papert's Eureka moment. He carries on describing how his thoughts wandered through several analogies, 'Previously I would have said that what was important about the program was that it represented a kind of thinking. Now I wanted to say that what counted was that it represented something the programmer does. It didn't matter that the something was thinking; it could just as well have been walking or drawing or whatever. In fact, maybe walking or drawing would be better than playing 21; children care more and know more about these activities'.

This led to Papert creating the idea of a Turtle[7] robot and adding to Logo a new geometry called Turtle Graphics. This geometry didn't depend on a Cartesian

[7] Grey Walter made robots Elmer and Elsie between 1947 and 1948. He called them tortoises, because of their shapes. Tortoise got translated from British English to the American Turtle. Grey Walter's work inspired Papert to use the name Turtle, but Grey Walter's robots had nothing to do with education. He made them as part of his studies in neuroscience – not education. Similarly, Braitenberg created his famous robots to explore neuroscience.

framework; it explained space from the robot's point of view. Like the child, the robot always faced forward, irrespective of whether that was north, south, east or west – or up, down, left and right on a computer screen (Ableson and diSessa 1981, pp. 11–16).

Danny Bobrow, who added the Turtle Graphics Code to Logo, told me that before he started coding they went to a school and got the children to 'play turtle'. Papert explains, 'The essential point about the Turtle is its role as transitional object that is a transitional between the body, the self and abstract mathematical ideas. [With] the Turtle you can identify with it, you can move your body in order to get how to command the Turtle. So it's related to you, to the body to the human and it's also related to mathematical ideas whose structure is such that it captures some extremely powerful geometric and physical ideas' (Papert 1983, p. Video 2 1:37–2:11).

As a graduate student Mike Paterson, who became Professor of Computing at Warwick University in England, visited BBN. In December 1969, Papert set him the task of specifying the Turtle robot (Paterson 1969). In January 1970 the MIT AI Lab built the first Turtle (See Fig. 1.1). Early Turtles included the Turtle Tot, Tasman Turtle, Jessop (Edinburgh) Turtle, the BBC Buggy (made from Fischertechnik) and the most popular the Valiant Turtle – which only stopped production in 2015. Meanwhile in January 1989, Valiant launched the Roamer, which contained a cut-down version of Turtle Graphics embedded in its chip. Children programmed Roamer directly using its on-board keypad. Later that year Swallow systems launched their version of Roamer called PIP and later a forerunner of BeeBot called Pixie.

Fig. 1.1 The first turtle. (Courtesy of Cynthia Solomon)

In the seventies, Logo development moved from BBN to MIT. Papert and his team started to play with Lego and their control boxes. These did the same tasks as the control boxes used to drive the 'Fleet Circus' automaton but using Lego bricks instead of maker materials. UK schools justified using control technology because both the design technology and the information and communications technology (ICT) curriculums mandated their use. Schools had a choice of control boxes. In late 1984 Kjeld Kirk Kristiansen, then the CEO of Lego (and grandson of its founder), saw a television programme with children explaining their programming skills and Papert explaining his educational philosophy. This inspired Lego to fund Papert's research team which in January 1998 resulted in the launch of Lego Mindstorms (Waterson 2015).

Other Early Robots

In his book Personal Robotics, Richard Raucci described several robots. Some of these had found their way into schools (Raucci 1999). Raucci asked: 'What Is a Robot, and What Isn't?' His answer – you can programme it and it must have sensors. This definition raises many questions and doesn't address a few important issues germane to education robots. Since around 2010 a flood of robots have appeared, many of them proclaiming their education credentials. To better answer Raucci's questions and give some semblance of order to the new robots, I worked with several researchers on producing a taxonomy for education robots.

EduRobot Taxonomy

In April 2018 we presented a provisional version of the taxonomy as a poster and a paper at the Robots in Education Conference held in Malta (Catlin et al. 2018a). Figure 1.2 shows the basic taxonomic structure. We surveyed and discussed the proposal with conference delegates, and with the aid of a few more colleagues, we presented a revised version at the Constructionism Conference in August 2018 (Catlin et al. 2018b).

We first identified two 'types' of robot, Build Bots, which the students had to build before they could do anything with them. The second we called User Bots – these you can take them out of the box and use them immediately. Figure 1.3 shows an example classification.[8]

---

[8]You can make a Turtle robot (taxonomy – User Bot: Turtle) from Lego. This doesn't make Lego a User Bot. The classification rule, the higher classification determines the choice. That is, Lego is first and foremost a Build Bot. Think of a platypus, which is a mammal despite its many reptile characteristics.

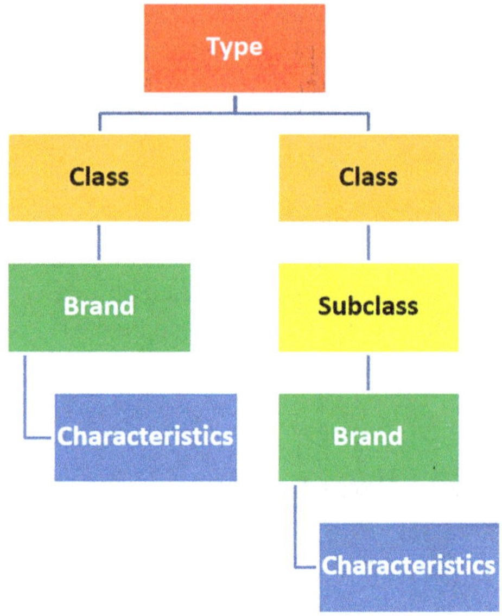

Fig. 1.2 EduRobot taxonomy. (Courtesy of Catlin, Kandlhofer and Holmquist)

Build Bots: Lego MindstormsEV3

Build Systems: Component Parts

A flexible system based on Lego building bricks with special robot parts. Suitable for K5 to University. Previous versions: RCX, NXT and NXT2.0.

Characteristic Tags:

Status: Commercial **Locomotion:** Wheels, Walking, Crawling, Tracked Drives, Static. **Power:** Batteries. **Command and Control:** Computer, Mobile Devices, Autonomous, Human-Robot Interaction, Robot to Robot. **Communication**: Interface Cable, Bluetooth, Wi-Fi, Infrared. **Sensors**: Digital, Analogue. **Outputs**: Digital, Analogue, Servomotor, Sound Effects, Screen. **Architecture**: Direct, Reactive, Deliberative, Hybrid. **Programming**: Block, Graphical, Text. **Modularity**: Expandable. **Morphology**: Personalised, Component Based.

Fig. 1.3 Example robot classification. (Courtesy of Catlin, Kandlhofer and Holmquist)

Following feedback in Malta, we changed the taxonomy to include Social Robots as a separate 'type'. We used the definition, 'Social Bots are autonomous robots that interact and communicate with students using HRI [Human Robot Interaction] technologies. The interactions aim to embody accepted social and cultural norms' (Li et al. 2011, p. 333).

Educational Robotic Application (ERA) Principle

In 2010 Mike Blamires and I wrote the ten ERA principles (Catlin and Blamires 2010a, b). We both started working in this field in the early 1980s and used different types of robots in thousands of schools spread over five continents. By following Papert's method of reflection we tried to make sense of our multiple experiences. This led us to draft ERA. The principles serve several purposes:

1. They reflect our past efforts with education robots.
2. They provide a set of heuristics for evaluating and thinking about education robots.
3. They provide a design specification to assist in the development of robots, their activities and learning environments.

Table 1.3 shows the ten principles and their axiomatic style definitions.

We can use ERA to better answer Raucci's questions: What is an education robot and must a robot have senses? First, we need to decide whether we should include physical and virtual robots.

The 'embodiment principle' refers to physical and not virtual robots. This doesn't mean virtual robots don't have educational value. It does mean the student experiences with physical and virtual robots aren't the same. The ERA paper justifies this based on the theory of embodiment, responses of teachers and some evidence from mathematicians on how children develop spatial understanding. Added to this I recently discovered more direct evidence from Sylvia Weir, 'The inventors of Logo treated the physical Turtle as much the same as the screen Turtle, but children do not' (Weir 1987, p. 155). She reports improvements gained by children using the floor Turtle over the screen version. This led the taxonomy group to decide virtual robots needed a taxonomy branch of their own – which is outside the scope of EduRobot.

Combining the 'intelligence, interaction and embodiment principles' provides a definition of an education robot. Put simply, they combine to say an education robot is a physical machine with enough intelligence to support a student's learning when they interact with it. This resolved a difficult question about whether you can classify toy robots as educational. We need to look at some subtle arguments to answer the sensor's question.

Defining a robot has always been tricky (Catlin et al. 2018a, pp. 4–6). The word 'robot' isn't the sole prerogative of engineers and scientists – after all the word came from the arts when Karel Capek entitled his play Rossum's Universal Robots.

Table 1.3 Educational robotic application (ERA) principles

Principle	Explanation
Technology focus	
Intelligence	'Educational robots can have a range of intelligent behaviours that enables them to effectively participate in educational activities'
Interaction	'Students are active learners whose multimodal interactions with educational robots take place via a variety of appropriate semiotic systems'
Embodiment	'Students learn by intentional and meaningful interactions with educational robots situated in the same space and time'
Student focus	
Engagement	'Through engagement educational robots can foster affirmative emotional states and social relationships that promote the creation of positive learning attitudes and environments, which improves the quality and depth of a student's learning experience'
Sustainable learning	'Educational robots can enhance learning in the longer term through the development of meta-cognition, life skills and learner self-knowledge'
Personalisation	'Educational robots personalise the learning experience to suit the individual needs of students across a range of subjects'
Teacher focus	
Pedagogical	'The science of learning underpins a wide range of methods available for using with appropriately designed educational robots to create effective learning scenarios'
Curriculum and assessment	'Educational robots can facilitate teaching, learning and assessment in traditional curriculum areas by supporting good teaching practice'
Equity	'Educational robots support principles of equity of age, gender, ability, race, ethnicity, culture, social class, life style and political status'
Practical	'Educational robots must meet the practical issues involved in organising and delivering education in both formal and informal learning situations'

Courtesy of Catlin and Blamires

The real issue lies with our focus: Raucci focuses on the technology – and I detect similar interests from many people behind new education robots. If you concentrate on the child's learning, you get a different perspective. Papert discussed Norbert Weiner and cybernetics and the idea of control (Papert 1993, pp. 179–204). In control you need to define the boundaries of your system. Figure 1.4 shows three control models: A has the system boundary around the technology, and B and C include the students as an active part of the setup. If the robot doesn't have sensors, it relies on the student's 'sensors' and creates a natural learning environment. In these circumstances, education robots may or may not have sensors.

Social Robots

The 'pedagogical principle' covers developmental theories sympathetic to education robots and a classification of activities used with them. The original ERA paper only outlined this classification and a later paper expanded on this

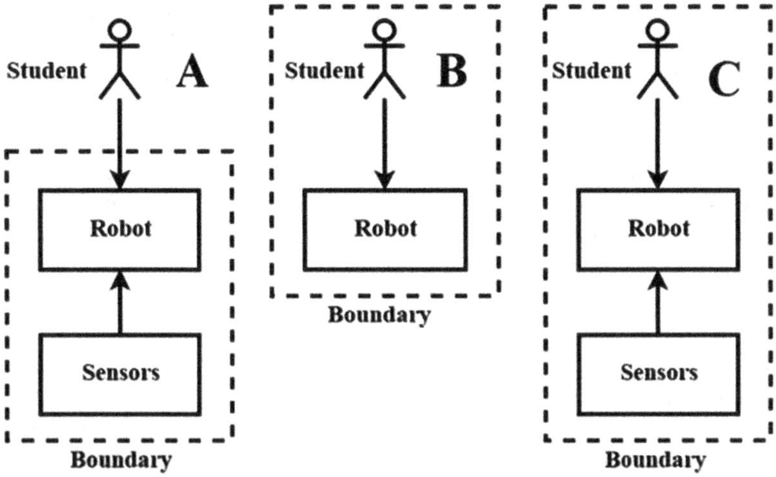

Fig. 1.4 Robot, student and sensor learning systems

(Catlin 2016a). Blamires and I believe constructionist theories best account for the use of education robots. But we recognise psychological theories view data from different standpoints. For example, instead of Piaget's cognitive ideas, you could look at how robots work from Gibson's perceptual development theory (Gibson 1969). If a particular theory helps create a better learning scenario, then it's worthwhile considering it.

The University of Hertfordshire developed the social robot Kaspar[9] to help autistic children. The robot reacts to children and the child gradually learns to adjust the way they respond to the robot. This helps them improve their social skills. You could understand this using Bandura's social learning theory which examines how we develop by copying the actions and behaviours of people around us. Sick, bedridden children have used a robot called Pebbles[10] to go to school for them. They control it and it enables them to take part in lessons which improves their morale and hastens their recovery. Although research is continuing, I believe these cases will comply with Papert's Paradigm. However, a new breed of education robot is starting to appear which doesn't.

Breaking Papert's Paradigm

'Robots will begin replacing teachers in the classroom within the next 10 years as part of a revolution in one-to-one learning, a leading educationalist has predicted. Sir Anthony Seldon, Vice-Chancellor of the University of Buckingham, said

[9] Taxonomy – Social Robot: human-like

[10] Taxonomy – Social Robot: telepresence

intelligent machines that adapt to suit the learning styles of individual children will soon render traditional academic teaching all but redundant. The former Master of Wellington College said programmes currently being developed in Silicon Valley will learn to read the brains and facial expressions of pupils, adapting the method of communication to what works best for them' (Bodkin 2017).

This wasn't news. A group from Carnegie Mellon University worked with Honda's Asimo robot, training it to tell stories. 'Engaging storytelling is a necessary skill for humanoid robots if they are to be used in education and entertainment applications. Storytelling requires that the humanoid robot be aware of its audience and able to direct its gaze in a natural way. In this paper, we explore how human gaze can be modelled and implemented on a humanoid robot to create a natural, human-like behaviour for storytelling' (Mutlu et al. 2006).

This paper appeared 11 years before Seldon's prediction. Then a school would need to pay $150,000 per month to hire Asimo. Now you can buy RoboThespian[11] for $75,000. This is a British robot, but you can find people developing this type of technology all over the world. And they're becoming more powerful and accomplished. Scientific American list '*Bots that Argue and Instruct*' (Meyrson 2018, p. 26) as one of the top ten emerging technologies of 2018. The article more or less claims the technology capable of passing the Turing Test![12] The article finishes with, 'The intelligent systems will be useful only for assembling existing knowledge, not for creating it… Still, as machines become more intelligent they raise the spectre of job losses. It behoves society to provide the next generation with the skills it needs to tackle problems that require human ingenuity to solve'.

We can challenge this last statement. Astronauts aboard the International Space Station had to shut down their robot CIMON[13] when it began to behave like Hal 9000 from 2001 Space Odyssey (Johnson 2018). In another incident Facebook shut down robots Bob and Alice when they stopped speaking English and started talking in a more efficient language they invented (Kenna 2017). Clearly, we can assign these glitches to teething problems, but they show the ingenuity of such machines and our inability to predict undesired outcomes.

How should robots with these skills work in the classroom? I'm sure many schoolchildren would enjoy pulling faces and doing their creative best to fool Mr. Robot Teacher. On a more serious note, this is the computer programming children. I think these developments are unavoidable – but their desirability is dubious. What's happening here revives debates between Papert and Patrick Suppes (who supported computer-aided instruction – CAI) except we now have an anthropomorphised computer on wheels or legs (Papert 1993, pp. 162–164) (Johnstone 2003, pp. 93–94). We need a strategy to manage this technology, and I suggest Papert's ideas is the place to start.

---

[11] Taxonomy – Social Robot: humanoid

[12] A test of machine intelligence. If a human can't distinguish the machine from another human by the replies to questions put to both, the machine is intelligent.

[13] Crew Interactive Mobile Companion

Fast Forward to the Future

Machine Learning and Human Robot Interaction

ERA anticipated machine learning (intelligence principle) and advances in human-robot (HRI) and human-computer interactions (HCI) (interaction principle). Robots like the latest Roamer[14] allow you to change their 'intelligence' to support specific activities. This isn't simply changing the programme, but how the robot works. You build the robot Cubelets[15] by assembling its modules in different arrangements (Fig. 1.5). How you do this decides the robot's behaviour and 'programmes' it. This is a modern version of the ideas expressed in David Miller's Scarecrow robot (Miller et al. 2006). Scarecrow represented an electromechanical robot – it didn't have a computer. Its mechanical arrangement determined what it did.

Programming is one way of interacting with a robot, but ERA imagined different ways for children to interact with education robots, for example, tangible computing – explained by Tangible Bits (Ishii and Ullmer 1997). The focus on coding distorts what we mean when we say 'programming'. It's clear the popular programme Scratch[16] (a descendant of Logo) is a coding system. You can interact with other new robots like Ozobot[17] by drawing lines: although the manufacturer calls it coding, I think interacting is a more accurate term.

Fig. 1.5 Cubelets. (Courtesy of Modular Robotics)

[14] Taxonomy – User Bot: Turtle

[15] Taxonomy – Build Bot: Build system; modular parts

[16] Characteristic tag: Block-based program

[17] Taxonomy – User Bot: Turtle

We live in a time experiencing rapid and significant advances in the technologies affecting the 'ERA interaction and intelligence principles'. This includes the Internet of things (IoT) which will enrich learning environments. We need careful research to develop robots with machine learning and HRI abilities and stay faithful to Papert's Paradigm.

Education, Policy, Schools and Teachers

It's my experience that a robot activity can work well with one teacher and be a failure with another. It appears from our earlier review that success and failure issues apply to school, researchers, teachers, school administrators and political policies.

Papert wasn't a fan of logical positivism. He reflected when he first met gears, 'If any "scientific" education psychologist had tried to "measure" the effects of this encounter, he probably would have failed. It had profound consequences but, I conjecture, only many years later. A "pre- and post" test at the age of two would have missed them' (Papert 1980, p. viii). A logical positivist would dismiss this experience as anecdotal and demand randomised control trials. I'm not against this sort of study, but I don't accept they occupy the pinnacle of quality research: we need to treat all research respectfully (Catlin and Blamires 2010a, b). However, our strategy here is not investigating why something didn't work, but why it worked.

We can't do better than starting with Costa Rica. In the 'ERA practical principle' we examine the conditions that foster systemic change. Success needs five conditions:

1. Vision (without it we get confused)
2. Participant buy-in (without it we resist change)
3. Participant skills (without appropriate skills we become fearful)
4. Resources (without resources we get frustrated)
5. Action plan (without a plan we dither)

An examination of the Costa Rican effort sees they met all these conditions. The projects' director Clotilde Fonseca explains the details of their project in a must-read paper which serves as a model for the application of Papert's ideas (Fonseca 2001). She explains their approach to evaluation. 'It is extremely useful to establish strategies and methodologies that facilitate the monitoring and formative evaluation... Unfortunately, daily practice reveals a lack of suitable methodologies for measuring the impact of technology and applying qualitative monitoring' (Fonseca 2001, p. 12).

Those who insist on quantitative assessments may hastily dismiss this approach. The Costa Rican government appointed IBM and Seymour Papert to set up the project. The head of their Latin American Education Group, Alejandrina Fernandez) summarised the impact of the effort (Fernandez n.d.):

- It inspired 13 Latin American to launch similar projects, seven of which evolved into larger national projects.
- One of the subprojects (Genesis) changed the way over two and a half million students in over 1000 schools learned and how they thought about themselves and their potential.
- It trained and transformed how over 12,000 K-12 teachers taught.
- The Costa Rican government officials credit the project for helping transform Costa Rica from an agricultural country to a technological one.
- Costa Rican leaders claim that the Costa Rican software industry emerged as a result of the exposure of millions of children to a new way of learning.

According to Johnstone the first 10 years of the project resulted in a fourfold increase in students studying computer science and persuaded Intel to locate its first Latin American assembly and testing plant in the country (Johnstone 2003, p. 132).

Costa Rica's success did attract high-ranking political visitors eager to learn from their approach. However, their reaction didn't always embrace the spirit of the enterprise. For example, the Minister of Finance of a Latin American country gave his views about the project. 'I find this programme of yours wonderful. It is a real asset. You install machines. The children learn. Teachers are dispensed with. The State payroll shrinks. And best of all you do away with strikes altogether' (Fonseca 2001, p. 2).

We need to laud the Costa Rican efforts, particularly their attitude to teacher training. I've sold many robots over the last 35 years. And I find it shocking how many schools, school districts and even whole countries fail to understand the importance of teacher training or the effort needed to do it properly. It's not simply a matter of learning how to programme the robot – it's about how to organise the use of robots and teach a constructionist lesson.

Going on a course or several courses is not enough, particularly if, as is the case with some countries climbing on to the coding bandwagon, the normal teaching style is drill and practice. Costa Rica committed to constructionism before selecting any technology. The issue with teacher training is normally economic and teachers' reluctance to change how they teach. This isn't a matter of obstinacy: teaching is both a science and intrinsic skill (Carr 2003). As we discussed programmers gain 'intuition' through practice. Irrespective of what talent a teacher starts with, they must climb Gordon's skill ladder shown in Fig. 1.6. I've found integrating, training (online) and research into activities will support teacher development (Catlin et al. 2015). This sets up a just-in-time, on-the-job training approach which gradually improves the teacher's knowledge and skills as they teach.

Robots and the Curriculum

Jim Howe of Edinburgh University did develop an alternative structured way of using Logo but Beryl Maxwell suggested a balanced mixture of structure and exploration (Howe and Maxwell 1983, p. Video 4 3:02–4:07). The law demands teachers

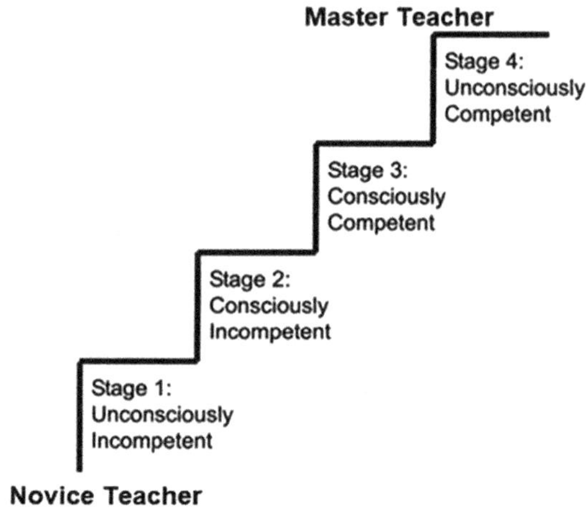

Fig. 1.6 Gordon's skill ladder

work to the curriculum. Maxwell's balanced approach satisfies the 'practical and the curriculum and assessment ERA principles'.

What you can do with robots varies according to its taxonomy. The 'pedagogy principle' identified 29 characteristics of education robot activities (Catlin 2016a, b). For example:

'Exploration We use the robot to explore and discover the knowledge hidden in a Microworld. This exercise adds excitement to primary school history lessons. For example, Roamer is an Archaeologist and it starts to explore an Ancient Roman Site. Pupils programme Roamer to explore the site. They discover artefacts and patterns that tell them whether the site was a marketplace, a barracks or a Roman bathhouse' (Catlin 2016a, b, p. 8).

This provoked a whimsical response from one reviewer who said: 'There were better ways of doing the activity: the robot served no purpose and you might as well use an electric toothbrush or a stone'.[18] We did a small test and asked a class of children which they preferred to play with (see Fig. 1.7).

I've two reasons for mentioning this. First, according to the United Nations, about 1.2 billion children go to school every day and in a year teachers cover every school topic. We can consider the number of lessons using robots as a percentage equates to zero. Somewhere, we can also assume, a teacher has found a fantastic way of presenting a topic. However, it's not a matter of finding those 'killer activities' and getting everyone to use them. It's a matter of finding what works for you [teacher] and your students. Robots based on Papert's Paradigm used by experienced teachers have a high chance of working.

A teacher devised the *Exploration* activity for a history lesson on the Romans. She lived near Lindum Colonia [Lincoln], the site of a legionary fortress where

[18] I reviewed a paper for the same conference. It focussed on using Lego and exploring poetry and it got the same sort of response from one of the other reviewers.

Fig. 1.7 Children's choice: do you want to play with the robot, toothbrush or stone. (Courtesy of Valiant Technology)

archaeologists were working on local digs and the children had watched a popular television programme called 'Time Team'. She could justify this activity because of the 'engagement and personalisation ERA principles'. It also gave students the chance to consolidate some maths and coding skills – 'curriculum and assessment principle'. Papert's Paradigm once again 'kicks in': all teachers know if you capture the students' interest they will learn.[19] The moral of this story is it's never about the technology (unless your subject is robotics), it's about the learning and what works for those involved.

In general, education authorities have gradually shoved schools closer to the 'school' Papert hated. The obsessive focus on high-stakes testing, league tables and

---

[19] The children who chose the stone and toothbrush were happy to play with the robot, but had specific interest in finding out how the toothbrush worked and writing a story about a magic stone.

accountability oppresses teachers and warps how we organise schoolwork. We need the courage of our convictions and back Papert's Paradigm: it works. This means discovering robot activities that connect to the curriculum and transform the way children learn. It also means finding opportunities to use robots throughout the school day.

Making Effective Use of Robots in Schools

Between September 1987 and December 1989, the National Council for Education Technology (NCET) ran a Turtle project in 21 school districts in the UK (Mills et al. 1989). As a result the use of *programmable toys* appeared in the first national curriculum for mathematics, ICT and design technology in England and Wales and has remained in all later revisions.

Papert talked about people 'doing Logo' but not entering into 'the spirit of Logo' (Papert 1999, pp. vi–vii). Valiant Service Desk often received calls from some schools saying: 'We've got Roamer scheduled into our teaching plan this month… or we have an Ofsted[20] inspection and we need to show pupils using Roamer'. These questions show schools 'doing' enough to meet curriculum demands but not embracing the spirit of using 'programmable toys'. Many of these teachers found it difficult to justify using a robot. Sometimes because they didn't know how, or the technology intimidated them, using it didn't match their teaching style or they were too set in their ways to even try. But in other cases, they simply didn't see the opportunities.

Opportunities to Use Robots

Everyday Lessons

You can use robots in everyday lessons. Even with one robot, you can engage a whole class. For example, in the robot rally activity (Fig. 1.8), students programme the robot to travel the course as fast as possible. They choose a route across different terrains. Working in pairs the whole class test the speed of the robot along the road, over the mountain or through the forest. Each team is responsible for timing the trials and using their data to chart a route. The class correlate the routes and then test them. Children do some scientific experiments, use mathematics and the data collected to decide a course and then test their idea.

You can find challenges to cover all subjects. Like Hollywood stars, robots can play a lead role, but they're also good playing cameo parts. I've used robots just to

[20]Office for Standards in Education – a government quango who inspect and report on school performance

Fig. 1.8 Robot rally race. (Courtesy of Valiant Technology Ltd)

introduce a topic or to get student's thinking and talking. We need to remind ourselves that the core of Papert's Paradigm is encouraging children to think, to get them to fall in love with learning. If that's all the robot does, it has done a good job.

The clarion call for STEM loudly drowns out a desperate need to enrich the cultural and moral life of our communities. An MIT experiment in Buenos Aries aimed to engage Jewish parents and their children to come together in a reflective way during the Jewish High Holidays. They aimed to build and programme Lego robots to explore values and identity (Bers and Urrea 2000, pp. 194–217). Another example, Roamer, signs the European Pledge to Peace. One peace event led a class of 30 children in Oldham to programme 6 robots to draw a line from one classmate to another. They sat in a circle and sent the robot to a 'person who has the same colour hair' or 'supported the same football team'. Their teacher proclaimed, 'They learnt more about each other in one hour than they had in the previous six months'. The task resulted in a network of lines representing what they had in common – not their differences. It's surprising how many arch enemies became friends when they did that (Catlin 2014a, b, c).

I don't hold with the opinion that arts are creative and somehow science, computing, engineering and mathematics aren't. Robots belong to both worlds and often combine the two. A mathematics and Islamic art project demonstrated this idea (Catlin 2018b). Other examples include art (Clayson 1988), robot performing arts (movies) (Catlin 2010) and dancing robots (Catlin 2017).

Prior Knowledge

You might find it useful to introduce a new topic with activities that engage students with the subjects' key ideas. Bransford and colleagues introduce the importance of prior knowledge and experience (Bransford et al. 2000, pp. 10–12). And of course Papert endorses the approach: 'By getting to know these Turtles as they get to know a person these children are learning to be mathematicians. This is Piaget's real message, knowledge built on experience' (Papert 1983, pp. Video 2 – 0:04–0:16).

Inspired by the work of Professors John Paulos (Paulos 1998) and Kieran Egan (Egan 1989), I ran a test in a high school with a group of average-ability students. Their teacher wanted to prepare them for a study of movement. He gave them a list of keywords about motion, like velocity, speed and distance, and asked them to write stories including the keywords. They had to programme the Roamer robot to act the story and then present it to their classmates. The exercise highlighted what students knew and what they misunderstood about the words they used in conversation. Their teacher thought this made them focus on the subject and helped him to know what confusions he had to correct.

It also revealed how we'd brainwashed students into views that limited their appreciation of the maths, their thinking and their creativity. 'We're meant to be doing maths. This is English: not maths!' Little wonder people don't enjoy the beauty and power of mathematics. We've locked them into such narrow-minded perspectives of what it is they fail to see how it surrounds us all. The more advanced our robots[21] become the more elegant and sophisticated we can make these activities.

Classroom Strategies

You can use User Bots: Turtles for revision by engaging students in tasks that help them with factual information and skills in a different context. The Biggest Number is an example which has proved successful in various situations (Catlin 2013; Hudson 2017). The students programme the robot to find a route from start to finish. They can only enter a square once and then use the operator and next number to amass a score. In Fig. 1.9 they can go from 12 + 4–2 and so on. The task tests and reinforces their grasp of arithmetic and arithmetic operations (including inverse operations): 759,942 is the highest to date.

If your normal teaching method hasn't helped the children understand a concept, then you could try a relevant robot task. A different approach often helps. Some teachers set up a special robot corner so a small group of children can work with a robot while the rest of the class do other activities. Others organise pull-out sessions and use the help of a teaching assistant. Both these methods help teachers to work with minimal equipment. Better equipped classrooms work with groups of five children with one robot.

[21] The 'ERA intelligence and interaction principle' predict we'll develop more natural interactions with robots.

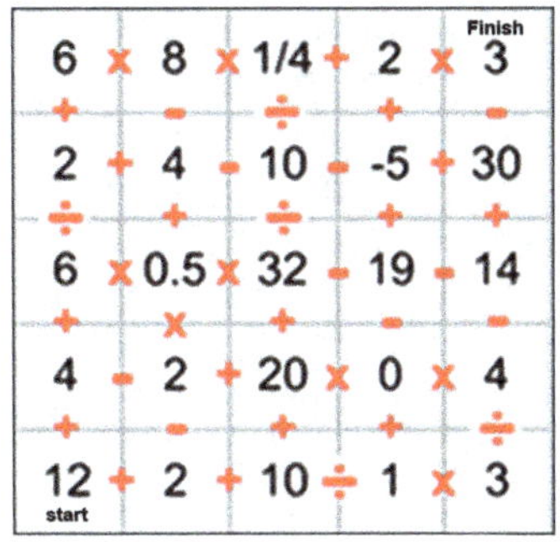

Fig. 1.9 Biggest number mat. You can change the mathematics in the puzzle to suit your students. (Courtesy of Valiant Technology)

Events

Teachers often use robots, especially Build Bots like Lego, in after-school clubs or they attend special events like Lego First League or Big Bang experiences. Robots used in special events always create enthusiasm and energy. What happens when the pupils return to schools – a return to the humdrum? Better planning harnesses the energy and makes sure it supports the curriculum. Some teachers find ways to set up events within their school. They often fit this into busy classroom schedules through cultural events and link it to the curriculum as prior knowledge or revision opportunities

Cultural Events

Some education robots have neutral designs, allowing the student to give them a personality that reflects their culture. The robot becomes a tool of culture and supports student's self-expression. Obvious cases include a project with Roamer Maori people which started in New Zealand and repeated by a small Squaxin tribe in Seattle (Catlin et al. 2012). The Squaxins used the robot in a summer camp during the canoe and potlatch event.[22] Students programmed Roamer to animate their traditional stories, simulate the canoe journey and perform traditional dances. Before the summer camp, the tribal elders approved the programme thinking it was a STEM project. When the children[23] started to approach them to find out how to do the dances and weave blankets to dress their Roamer up, they realised it was a cultural project.

[22] West Coast Native American used ocean-going canoes to travel up and down the US Pacific Coast. Celebrating this tradition is now an annual adult-only event. Native Americans revived potlatch festivals (banned in the nineteenth century) which traditionally brought tribes together to share wealth, news, food, music and dance.

[23] Children attend the summer camp voluntarily and those involved were of all ages and abilities.

The 'ERA equity and personalisation principles' provide more understanding of how robots grounded in Papert's Paradigm engage children no matter their circumstance. While we normally cite gender, ethnicity and cultural factors, we shouldn't forget poverty (Catlin and Robertson 2012). Again we cite the Costa Rican project. 'Perhaps one of the more valuable contributions that the Computers in Elementary Education Program has made to the international community is showing that it is possible to obtain significant results from introducing new technological and educational opportunities to children and teachers from deprived communities' (Fonesca 1999, p. 13).

You can also use the energy and cultural interests aroused by festivals from the Rio Carnival to Chinese New Year: or major international events like the Olympics (Catlin 2012b).

Robots Versus Computers

People, heeding Wing's call for coding to touch all subjects, have found ways to, for example, engage children with Newton's laws of motion. Some teaching resources simply substitute the traditional CAI student-computer interactions with students writing trivial bits of code. Scratch shows its Logo ancestry by challenging children to code simulations to explain the laws. Simulation plays an important role in modern sciences, like astrophysics and cosmology. But robots live in the concrete world and provide tangible experiences of Newton's laws giving students the chance to form and test theories. For example, in the World Cup penalty shoot-out, pupils programme a robot to run at and hit a ball into the goal (Catlin 2018a). At normal speed, the robot will not score. How do you make sure it does? Better follow through? Hit the ball harder – how do you do that – why does it work? The problem throws up many relevant questions needing concrete, testable answers.

Assessment

Teachers around the world bemoan assessment and high-stakes testing. 'When they feel they are successful they feel it is despite the undermining opposition, they feel they triumph despite obstacles … But the sense of being undermined comes about because they feel their voice is ignored. The curriculum and its assessment is thrust upon them from outside. The sense of personal involvement and freedom to be inspired are constrained' (Cullingford 1997, p. 266). I believe the political drive for accountability has significantly worsened since Cullingford wrote those words.

Although subject to many pressures and sometimes politically motivated interference, teachers still, more or less, control what goes on in their classroom. We need to capture Trevor Thomson's expertise. And, more importantly, we need a way

Table 1.4 Components of assessment for learning

Component	Meaning
Learning intentions	The student's perspective on what they're learning
Success criteria	How students recognise when they've succeeded
Peer assessment	Students interacting, discussing, challenging and criticising each other's efforts
Quality feedback	Ways for teachers to help students overcome difficulties and guides improvements to their understanding

of helping teachers improve their skills and climb on to the top step of Gordon's skill ladder. Assessment for learning (AfL), properly used, is a set of heuristics that can support these ambitions. More importantly, AfL provides the ideal way of managing lessons with education robots (Catlin 2012a).

Black and Wiliam's seminal paper, Inside the Black Box, outlines a formative assessment approach to managing lessons (Black and Wiliam 1990) (see Table 1.4).

Effectively, AfL codifies good teaching practice, and you'll find when used properly, it reflects Papert's Paradigm. A review of the report of the 1989 NCET Turtle project referred to earlier shows a natural correlation with AfL ideas and the use of education robots. AfL is a formative assessment method; it's a set of techniques to help teachers understand and manage the dynamics of a lesson and maximise student learning. Unfortunately, bureaucrats are increasingly trying to control the classroom and the way teachers teach. Wiliam cites the example of teachers forced to write learning 'intention' on the blackboard and even making children copy it into their exercise books. Teachers know this is nonsense (Wiliam 2011, p. 56).

A learning intention is, '*What the children think they're learning*'. You should make sure this is the children telling you: not simply students paraphrasing your lesson objectives. I argue you can only set up learning intentions once you've engaged the pupils in the lesson (Catlin 2016a, b). Dylan Wiliam discusses many exceptions to the bureaucratic constraints insisting AfL guidelines become 'must follow rules' (Wiliam 2012).

Conclusion

Jeannette Wing's coding revolution coincided and perhaps promoted a surge in new education robots. However, her rationale encouraging schools to take up coding and its promised education and economic rewards lack long-term conviction. It's true there's an immediate need for more programmers. But it's hard to believe this is a long-term issue: advances in machine learning technologies will resolve the problem. While her idea about computational thinking has merit, it's not something we should impose on children. Instead, we should encourage its natural development through experience while allowing children to cultivate their own styles. We need to remember that computational thinking isn't the only professional thinking method that works, and it's an emergent process that comes from experience.

This doesn't mean we should abandon the idea of children learning to code, but it changes the reason we're doing it and so how we organise it and assess its benefits.

We can see sophisticated robots taking their first baby steps towards becoming teachers. What they do now is mechanical and predictable. However, we can't rule out developments in AI leading to robots that compete for jobs with human teachers. We've seen how some politicians see teachers as part of the education problem – not the solution. They'd like to get rid teachers as far as possible and no doubt would like to replace them with robots. I recognise this is an extreme, even fanciful viewpoint, but it does indicate a tendency among some people. If we adopt this approach, computers will teach children, which runs counter to the constructionist way we learn.

I presented two examples where constructionism lay at the heart of school education. One in the classroom (Trevor Thomson) and national programme (Costa Rica). Both of these embraced Papert's ideas by showing how we can achieve a balance between developing student talents and meeting the education expectations of society. They prove both the worth and practicality of Papert's Paradigm.

Robots can and should play a part in such education as teaching aides and not as teacher replacements. I've outlined some ways teachers can incorporate their use in everyday schooling. The question developers must tackle is how to include powerful machine learning technology into education robots compliant with Papert's Paradigm. This will still keep children in charge of their own learning while giving them the experiences needed to grow their talents and gain knowledge. Students will still programme robots, but they'll also interact with them in more natural ways offered by HRI.

References

Ableson, H., & diSessa, A. A. (1981). *Turtle geometry*. Cambridge, Massacheusetts: London: MIT Press.

Akermann, E. K. (n.d.). *Piaget's constructivism, papert's constructionism: What's the difference*. (MIT Media Lab) Retrieved November 15, 2018, from Future Learning Group: http://learning.media.mit.edu/content/publications/EA.Piaget%20_%20Papert.pdf

Angulo, C., Canas, A. J., Castro, A. G., Munoz, L., & Zamora, N. (2018). Think, create and program: Evolving to a K-9 nationwide computational thinking program in Costa Rica. In V. Dagienė, & E. Jasutė, *Constructionism 2018*, (pp. 515–520). Vilnius. Retrieved October 25, 2018, from http://www.constructionism2018.fsf.vu.lt/file/repository/Proceeding_2018_Constructionism.pdf

Bers, M. U., & Urrea, C. (2000). Technological prayers: Parents and children exploring robotic values. In A. Druin, J. Hendler, A. Druin, & J. Hendler (Eds.), *Robots for kids* (pp. 194–217). San Diego, London: Academic Press.

Black, P., & Wiliam, D. (1990). *Inside the black box: Raising standards through classroom assessment*. GL assessment Limited. Retrieved November 30, 2018, from https://weaeducation.typepad.co.uk/files/blackbox-1.pdf

Bodkin, H. (2017). 'Inspirational' robots to begin replacing teachers within 10 years. *Telegraph online*. Retrieved November 15, 2017, from https://www.telegraph.co.uk/science/2017/09/11/inspirational-robots-begin-replacing-teachers-within-10-years/

Bradly, K. (2017). *Creative industries record contribution to UK economy.* Department for Digital, Cultural, Media and Sport. UK Government. Retrieved December 7, 2018, from https://www.gov.uk/government/news/creative-industries-record-contribution-to-uk-economy

Brand, S. (1972). *Spacewar*. Retrieved November 8, 2018, from Rolling Stone: http://wheels.org/spacewar/stone/rolling_stone.html

Bransford, J. D., Brown, A. L., & Cocking, R. R. (Eds.). (2000). *How people learn: Brain, mind, experience and school* (Vol. Expanded Edition). Washington, DC: National Academy Press.

Carr, D. (2003). *Making sense of education*. London and New York: Routledge Falmer, Taylor & Francis Group.

Catlin, D. (2010). Robotic performing arts project. *Constructionism 2010.* Paris. Retrieved December 6, 2018, from http://research.roamer-educational-robot.com/2010/08/16/robotics-performing-arts-project/

Catlin, D. (2012a). Maximising the effectiveness of education robotics through the use of assessment for learning methodologies. *TR-TWR conference.* Riva del Garda, Italy. Retrieved November 30, 2012, from http://research.roamer-educational-robot.com/2012/04/20/maximising-the-effectiveness-of-educational-robotics-through-the-use-of-assessment-for-learning-methodologies/

Catlin, D. (2012b). *Roamer olympics*. Retrieved November 22, 2012, from Tumblr: http://roamer-robot.tumblr.com/post/27185616533/roamer-olympics

Catlin, D. (2013). *The big bang experience*. Retrieved November 22, 2018, from Tumblr: http://roamerrobot.tumblr.com/post/45871836724/roamer-a-big-bang-experience

Catlin, D. (2014a). *Roamer makes friends*. Retrieved November 29, 2018, from Tumblr: http://roamerrobot.tumblr.com/post/158920999823/roamer-makes-friends

Catlin, D. (2014b). Using peer assessment with education robots. *1st international conference on peer review, peer assessment and self assessment.* Tallinn, Estonia. Retrieved December 1, 2018, from http://research.roamer-educational-robot.com/2014/08/21/using-peer-assessment-with-educational-robots/

Catlin, D. (2014c). Using peer assessment with educational robots. *1st International Conference on Peer Review, Peer Assessment and Self Assessment.* Tallinn, Estonia: PRASAE. Retrieved November 30, 2018, from http://research.roamer-educational-robot.com/2014/08/21/using-peer-assessment-with-educational-robots/

Catlin, D. (2016a). 29 effective ways you can use robots in the classroom: An explanation of ERA pedagogical principle. *Edurobotics 2016.* Athens. Retrieved November 15, 2018, from http://research.roamer-educational-robot.com/2016/12/23/29-effective-ways-you-can-use-robots-in-the-classroom/#more-368

Catlin, D. (2016b). Learning intentions and educational robots. *Constructionism 2016.* Bangkok. Retrieved November 30, 2018, from http://research.roamer-educational-robot.com/2016/02/05/learning-intentions-and-educational-robots/

Catlin, D. (2017). 5 things that made roamer a terrific dancer: How student's loved roamer's creative coding challenge. *GO Magazine*. Retrieved December 6, 2018, from http://go.roamer-educational-robot.com/2017/10/16/5-things-that-make-roamer-a-terrific-dancer

Catlin, D. (2018a). *How did world cup fever win in essex.* Retrieved November 29, 2018, from GO Magazine: http://go.roamer-educational-robot.com/2018/07/19/how-did-world-cup-fever-win-in-essex/

Catlin, D. (2018b). How to engage 14 year-old students with mathematics: Using roamer to explore islamic art. *GO Magazine*. Retrieved December 6, 2018, from http://go.roamer-educational-robot.com/2018/05/18/how-to-engage-14-year-old-students-with-mathematics/

Catlin, D., & Blamires, M. (2010a). The e-robot project: A longitudinal on-line research collaboration to investigate ERA principles. *Simpar (TRTWR).* Darmstadt. Retrieved November 22, 2018, from http://research.roamer-educational-robot.com/2010/11/15/the-e-robot-project/

Catlin, D., & Blamires, M. (2010b). The principles of educational robotic applications (ERA):A framework for understanding and developing educational robots and their activities. *Constructionism 2010.* Paris: Proceedings of Constructionism 2010. Retrieved May 15, 2018, from goo.gl/N7z84k.

Catlin, D., & Blamires, M. (2018). Designing robots for special needs education. In L. Daniela & M. D. Lytras (Eds.), *Technology, knowledge and learning special issue on education robotics for inclusive education*. Netherlands: Springer. https://doi.org/10.1007/s10758-018-9378-8.

Catlin, D., & Robertson, S. (2012). Using educational robots to enhance the performance of minority Students. *TR-TWR*. Riva del Garda, Italy. Retrieved November 22, 2018, from http://research.roamer-educational-robot.com/2012/04/12/74/Using_Educational_Robots_to_Enahnce_the_Performance_of_Minority_Students

Catlin, D., & Thomson, T. (1998). *Fleet circus of imagination*. Retrieved November 20, 2018, from GO Magazine: http://go.roamer-educational-robot.com/1998/11/21/fleet-circus-of-imagination/

Catlin, D., & Woollard, J. (2014). Educational robots and computational thinking. *Teaching Robotics Teaching with Robotics (TRTWR) and Robotics in Education (RIE) 2014 Conference, Padova, Italy Teaching Robotics Teaching with Robotics (TRTWR) – Robotics in Education (RIE) 2014 Conference, Padova, Italy, Italy. 8 pp.* Padova, Italy. Retrieved October 29, 2018, from http://research.roamer-educational-robot.com/2014/07/22/educational-robots-and-computational-thinking/

Catlin, D., Smith, J. L., & Morrison, K. (2012). Using educational robots as tools of cultural expression: A report on projects with indigenous communities. *3rd International Conference Robots in Education*. Prague. Retrieved November 29, 2018, from http://research.roamer-educational-robot.com/2012/09/13/using-educational-robots-as-tools-of-cultural-expression/

Catlin, D., Csizmadia, A. P., Omera, J., & Younie, S. (2015). Using education robotic research to transform the classroom: Establishing a robotics community of evidenced-based practice using MESH guides and the TACTICS framework. *6th International Conference on Robots in Education*. Yverdon-les-Bains. Retrieved November 22, 2018, from http://research.roamer-educational-robot.com/2015/05/21/using-educational-robotics-research-to-transform-the-classroom/

Catlin, D., Kandlhofer, M., & Holmquist, S. (2018a). EduRobot taxonomy a provisional schema for classifying educational robots. *9th International Conference Robotics in Education*. Qwara, Malta. Retrieved November 12, 2018, from https://www.researchgate.net/publication/324786505_EduRobot_Taxonomy_A_Provisional_Schema_for_Classifying_Educational_Robots

Catlin, D., Kandlhofer, M., Holmquist, S., CsIzmadia, A. P., Angel-Fernandez J., & Cabibihan, J. (2018b). Edurobot taxonomy and Papert's paradigm. In V. Dagienė, & E. Jasutė (Ed.), *Constructionism 2018: Constructionism, computational thinking and innovation*, (pp. 151–159). Vilnius, Lithuania. Retrieved November 12, 2018, from http://www.constructionism2018.fsf.vu.lt/file/repository/Proceeding_2018_Constructionism.pdf

Clayson, J. (1988). *Visual Modeling with logo*. Cambridge, MA: MIT Press.

CodeWeek EU. (2018). Retrieved October 25, 2018, from CodeWeek EU: https://codeweek.eu/about/

Computing Programmes of Study. (2013). *Statutory guidance: National curriculum in England*. Department of Education. Retrieved October 25, 2018, from https://www.gov.uk/government/publications/national-curriculum-in-england-computing-programmes-of-study/national-curriculum-in-england-computing-programmes-of-study

Cooper, D. E. (2001). Plato. In J. A. Palmer (Ed.), *Fifty major thinkers on education* (pp. 10–14). London and New York: Routledge.

Crevier, D. (1993). *AI: The tumultuous history of the search for artificial intelligence*. New York: Basic Books.

Cribben, J. (2013). *Computing with quantum cats: From Alan turing to teleportation* (Black Swan Edition 2014 ed.). London: Bantum Press.

Cullingford, C. (1997). Conclusion: The personal effects of assessment. In C. Cullingford (Ed.), *Assessment versus evaluation*. London: Cassell.

Dennett, D. C. (1993) Review of papert, the children's machine. *New Scientist*, pp. 45–46. Retrieved November 16, 2018, from https://ase.tufts.edu/cogstud/dennett/papers/papert.htm

Egan, K. (1989). *Teaching as story telling*. Chicago: Chicago University Press.

Fernandez, A. (n.d.). *Masterpiece: Education technology industry: Projects genisis and quorum.* Retrieved December 22, 2018, from Breakthrough Design Lab: http://www.breakthrough-designlab.com/new-page-2/

Fonesca, C. (1999). The Computer in Costa Rica: A new door to educational and social opportunities. In *Logo philosophy and implementation* (pp. 3–21). LCSI. Retrieved November 16, 2018, from http://www.fod.ac.cr/aplicacionesFOD/ARCHIVOS/RECURSOS/thecomputerin-costarica.pdf

Fonseca, C. (2001). Fallacies and objectives regarding the uses of new technologies in education. *New Technologies in Education: Prospects,* XXXI(3), 399–413. Retrieved December 22, 2018, from http://www.fod.ac.cr/aplicacionesFOD/ARCHIVOS/RECURSOS/fallaciesandob-jectivesregardingtheusesofnewtechnologiesineducation.pdf

Furber, S. (2012). *Shut down or restart? The way forward for computing in UK schools.* London: Royal Society. Retrieved October 24, 2018, from https://royalsociety.org/~/media/education/computing-in-schools/2012-01-12-computing-in-schools.pdf

Garvin, J. (2016). *Software developers worry they will be replaced by AI.* Retrieved October 28, 2018, from https://evansdata.com/press/viewRelease.php?pressID=231

Gibson, E. J. (1969). *The principles of perceptual learning and development.* New York: Prentice Hall.

Hayes, B. (2014). Programming yur quantum computer. *American Scientist, 102*(1), 22. https://doi.org/10.1511/2014.106.22.

Hemmendinger, D. (2010). A plea for modesty. *Inroads, 1*(2), 4–7.

Hobson, P. (2001). Aristotle. In J. A. Palmer (Ed.), *Fifty major thinkers on education* (pp. 14–19). London and New York: Routledge.

Howe, J., & Maxwell, B. (1983). Talking turtle. *Horizon*, Part 1–05:30 to 05:55. BBC and the Open University. Retrieved November 5, 2018, from http://research.roamer-educational-robot.com/1983/12/14/talking-turtle/

Hoyles, C., & Sutherland, R. (1989). *Logo mathematics in the classroom* (Revised ed.). London and New York: Routledge.

Hudson, K. R. (2017). *How did roamer do at amazing robot exhibition.* Retrieved November 30, 2018, from GO Magazine: http://go.roamer-educational-robot.com/2017/11/14/how-did-roamer-do-at-amazing-robot-exhibition/

Hutchinson, D. S. (1995). Ethics. In J. Barnes (Ed.), *The Cambridge companion to aristotle* (pp. 195–232). Cambridge: Cambridge University Press.

Ishii, H., & Ullmer, B. (1997). Tangible bits: Towards seamless interfaces between people, bits and atoms. *Conference on Human Factors in Computing Systems.* Atlanta, GA. Retrieved November 22, 2018, from http://alumni.media.mit.edu/~ullmer/papers/tangible_bits/

Johnson, A. (2018). *'Don't be so mean': CIMON the European space robot cops an attitude.* Retrieved December 22, 2018, from NBC News: Innovation: https://www.nbcnews.com/tech/innovation/don-t-be-so-mean-cimon-european-space-robot-cops-n943336

Johnstone, B. (2003). *Never mind the laptops.* New York: IUniverse Inc.

Kenna, S. (2017). *Facebook shuts down AI robot after it creates its own language.* Retrieved December 22, 2018, from Huff Post: Innovation: https://www.huffingtonpost.com.au/2017/08/02/facebook-shuts-down-ai-robot-after-it-creates-its-own-language_a_23058978/?guccounter=1&guce_referrer_us=aHR0cHM6Ly93d3cuZ29vZ2xlLmNvbS8&guce_referrer_cs=wDg16blW8AFS0-SOeZT0ow

Klahr, D., & Carver, S. M. (1988). Cognitive objectives in a LOGO debugging curriculum: Instruction learning and transfer. *Cognitive Psychology,* 20(3), 362–404. Retrieved November 11, 2018, from http://psycnet.apa.org/doi/10.1016/0010-0285(88)90004-7

Korzonek, T. (2016). *Is AI on its way to disrupting software development?* Retrieved October 28, 2018, from https://blog.codebeat.co/: https://blog.codebeat.co/is-ai-on-its-way-to-disrupting-software-development-e7933fc0a75

Kuhn, T. S. (1996). *The structure of scientific revolutions* (3rd ed.). Chicago: University of Chicago Press.

Lave, J., & Wenger, E. (1991). *Situated learning: Legitimate peripheral participation*. Cambridge, England: Cambridge University Press.

Li, H., Cabibihan, J., & Tan, Y. K. (2011). Towards an effective design of social robots. *International Journal of Social Robots, 3*, 333–335.

Luria, A. R. (1976). In M. Cole (Ed.), *Cognitive development: Its cultural and social foundations*. Caambridge Massachusetts and London: Harvard University Press.

Mayer, R. E. (2004). Should there be a three-strike rule against pure discovery learning. *American Psychologist, 59*(1), 14–19. https://doi.org/10.1037/0003-066X.59.1.14.

McCorduck, P. (2004). *Machines who think*. Natick, MA: A.K. Peters.

Meyrson, B. S. (2018). Bots that argue and instruct. *Scientic American*. Scientific American, New York.

Miller, D. P., Milstein, J. Q., & Stein, C. (2006). *Scarecrow: If I only had AI*. Retrieved December 1, 2018, from Smantic Scholar: https://pdfs.semanticscholar.org/dc0b/216f5a99a84d57ed8f7303b1240b57644b18.pdf

Mills, R., Staines, J., & Tabbeerer, R. (1989). *Turtling without tears*. Coventry, UK: National Council for Educational Technology.

Mubeen, J. (2017). *Seymour papert dreamed of a learning revolution -Why hasn't it happened?* Retrieved November 16, 2018, from Medium: https://medium.com/@fjmubeen/seymour-papert-dreamed-of-a-learning-revolution-why-hasnt-it-happened-b280f542e4ff

Mutlu, B., Forlizzi, J., & Hodgins, J. (2006). A storytelling robot: Modeling and evaluation of. In *6th IEEE-RAS International Conference on Humanoid Robots Human-like Gaze Behaviour*. Genoa, Italy: IEEE.

Ng, D. (2017). *Where seymour papert got it wrong*. Retrieved November 16, 2018, from Medium.Com: https://medium.com/vertical-learning/where-seymour-papert-got-it-wrong-6203f94149d1

Noss, R. (1995). Thematic chapter: Computers as commodities. In A. DiSessa, C. Hoyles, & R. Noss (Eds.), *Computers and exploratory learning* (pp. 363–382). Berlin: Springer Verlag.

Papert, S. (1969). Programming and problem solving: The logo programming language. In W. Feurzeig, *Computer systems for teaching complex concepts: Final report* (Final Report ed., Vol. Report 1742, pp. 94–108). Cambridge, MA, USA: Bolt, Beranek and Newman. Retrieved 11 7, 2018, from http://research.roamer-educational-robot.com/1969/03/10/programming-and-problem-solving-the-logo-programming-language/

Papert, S. (1980). *Mindstorms: Children, computers and powerful ideas*. New York: Basic Books.

Papert, S. (1983). Talking turtle. *Horizon*, Part 1–02:48 to 03:10. BBC and the Open University. Retrieved November 5, 2018, from http://research.roamer-educational-robot.com/1983/12/14/talking-turtle/

Papert, S. (1993). *The children's machine*. New York: Basic Books.

Papert, S. (1999). Introduction. In *Logo philosophy and implementation* (pp. v–xvi). LCSI. Retrieved November 16, 2018, from http://www.fod.ac.cr/aplicacionesFOD/ARCHIVOS/RECURSOS/thecomputerincostarica.pdf

Papert, S. (2005). Teaching children thinking. *CITE Journal: Contemporary Issues in Technology and Teacher Education,* 5(3). Retrieved November 9, 2018, from https://www.citejournal.org/volume-5/issue-3-05/seminal-articles/teaching-children-thinking/

Papert, S., & Harel, I. (1991). Situating constructionism. New York: Ablex Publishing. Retrieved from http://web.media.mit.edu/~calla/web_comunidad/Reading-En/situating_constructionism.pdf

Parnas, D. L. (1985). Software aspects of strategic defense systems. *Communications of the ACM,* 28(12), 1326–1335. Retrieved October 27, 2018, from https://web.stanford.edu/class/cs99r/readings/parnas1.pdf

Paterson, M. (1969). *LOGO ambulatory executor – Turtle robot*. Scientific American. Retrieved May 1, 2018, from goo.gl/VK2DMv.

Paulos, J. A. (1998). *Once upon a number*. New York: Basic Books.

Pea, R. D. (1984). Symbol systems and thinking skills: Logo in context. In *Pre-proceedings of the 1984 National Logo Conference* (pp. 55–66). Cambridge, MA: Laboratory of Computer Science, Massachusettes Institute of Technology.
Pea, R. D., & Kurland, M. (1984). On the cognitive effects of learning computer programming. *New Ideas in Psychology,* 2, 138. Retrieved from On the Cognitiive Effects of Learning Computer Programming.
Plowden, B. (1967). *Children and their primary schools.* Her Majesty's Stationary Office. Retrieved Nov 10, 2017, from http://www.educationengland.org.uk/documents/plowden/plowden1967-1.html
Polya, G. (1990). *How to solve it* (New Edition ed.). London and New York: Penguin Books.
Price, B. D. (n.d.). *The con in constructivism.* Retrieved November 24, 2018, from Improve Education: http://www.improve-education.org/id55.html
Raucci, R. (1999). *Personal robotics.* Natick, MA: A K Peters.
Resnick, M. (2016). *Mitch resnick: Designing for wide walls.* Retrieved November 6, 2018, from Design Blog: https://design.blog/2016/08/25/mitchel-resnick-designing-for-wide-walls/
Rice University. (2018). Retrieved from Pliny: http://pliny.rice.edu/
Southampton University. (2013). Computational thinking: The developing definition. Southampton. Retrieved October 29, 2018, from http://eprints.soton.ac.uk/356481
Stewart I. (1990). Foreword G. Polya, *How to solve it* (New Edition ed.). London and New York: Penguin Books.
Stonier, T. (1983). *The wealth of information: A profile of the post-industrial economy.* London: Thames Methuen.
Turing, A. M. (1947). *Lecture to the London mathematical society.* London: London Mathematical Society. Retrieved October 27, 2018, from https://www.vordenker.de/downloads/turing-vorlesung.pdf
Turkle, S. (2005). *The second self: Computers and the human spirit.* (20th Anniversary Edition ed.). MIT Press, Cambridge, MA.
Vanderbilt, C. A. (1996). Looking at technology in context: A framework for understanding technology in context. In D. C. Berliner & R. C. Calfee (Eds.), *The handbook for educational psychology* (pp. 807–840). New York: Simon and Schuster-MacMillan.
Waterson, A. (2015). *Lego mindstorms: A history of educational robots.* Retrieved November 22, 2018, from Hack Education: http://hackeducation.com/2015/04/10/mindstorms
Weir, S. (1987). *Cultivating minds: A logo case book.* New York: Harper Row.
Wiliam, D. (2011). *Embedded formative assessment.* Bloomington, ID: Solution Tree.
Wiliam, D. (2012). *Unpacking formative assessment.* (Northwest Evaluation Association) Retrieved November 30, 2018, from You Tube: https://www.youtube.com/watch?v=kPf0nQFfv50
Wing, J. M. (2006). Computational thinking. *Communications of the ACM, 49*(3), 33–35.

Chapter 2
Educational Robotics for Reducing Early School Leaving from the Perspective of Sustainable Education

Linda Daniela and Raimonds Strods

Abstract Early School Leaving (ESL) is a problem for many countries and some have pledged to reduce the number of children leaving school early to below 10% by 2020. Between October 2015 and September 2017, Italy, Greece and Latvia implemented an Erasmus+ project that used robotics to reduce the risk of ESL. The effectiveness of the teaching and learning materials developed during the project and the pedagogical strategies used were examined in groups at high-risk of ESL and in the work of the teachers participating in the project. In this paper, the use of robotics to reduce the risks of early school leaving is analysed from the perspective of sustainable education. Mixed methods were used to evaluate the project, and several tools were developed to gather qualitative and quantitative data. Preliminary evaluation of the project was based on action research principles.

It was concluded that the use of robotics enhanced the motivation to learn in students at high-risk of ESL and encouraged them to construct knowledge actively and independently, thus reducing their risk of ESL and in the long-term ensuring the 4th SDG was reached, particularly sustainable education. Analysis of teachers' responses also supported the conclusion that the use of robotics improved the students' attitude towards learning, motivation and ensured active participation in the learning process.

Keywords Educational robotics · Robotics-based learning · Sustainable education Early school leaving

L. Daniela (✉)
Faculty of Education, Psychology, & Art, University of Latvia, Rīga, Latvia
e-mail: linda.daniela@lu.lv

R. Strods
University of Latvia, Rīga, Latvia

L. Daniela (ed.), *Smart Learning with Educational Robotics*,
https://doi.org/10.1007/978-3-030-19913-5_2

Introduction

Many countries of the world are trying to reduce rates of Early School Leaving (ESL). A low population level of education has a negative effect on a country's competitiveness because of the waste of human capital, but it also has a social impact, which in turn influences the different aspects of sustainable development. People with little education may be unable to provide for themselves economically and require social assistance, thus creating a burden for the economy and potentially endangering the safety of other people, etc. People with low levels of education may not be able to take responsible decisions about their families, they are not ready for innovations and so on. In 2015, countries agreed on Sustainable Development Goals (SDG), where the 4th goal is devoted to quality education for everyone; thus, we can understand that activities aimed to reduce risks of early school leaving ensures that everyone can realise his/her potential. However, educational robotics can support reaching not only educational objectives but also can support reaching other goals of sustainable development, for example, support reducing the gender gap in the field of ICT, support innovations to improve the industry, help to develop sustainable cities and societies and so on (United Nations Development Programme 2015). The focus of this paper is on outcomes of the use of educational robotics for students at risk of ESL, but these outcomes will be analysed trough the lenses of SDGs. The sustainable development goal for education in broad understanding is perceived as education that is inclusive and available for everyone without borders or other restrictions, such as gender, educational availability and so on. In this paper, we look at this goal from another perspective — how to ensure that children stay in education and reach higher objectives — and we believe that this perspective is also important for reaching sustainable development goals.

Rates of ESL in the countries involved in the project were rather high during the preparation phase: in 2014 they were 9% in Greece, 15% in Italy and 8.5% in Latvia. Hence, the aim of the project was to reduce rates of ESL through use of educational robotics (ER) (Eurostat 2016).

The risk factors for ESL are often complex and combined with other risk factors (Bhowmik 2017; Daniela et al. 2014; Nevala et al. 2011). Family-related risk factors include lack of social and emotional support, but other risk factors are related to the teaching and learning process or the student's special needs (health issues, learning disabilities, etc.) (Melkevik et al. 2016; Downes 2016). Risk of ESL is also increased by lack of emotional attachment to one's school, feeling rejected by the school and unsupported by the teachers. Fredrick and colleagues concluded that school attachment was influenced by factors in three broad categories: behaviour, i.e. factors relating to involvement in learning and social activities; emotion, i.e. relationships with teachers and other students and attitude towards school in general; cognition, i.e. factors relating to one's willingness and readiness to devote one's energy and intellectual capacity to performing learning tasks of varying complexity (Fredricks et al. 2004). ESL risk is also related to motivation to learn and overcome difficulties and hence can be related to problems with the teaching process, such as neglect of students' individual needs or use of lessons and activities that are perceived as boring or not relevant to real life. There are also many factors outside the school environment that can cause ESL: the ease with which information can be accessed today and the availability of opportunities to participate in different activities in or out of the school environment can give students the impression that time spent at school is lost time.

The fast tempo of contemporary life has created a paradoxical situation: on the one hand, the need for educated, creative and innovative people has increased, yet at the same time young people are looking more critically at the potential opportunities opened up by education. The ever-increasing pace of technological progress and urbanisation mean there is an urgent need for people who are able to think creatively, to solve problems, to make prompt decisions and take responsibility for their actions. This poses a challenge to educational systems on many levels, because it is necessary to ensure that the education process is interesting and exciting for students and also equips students with the skills required to take responsibility for constructing their knowledge. Boring teaching and learning processes are often mentioned as the cause of the ESL problem, but they represent only one facet of ESL risk. If the emphasis is on ensuring that lessons and activities are always interesting and exciting, it encourages students to focus their attention on what is new, attractive and exciting and makes the education system hostage to the need for materials and activities that are new and interesting and reduces students' attention span. We have reached the point where learners have become *external experts* assessing whether the teaching and learning process is interesting enough. Teachers have become service providers and have to take responsibility for ensuring that the learning process is interesting yet also equips students with the skills to analyse information critically and to take a broad view of areas of knowledge rather than simply considering units of information individually. We do not deny the importance of making teaching and learning interesting, but it should be taken into account that making this the sole guiding principle of pedagogy promotes the development of a short attention span, because students' attention is continuously shifting to the next interesting stimulus. This phenomenon affects learners' ability to memorise information in order to analyse it and it also affects the development of metacognitive processes, which influence learners' ability to analyse and synthesise acquired knowledge to generate new levels of competence. Hence, pedagogical science must seek new ways of facilitating students' learning, assessing the knowledge and skills they have acquired and preparing them to collaborate with processes taking place in the urban world. It also brings the necessity for innovative pedagogies to a new level (Daniela 2018) and shows the necessity to analyse sustainable education not only from the perspective of how to ensure that everyone can access education but also how to ensure that students are not leaving the education system. The next generation must be equipped to be responsible participants in the teaching and learning process and heed should be paid to Papert's argument that knowledge should be acquired in a complex way and fragmentation should be avoided (Papert 1980; Daniela et al. 2017; Alimisis 2014).

Project participants tried to take into consideration both the need to make the teaching and learning process interesting and exciting and the need to allow students to acquire new knowledge through *hands-on* activities. The latter is necessary if learners are to link the exciting robotics activities with other knowledge that is acquired in the compulsory teaching and learning process. The aim was to make the learning process more interesting and show students that it is possible to learn differently and acquire knowledge through practical activities. LEGO Mindstorm robots were the tool chosen to realise this aim. Although the idea of using robotics in the teaching and learning process has been around since 2006 (Eguchi 2014), in the majority of cases these principles of *hands-on* activities by using robotics are used in the acquisition of mathematics, sciences and engineering knowledge (Benitti and Spolaōr 2017; Alimisis 2013). During the past decades some studies have appeared on the use of robotics with students who have special needs (Tweddle 2008; Lund and Marti 2005), and there are still few studies examining the use of educational robotics as a means of reducing risk of ESL (Moro et al. 2018; Alimisis 2014; Karampinis 2018; Karkazis et al. 2018).

There have also been studies where it is concluded that students of different ages and both genders should be able to achieve the same level of Computational Thinking (CT) skills, but development of CT is affected by the time devoted to relevant activities, students level of proximal development and attitude towards the tasks that need to be done and by gender. The influence of gender can, however, be overcome by use of appropriate teaching and learning methods (Atmatzidou and Demetriadis 2016). The evaluation of this project yielded a similar conclusion, namely that girls are equally capable of developing their digital competence if they are supported (Daniela et al. 2017).

The aim of this study was to determine whether the activities designed to reduce the risk of ESL that were developed as part of our Erasmus+ robotics project do, in fact, reduce the risk of ESL in students who are at high-risk, thereby reaching the 4th goal of SDG (United Nations Development Programme 2015).

Didactical Model of Activities

It was decided that work with students at high-risk of ESL would take the form of after-school, extracurricular activities in participating schools. Ten programmes of progressively increasing difficulty were developed by the project experts to guide students' learning. The programmes were delivered by teachers who volunteered to participate in the project (from a variety of subjects, including mathematics, physics, information and communication technology (ICT), home economics, philosophy and English); we did not seek teachers of specific school subjects. Before they started to deliver the new programmes, teachers participated in a number of training activities provided by project experts.

The ESL risk evaluation tools developed as part of the project (Daniela 2016) were used to select students at high-risk of ESL. These students were offered the opportunity to participate in extracurricular activities and learn to work with LEGO Mindstorm robots. Separating high-risk groups from the rest of a population and organising special activities for them risks causing social exclusion (Midgley 2000; Migdley and Urdan 2001; Daniela et al. 2014); nevertheless, this was the approach we adopted because robotics activities are not part of the standard curriculum. Therefore, it was necessary to verify the developed teaching/learning curricula organising extracurricular activities, which, on the one hand, created several risks — even greater marginalisation of ESL students, possible unwillingness of these students to participate in additional activities that require cognitive effort and the necessity to stay for longer at school, etc. Yet, on the other hand, a positive effect was anticipated if the teaching/learning was organised in that way because students subjected to ESL were collaborating with students who also had poor social links with their classmates and formed new social links that facilitated their willingness to learn and to cooperate with others. Secondly, students' knowledge and academic achievement was not assessed summatively; we opted to use formative assessment instead in order to reduce the possible stress associated with being assessed. It was decided to use ER to reduce risk of ESL and to introduce the activities outside compulsory lessons so that the target group (students at high-risk of ESL) could work without worrying about the learning speed or academic achievement of other students and would not be competing with students who were highly motivated to learn and had faster cognitive processes. This might have been off-putting for the target participants, who had lower self-esteem, slower cognitive processes and lower motivation to learn, developed avoidance motivation and so on.

Classes with ER were organised in two rounds (Moro et al. 2018). During the first round, students tried out five programmes, working alongside their teachers. Students' achievements and the teaching and learning materials were evaluated after each round. A new group of students at high-risk of ESL was involved in the second round and worked with 10 ER programmes. In both the rounds, the programmes were delivered over a 3-month period and ESL risk was assessed before and after the programmes to find out the positive outcomes from ER activities.

Delivery of the programmes was based on *hands-on* learning principles and, to a certain extent, smart learning principles (students had to use LEGO software to program the robots). Samra and colleagues reported that *hands-on* learning provides immediate feedback and allows students to choose the learning content and assess their own achievements, thus promoting willingness to participate (Samra et al. 2017). Using the LEGO Mindstorms robots also enabled us to apply pedagogical techniques, such as peer learning and collaborative learning — students worked in groups in order to teach the robot to do thing; active learning — students programmed the robots and then tested whether their code worked; blended learning strategies — students had to use the e-environment as well as asking the teachers for support and assistance.

Methodology

Students at risk of ESL were selected to participate in the project by using methodology developed as part of the project (Daniela 2016).

Several original instruments were developed to evaluate the results of the extracurricular programmes and detailed information on progress evaluation instruments is given in Chap. 10.

- A questionnaire evaluating risk of ESL. This was completed by teachers before students started participating in the extracurricular ER activities and was used only to select the students of target group. This article does not present detailed data on the risk of ESL.
- A student evaluation questionnaire. This was completed by the students who participated in the programmes and by their regular teachers. The questionnaire captured data on several educational environment-related ESL risk variables.
- A questionnaire consisting of open questions about the delivery and outcomes of the ER programmes. This was completed by the participating teachers and included questions on the programmes and students' participation and engagement.

Two rounds of programme activities were organised, in accordance with action research principles (Orland-Barak and Becher 2011; Corey 1954). In the first round, the educational programmes were delivered to students at high-risk of ESL who had been involved in the process of developing the programmes. The teachers' evaluations of the results of the first round were then used to improve the materials for the second round. In the second round, the improved programmes were delivered to a new group of students at high-risk of ESL.

Results

Risk of ESL is associated with low socio-economic status, low motivation to learn, learning difficulties and special educational needs. In our analyses we did not distinguish between groups on the basis of ESL risk or special needs.

We compared students' opinions of learning motivation, attitude towards learning, perception about educational robotics and teachers' opinions of students learning motivation, attitude towards learning, and behaviour in school before and after the ER programmes. The average age of participating students was 15 years, but the range was 11–19 years, which created some implementation challenges which will be explained in the chapter. The results from the first round are based on data from 62 students and the results of the second round are based on data from 80 students.

Initially, we summarised and compared students' answers to questions about participation in project activities in the first and second rounds; these results are summarised in Table 2.1. There were mean changes in the positive direction in all the

Table 2.1 Students' opinions of the impact of the activities in the first and second rounds

	First round		Second round		
	N	Mean	*N*	Mean	Difference
Learning by using robots was exciting	62	4.13	80	4.40	0.27
I have learned how to program robots	62	3.52	80	3.89	0.37
I liked working in groups on assignments with the robots	62	3.68	80	4.34	0.66
I liked doing calculations while programming	62	2.98	80	3.50	0.52
I can use this knowledge in other activities	62	2.97	80	3.72	0.75
I liked solving programming problems by myself	62	2.79	80	3.44	0.65
I liked that others helped me to solve programming problems	62	3.68	80	4.05	0.37
I liked looking for the extra information needed to use robots	62	2.77	80	3.48	0.71
Improved knowledge in Maths	62	2.81	80	3.21	0.4
Improved knowledge in Physics	62	2.55	80	3.05	0.5
Improved knowledge in ICT	62	3.66	80	4.06	0.4
Improved attitude towards learning	62	3.24	80	3.66	0.42
Improved cooperation with classmates	62	3.69	80	4.10	0.41
Improved cooperation with teachers	62	3.50	80	4.25	0.75
	Average first round	3.28	Average second round	3.8	

criteria included in the questionnaire. In the first round, the mean score for all the variables was 3.28 (all the variables were evaluated by using the Likert scale from 1 to 5, where 1 was *completely disagree* and 5 was *completely agree*), but in the second round the mean was 3.8, indicating that the programmes had a positive impact. The greatest changes relate to the following items: "I can use this knowledge in other activities" ($\Delta M = 0.75$); "cooperation with teachers" ($\Delta M = 0.75$); "I liked looking for the extra information needed to use the robots" ($\Delta M = 0.71$); "I liked working in groups on assignments with the robots" ($\Delta M = 0.66$). Not all these changes were statistically significant, but it should be remembered that the results relate to 3-month programmes. The results indicate that the second round of programmes had a greater impact than the first round of programmes. There are several possible explanations for this, e.g. the participating teachers became better at working with the target group. The students in both rounds indicated that the programmes had an impact on their ICT knowledge. The greatest increase in positive direction is in the variable about the knowledge in physics, where the mean increase is by 0.5 points.

Table 2.2 Factor analysis: first round. Rotated component matrix[a]: first round

Activities with robotics		Digital and real world interaction		Positive learning environment		Stem knowledge	
0.774	I liked working in groups on assignments with the robots	0.808	Cooperation with teachers	0.760	Learning by using robots was exciting	0.846	Knowledge of Physics
0.768	I have learned how to program robots	0.764	Knowledge of ICT	0.663	I liked working in groups on assignments with the robots	0.782	Knowledge of Maths
0.707	I liked looking for the extra information needed to use the robots	0.666	Cooperation with classmates	0.658	I liked that others helped me to solve programming problems		
0.648	I liked solving programming problems by myself	0.618	Attitude towards learning	0.620	I can use this knowledge in other activities		

Extraction Method: Principal Component Analysis
Rotation Method: Varimax with Kaiser Normalisation
[a]Rotation converged in six iterations

Next, separate factor analyses of data from the first and second rounds were performed, where for Extraction Method: Principal Component Analysis was used; for Rotation Method: Varimax with Kaiser Normalisation. Results of the first round were split into 4 factors (see Table 2.2). The most expressive dimension was named "Activities with robotics" because all the variables were those indicating that students enjoyed such activities. This indicated that, when using the didactic model developed in the project involving Lego Mindstorms robots in the teaching process of students subjected to the ESL risk, the highest results were in the robotics activities. The second most important factor was "Digital and real world interaction", followed by "Positive learning environment" and finally "Science, Technology, Engineering and Mathematics (STEM) knowledge". These results confirm that purposeful use of ER affects not only students' knowledge about programming robots and willingness to learn in a self-directed way but also their perceived knowledge of physics and mathematics, which in turn influences their self-assessment and can have a favourable impact on their motivation to learn.

The data from the second round were also subjected to factor analysis (see Table 2.3), and three factors emerged: "Synergy between attitude and digital knowledge", "Knowledge improvements" and "Relationships relevant to the learning process". Comparison of the factor analyses of the first and second rounds revealed that factors obtained after the second round had become reciprocally complementary, because, for example, knowledge of mathematics and physics no longer constituted a separate factor but contributed to the Knowledge improvement factor.

Table 2.3 Factor analysis: Second round. Rotated component matrix[a]: Second round

Synergy between attitude and digital knowledge		Knowledge improvement		Relationships relevant to the learning process	
0.810	I liked working in groups on assignments with the robots	0.820	I liked solving programming problems by myself	0.868	I liked that others helped me to solve programming problems
0.762	Learning by using robots was exciting	0.741	I liked looking for the extra information needed to use the robots	0.625	Cooperation with teachers
0.754	Cooperation with classmates	0.653	I can use this knowledge in other activities		
0.679	Attitude towards learning	0.611	I liked working in groups on assignments with the robots		
0.475	Knowledge of ICT	0.573	I have learned how to program robots		
		0.528	Knowledge of Physics		
		0.491	Knowledge of Maths		

Extraction Method: Principal Component Analysis
Rotation Method: Varimax with Kaiser Normalisation
[a]Rotation converged in four iterations

The data shows that the developed model of working with students at risk of ESL not only supports development of knowledge about robotics and programming, which affects knowledge of mathematics, physics and ICT, but also influences student–teacher cooperation, the development of a positive attitude towards learning and self-directed learning. Hence, we can infer that students' risk of ESL decreased as a result of their participation in the programmes.

The next step in the analysis was the calculation of correlations using Spearman's rho, which was chosen because Kolmogorov–Smirnov tests indicated that the data distributions were non-parametric. Correlations were calculated separately for the first and second rounds and only those results proving high correlation are presented in Table 2.4; results having no mutual correlation between variables were not included in the table. Correlations in the first round are shown in white and correlations in the second round in grey.

The obtained results correspond to the results of the factor analysis and the correlations between variables were higher in the second round, indicating that the ER-based programmes were more effective in the second round. In the first round, the highest correlations between variables were:

- "I liked working in groups on assignments with the robots" and "Improved cooperation with classmates": −0.569;
- "Improved knowledge in ICT" and "Improved cooperation with classmates": −0.563;

Table 2.4 Correlations

Spearman's rho	Round	Learning by using robots was exciting	I have learned how to program robots	I liked working groups on assignments with the robots	I liked doing calculations while programming	I can use this knowledge in other activities	I liked solving programming problems by myself	I liked that others helped me to solve programming problems	I liked looking for the extra information needed to use the robots	Improved knowledge in Math	Improved knowledge in Physics	Improved knowledge in ICT	Improved attitude to learning	Improved cooperation with classmates	Improved cooperation with teachers
Learning by using robots was exciting	1st	1				0.339**									
	2nd	1	0.487**	0.494**	0.549**	0.365**				0.335**	0.402**	0.341**	0.575**	0.396**	0.363**
I have learned how to program robots	1st		1		0.479**				0.527**						
	2nd	0.487**	1	0.521**	0.560**	0.656**	0.332**	0.310**	0.465**	0.400**	0.392**	0.496**	0.527**	0.382**	0.356**
I liked working groups on assignments with the robots	1st			1	0.348**	0.335**						0.439**		0.569**	
	2nd	0.494**	0.521**	1	0.463**	0.350**			0.331**		0.300**	0.306**	0.435**	0.550**	
I liked doing calculations while programming	1st		0.479**	0.348**	1		0.394**		0.528**				0.392**		
	2nd	0.549**	0.560**	0.463**	1	0.482**	0.432**		0.498**	0.414**	0.538**	0.422**	0.585**	0.416**	0.318**
I can use this knowledge in other activities	1st	0.339**		0.335**		1		0.374**		0.338**					
	2nd	0.365**	0.656**	0.350**	0.482**	1	0.492**	0.289**	0.450**	0.409**	0.338**	0.560**	0.521**	0.403**	0.352**
I liked solving programming problems by myself	1st				0.394**		1								
	2nd		0.332**		0.432**	0.492**	1		0.507**	0.316**	0.392**	0.375**	0.361**		
I liked that others helped me to solve programming problems	1st					0.373**		1							
	2nd		0.310**			0.289**		1				0.313**			0.336**
I liked looking for the extra information needed to use the robots	1st		0.527**						1						
	2nd		0.465**	0.331**	0.498**	0.450**	0.507**		1	0.391**	0.433**	0.345**	0.412**	0.338**	0.352**
Improved knowledge of Maths	1st					0.338**				1	0.540**	0.373**	0.393**	0.431**	
	2nd	0.335**	0.400**		0.414**	0.409**	0.316**		0.391**	1	0.483**	0.420**	0.548**	0.388**	0.342**
Improved knowledge of Physics	1st									0.540**	1		0.370**		
	2nd	0.402**	0.392**	0.300**	0.538**	0.338**	0.392**		0.433**	0.483**	1	0.519**	0.528**	0.405**	0.344**
Improved knowledge of ICT	1st			0.439**						0.373**		1	0.476**	0.563**	0.381**
	2nd	0.341**	0.496**	0.306**	0.422**	0.560**	0.375**	0.313**	0.345**	0.420**	0.519**	1	0.523**	0.508**	0.537**
Improved attitude to learning	1st								0.385**	0.393**	0.370**	0.476**	1	0.501**	0.476**
	2nd	0.575**	0.527**	0.435**	0.586**	0.521**	0.361**		0.412**	0.548**	0.528**	0.523**	1	0.680**	0.612**
Improved cooperation with classmates	1st			0.569**						0.431**		0.563**	0.501**	1	0.500**
	2nd	0.396**	0.382**	0.550**	0.416**	0.403**		0.302**	0.338**	0.388**	0.405**	0.508**	0.680**	1	0.701**
Improved cooperation with teachers	1st											0.381**	0.476**	0.500**	1
	2nd	0.363**	0.356**		0.318**	0.352**		0.336**	0.352**	0.342**	0.344**	0.537**	0.612**	0.701**	1

** Correlation is significant at the 0.01 level (two-tailed)

- "Improved knowledge of Physics" and "Improved knowledge of Maths": −0.540.

In the second round, the highest correlations between variables were:

- "Improved cooperation with teachers" and "Improved cooperation with classmates": −0.701
- "Improved attitude towards learning" and "Improved cooperation with classmates": −0.680
- "I have learned how to program robots" and "I can use this knowledge in other activities": −0.656.

In the first round, the variable "Improved attitude towards learning" was correlated with the greatest number of other variables (six) and four variables were correlated with only one other variable. In the second round, there were three variables that were correlated with all 13 other variables. The greatest increase in correlations applied to the variable "I have learned how to program robots", which was correlated with 2 variables in the first round and 13 in the second, indicating that students who participated in the second round had learned how to apply what they had learned from using robotics to other areas and aspects of learning.

The results also confirmed that students' self-directed learning had developed considerably, because the variable "I liked looking for the extra information needed to use the robots" was correlated with only 1 variable ("I have learned how to program robots") in the first round, but with 11 in the second. The same conclusion is confirmed by the change in the correlations of the variable "I can use this knowledge in other activities", which was correlated with 4 variables in the first round and 13 in the second.

These data also confirm that the programmes delivered greater benefits in the second round than the first. This indicates that the main factor in the greater efficacy of the programmes in the second round was an increase in teacher competence, as different students were involved in rounds one and two, whereas the teachers did not change. As the teachers gained experience with the programmes, the programmes became more effective in reducing the risk of ESL.

The next step in the analysis was to compare teachers' opinions of the impact of the programmes in the first and second rounds; these results are summarised in Table 2.5. After the first round, 203 teachers expressed an opinion about the impact of the programmes on participating students, and after the second round, this figure increased to 278 teachers. As already noted, the teacher's evaluation questionnaire was completed by teachers who worked with the participating students during compulsory school lessons. A part of the criteria included in the evaluation questionnaire had been formulated so that positive tendencies were observed if teachers had chosen higher variables on the scale from 1 to 5; other criteria, in their turn, were inversely proportional and positive tendencies appeared if indicators were lower. These criteria (which should be read inversely) in the table are indicated with grey background. Common changes in mean indicators were from 2.88 in the criteria, where a higher assessment level was valued positively (the criteria had to be evaluated on the scale from 1 to 5, where 1 was for *never* and 5 for *always*), and in the

Table 2.5 Teachers' opinion after activities: First and second rounds

	1st round		2nd round		difference
	N	Mean	N	Mean	
Preparation of homework assignments	203	2.70	278	3.58	0.88
Positive cooperation with teachers	203	3.07	278	3.92	0.85
Positive cooperation with classmates	203	2.96	278	3.71	0.75
Readiness for work in lessons	203	2.76	278	3.68	0.92
Understanding of the connection between learning and achievements	203	2.85	278	3.68	0.83
Readiness to do extra assignments to improve achievements	203	2.74	278	3.46	0.72
Following of behavioural rules in the classroom	203	3.05	278	4.16	1.11
Readiness participate in out-of-class or -school activities with classmates	203	2.82	278	3.79	0.97
Readiness to join activities led by other classmates	203	2.77	278	3.58	0.81
Motivation to learn the subject you teach	203	2.86	278	3.73	0.87
Motivation to understand his/her mistakes and correct them	203	2.90	278	3.70	0.8
Motivation to improve achievements	203	2.80	278	3.64	0.84
Motivation to overcome difficulties in learning	203	2.81	278	3.62	0.81
Readiness to work hard to achieve an aim	203	2.70	278	3.55	0.85
Being late for lessons	203	2.25	278	1.62	0.63
Problematic behaviour during recess (break)	203	2.22	278	1.53	0.69
Aggression towards other students	203	2.28	278	1.42	0.86
Aggression towards teachers	203	2.26	278	1.21	1.05
Using rude language with classmates	203	2.19	278	1.49	0.7
Using rude language with teachers	203	2.18	278	1.22	0.96
Refusing to do assignments during lessons	203	2.26	278	1.59	0.67
Aggression during conflicts	203	2.30	278	1.58	0.72
Solves learning problems by himself/herself	203	2.93	278	3.50	0.57
Asks for help from teachers	203	2.87	278	3.30	0.43
Resolves conflict calmly	203	3.37	278	4.20	0.83
Readiness to reach learning aims	203	2.87	278	3.45	0.58

second round, it was 3.68. When assessing the second part of the criteria, improvements of situations were assessed positively if teachers had assessed them with a lower value on the scale from 1 to 5. The mean variable in these criteria in the first round was 2.24 and 1.46 in the second round. Overall, the results indicate that teachers perceived that the programmes had a positive impact on students which are at risk of ESL, and it confirmed that such activities can support reaching the 4th SDG from the perspective that students are not dropping out from education.

Teachers perceived that the after-school programmes' greatest impact was on behaviour, thus confirming that the programmes increased students' involvement in the learning process and improved their attitude towards learning. Teachers were more sceptical about changes in students' attitude after the first round (Daniela 2016; Daniela et al. 2017), but the results of the second round confirmed that teachers perceived positive changes in students' behaviour and attitudes following their participation in the after-school programmes. There were significant changes in mean scores for the following variables "Following of the behavioural rules in the classroom" ($\Delta M = 1.11$); "Readiness to participate in out-of-class or -school activities with other classmates" ($\Delta M = 0.97$); "Aggressiveness to teachers" ($\Delta M = -1.05$); "Using rude language with teachers" ($\Delta M = -0.96$) (note that in the case of the latter two variables a decrease in score indicates an improvement in student performance).

These data confirm that in the opinion of their teachers, students' ability to follow instructions improved markedly as a result of the programmes. This was probably connected to changes in students' motivation, because the use of LEGO robots made the teaching and learning process more interesting and in consequence students performed set tasks more willingly. These changes for the students subjected to ESL risks are vital, as this group of students is usually passive and does not participate in extracurricular activities. Developing new friendships and strengthening existing friendships through participation in the after-school ER programmes may also have increased the target group's motivation to stay in education. One can conclude from these results that the ER programmes improved students' behaviour and attitude towards their teachers. Empirical research has shown that students at high-risk of ESL often have a negative attitude towards teachers. We have shown that the techniques and content of the ER programmes used in this project reduced the risk of ESL, because they improved students' attitude. The scores teachers gave for the item "Is ready for work in lessons" (post-1st round $M = 2.76$; post-2nd round $M = 3.68$; $\Delta = 0.92$) confirm that in general students were more motivated to participate in education after being involved in the after-school ER programmes.

Problem-solving was the area in which teachers perceived the least improvement in students' performance, this can be seen from score for the following items: "Asks for help from teachers" (post-1st round $M = 2.87$; post-2nd round $M = 3.30$; $\Delta = 0.43$) and "Solves the learning problems by himself/herself" (post-1st round $M = 2.93$; post-2nd round $M = 3.50$; $\Delta = 0.57$). We can conclude from these results that students involved in the project used various evasion strategies instead of solving their problems. Teachers' assessments indicated that the item "Is ready to do extra assignments to improve achievements" (post-1st round $M = 2.74$; post-2nd round $M = 3.46$; $\Delta = 0.72$) saw the third smallest improvement of all the items measured.

We also analysed teachers' responses to a specially developed questionnaire about the ER programmes. The questionnaire contained questions about the respondent's country, the school in which they were working and which of the ER programmes they had used, and the questionnaire was anonymous, allowing teachers to give an honest assessment of their work in the project. These were followed with open-ended questions, e.g. "Please briefly describe your students' interest in activities provided during all Robotics lessons"; "Which activities did they like the most?"; "Which activities did they not like?"; "Which activities were challenging for them? How did you deal with this?"; "What worked well? Why did it work well?"; "What did not work well? How did you cope with this?"; "Did you observe some signals which showed that students' motivation to learn rose during the project? What kind of signals can you name?"; "What kind of difficulties did you face regarding your students during the Project?"; "What kind of difficulties did you face with regard to cooperation among students during the project?"

Teachers gave positive responses that indicate the strengths of the project, for example:

Students' interest increased continuously… The vast majority did more work than expected, solved simple problems and in some cases displayed an innovative approach.

Interest was generally high. In the case of several students this was quite remarkable because they showed very little interest in normal school activities.

The majority of our activities worked well. The preparation helped a lot. One reason that the programme went well is that we avoided lots of theory and focused on practical activities.

My students were highly motivated. They showed skills that aren't always evident in compulsory school activities.

Students were asking for more information (mainly about robotics, competitions, further development, etc.).

The oldest students helped and taught the youngest. They all were very friendly. They always wanted to complete the project in one session.

Many teachers commented positively on the benefits of the project. We concluded from teachers' responses that they had noticed improvements in the following areas of students' performance: interest in learning had increased; students worked more diligently than previously; students' motivation to persevere with more complicated tasks had improved; it was indicated that students demonstrated skills that teachers had not anticipated before; students wanted to find additional information and take responsibility for their own learning; students worked together and helped each other in order to complete the tasks more quickly and achieve better results. These and many other previously described benefits certainly diminish the risk of ESL. However, various challenges emerged during the project that may constitute new risk factors and it was important to analyse them so that we could explore ways of reducing or eliminating them in future work.

Examples of responses that indicate weaknesses of the ER programmes are given below:

As soon as the task became more difficult the students lost interest.

The children didn't like tasks where they had to make calculations or tasks where great accuracy was needed.

Not all students were able to complete the tasks in the specified time and according to the plan.

Difficulties were caused by the fact that some children worked at a slow pace. Some students wanted to do the task individually, using the robot on their own. We tried to allow this as much as possible. We took a personalised approach.

The difficulty was that the robotics activities were after-school lessons and at the end of the day pupils are tired and unable to perceive and process information.

Teachers' comments on the weaknesses of the project did not mention poor choice of learning approaches or methods or poorly structured teaching activities; instead there were statements describing rather typical characteristics of students at high-risk of ESL. The teachers' comments on the weakness of the programmes as delivered can be summarised as follows: students' motivation drops if the tasks are too complicated; they find the mathematical calculations difficult and there is not enough time for all students to complete all the tasks because they work at different paces. Knowing about these weaknesses should enable us to devise ways of tackling them.

Teachers also noted various threats to the success of the programmes:

> They [students] didn't like the tasks they failed at or tasks which had logical conclusions.
> The students didn't like the theory, they wanted to experiment themselves.
> Students experimented and used the Internet to find solutions.
> Sometimes there was not enough time to complete all the activities and explain all the 'solutions'. We asked the students to email us about their difficulties, so that we could go over a solution in the next session, but none of them did.

These teachers' comments highlight factors to which more attention should be paid in future work. For instance, the programmes should be adjusted so that they promote logical thinking and thus help students to find solutions to problems without seeking concrete instructions. Students showed willingness to experiment and search for creative solutions to problems. They also looked for answers on the Internet; this means that teachers have to give students a possibility to work independently if there is such need. Considerable emphasis was placed on improving students' cooperation and group work skills, and the teachers' feedback indicated that they had to devote considerable effort to this aspect of students' performance in order to achieve the desired result.

Participating in the project gave the teachers new experience and they commented on various opportunities the project presented:

> There is a need to change the compulsory learning process by combining various types of learning activities to enhance students' in-depth understanding and willingness to use emerging technologies to learn.
> It is necessary to train teachers to use the modern pedagogical strategies and teaching methods that are central to these programmes, which are radically different from normal teaching and learning activities.
> It is necessary to train teachers to work with students at risk of ESL.

Conclusions

Summarising and analysing the data from the project enables us to conclude that:

- During the project, teachers acquired new knowledge and gained experience in applying it in their teaching. Their perception of the benefits of the programmes is so great that they recommended that all teachers be trained in the learning methods used in the programmes, because these methods enabled students to acquire deep knowledge of a subject and avoid fragmentary mastery of content.
- It is vital to ensure that teachers delivering programmes such as those used in our project to students at high-risk of ESL are trained to work with this specific student population, which typically displays low motivation to learn and a negative attitude towards learning. This means that teachers' support and assistance is crucial to the success of the programmes, although the programmes are structured so that the teacher is not in the centre of the teaching and learning process. This is confirmed by the finding that the programmes delivered better results in

the second round than the first; a result we attribute to the teachers' increased competence in working with students at high-risk of ESL. The approach to train teachers for using robotics can help to reach the 4th SDG (United Nations Development Programme 2015), where it is stated that adequately trained teachers can help to ensure quality education.

- The teacher's questionnaire enabled teachers to reflect on their experiences of developing and delivering the programmes and offer detailed and considered comments on the strengths and weaknesses of the programmes in the format used in this project. We can conclude from all the data analysed that the greatest benefit of the ER programmes is that they increased students motivation and engagement in the learning process, encouraging them to construct knowledge themselves and work together to complete tasks and achieve good results.
- Participating in planned ER activities reduces risk of ESL in high-risk students because it leads to improvements in several ESL risk indictors, such as learning motivation, attitude towards learning, behaviour in lessons and cooperation with classmates and teachers, and it proved that such innovative and fun activities support the objective defined in SDG to ensure quality education from the perspective that children are not leaving the educational system.
- Although it has been argued that students subjected to social risk should not be separated from other students in order to avoid the risks associated with further social exclusion (Midgley 2000; Migdley and Urdan 2001), the didactic model employed in the project demonstrated that delivering tailored activities to a target group — students at high-risk of ESL who were selected using well-defined criteria — helped this group to experience the joy of learning and to feel a sense of belonging to the group of students that participated in the teaching and learning process, which should, in the long-term reduce their risk of ESL. It should be noted, however, that we did not assess the impact of ER activities on other students who might have wanted to participate but were ineligible for the programmes run under this project as they were not considered to be at high-risk of ESL.
- Use of active learning principles and emphasis on *hands-on* activities in the programmes had a positive influence on participants' knowledge of mathematics, physics and information technologies as well as improving their cooperation skills.
- The nature of the ER activities meant that students received immediate feedback on their work, because if their calculations or programming were incorrect the robot did not perform as intended. This prompted the students to look for their mistakes and try to eliminate them and such activities also support reaching the 9th SDG, which states that transport, irrigation, energy and information and communication technology are crucial to achieving sustainable development (United Nations Development Programme 2015), and robotics supports the development of ICT competence, by supporting innovative thinking to obtain solutions for problems connected with sustainable development (Moro et al. 2018; Karampinis 2018; Karkazis et al. 2018).

- Through working with ER, the students involved in the programmes became interested in compulsory lessons, leading to the improvements in their academic achievement that were reported by their regular class teachers.
- Delivering ER activities as an after-school activity led to a 'digital divide' between students who had the opportunity to participate and those who did not. High-achieving students were excluded from our programmes because they were targeted at students at high-risk of ESL, but students who have other after-school activities to attend would also be unable to attend project activities. It is therefore necessary to include ER activities in the compulsory curriculum in order to avoid creating a digital divide.
- It is important to continue research into the impact of ER on different groups in order to determine which kinds of activity are best suited to specific groups. Questions that should be addressed include whether ER activities are suitable for all children with special educational needs, what should be taken into account and how the learning process should be organised for students who are introverted and do not want to cooperate.

In this chapter, we analysed how the ER activities can reduce the risks of ESL from the perspective that not only the access to education is important but also actions should be taken to ensure that children stay at educational institutions. We believe that by reducing the risks of ESL, reaching the 4th SDG is supported because students stay at educational institutions. By using innovative learning methods with ER, the development of computational thinking is ensured and it supports reaching the 9th SDG. Our activities were aimed at children who are at risk of ESL without any other limitations, and they ensured that children of different ages and girls and boys were working together, thus helping to reach the 6th goal because equal opportunities were provided for girls to participate in ER activities.

References

Alimisis, D. (2013). Educational robotics: Open questions and new challenges. *Themes in Science and Technology Education, 6*(1), 63–71.

Alimisis, D. (2014). Educational robotics in teacher education: An innovative tool for promoting quality education. In L. Daniela et al. (Eds.), *Teacher of the 21st century* (pp. 28–39). Newcastle upon Tyne: Cambridge Scholars Publishing.

Atmatzidou, S., & Demetriadis, S. (2016). Advancing students' computational thinking skills through educational robotics: A study on age and gender relevant differences. *Robotics and Autonomous Systems, 75*, 661–670.

Benitti, F. B. V., & Spolaõr, N. (2017). How have robots supported STEM teaching? In M. Khine (Ed.), *Robotics in STEM education* (pp. 103–129). Cham: Springer. https://doi.org/10.1007/978-3-319-57786-9_5.

Bhowmik, M. K. (2017). Dropping out: Why students drop out and what can be done about it. *Educational Review, 70*(4), 527. https://doi.org/10.1080/00131911.2017.1336276.

Corey, S. M. (1954). *Action research to improve school practices*. New York: Bureau of Publications, Teachers College, Columbia University. https://doi.org/10.1002/sce.37303805127.

Daniela, L. (2016). Preliminary results of the project "robotics-based learning interventions for preventing school failure and early school leaving. *ICERI 9th annual international conference of education, research and innovation proceedings*. https://doi.org/10.21125/iceri.2016.

Daniela, L. (2018). Smart pedagogy for technology enhanced learning. In *Didactics of smart pedagogy: Smart pedagogy for technology enhanced learning* (pp. 1–20). Cham: Springer. ISBN 978-3-030-01550-3.

Daniela, L., Nīmante, D., & Kraģe, G. (2014). Development of support system for decreasing social exclusion. In *Teacher of the 21st century: Quality education for quality teaching* (pp. 1–12). Newcastle upon Tyne: Cambridge Scoolar Publishing. ISBN (13): 978-1-4438-5612-6.

Daniela, L., Strods, R., & Alimisis, D. (2017). Analysis of robotics-based learning interventions for preventing school failure and early school leaving in gender context. *9th international conference on education and new learning technologies (Edulearn17)* (pp. 810–818), 3–5 July. Barcelona: Conference Proceedings Barcelona. https://doi.org/10.21125/edulearn.2017.1176.

Downes, P. (2016). Developing a framework of system change between diametric and concentric spaces for early school leaving prevention. *Educational Philosophy and Theory, 48*(9), 899–914. https://doi.org/10.1080/00131857.2015.1079517.

Eguchi, A. (2014). Educational robotics for promoting 21st century skills. *Journal of Automation. Mobile Robotics & Intelligent Systems, 8*(1), 5–11. https://doi.org/10.14313/JAMRIS_1-2014/1.

Eurostat. (2016). *Early leavers from education and training by sex*. Retrieved from: http://ec.europa.eu/eurostat/tgm/table.do?tab=table&plugin=1&language=en&pcode=t2020_40

Fredricks, J. A., Blumenfeld, P. C., & Paris, A. H. (2004). School engagement: Potential of the concept, state of the evidence. *Review of Educational Research, 74*(1), 59–109. https://doi.org/10.3102/00346543074001059.

Karampinis, T. (2018). Activities and experiences through RoboESL project opportunities. *International Journal of Smart Education and Urban Society (IJSEUS)*. https://doi.org/10.4018/IJSEUS.

Karkazis, P., Balourdos, P., Pitsiakos, G., Asimakopoulos, K., Saranteas, I., Spiliou, T., & Roussou, D. (2018). "To water or not to water". The Arduino approach for the irrigation of a field. *International Journal of Smart Education and Urban Society (IJSEUS)*. https://doi.org/10.4018/IJSEUS.

Lund, H. H., & Marti, P. (2005). Designing manipulative technologies for children with different abilities. *Artificial Life and Robotics, 9*(4), 175–187. https://doi.org/10.1007/s10015-005-0351-y.

Melkevik, O., Nilsen, W., Evensen, M., et al. (2016). Internalizing disorders as risk factors for early school leaving: A systematic review. *Adolescent Research Review*. https://doi.org/10.1007/s40894-016-0024-1.

Midgley, G. (2000). *Systemic intervention: Philosophy, methodology and practice*. New York: Kluwer Academic/Plenum. 447 p.

Migdley, C., & Urdan, T. (2001). Academic self-handicapping and performance goals: A further examination. *Contemporary Educational Psychology, 26*(1), 61–75. https://doi.org/10.1006/ceps.2000.1041.

Moro, M., Agatolio, F., & Menegatti, E. (2018). The development of robotic enhanced curricula for the RoboESL project: Overall evaluation and expected outcomes. *International Journal of Smart Education and Urban Society (IJSEUS), 9*(1), 48–60. https://doi.org/10.4018/IJSEUS.

Nevala, A.M., Hawley, J., Stokes, D., Slater, K., Souto-Otero, M., Santos, R., et al. (2011). *Reducing early school leaving in the EU*. Retrieved from: http://www.europarl.europa.eu/RegData/etudes/etudes/join/2011

Orland-Barak, L., & Becher, A. (2011). Cycles of action through systems of activity: Examining an action research model through the lens of activity theory. *Mind, Culture, and Activity, 18*(2), 115–128. https://doi.org/10.1080/10749039.2010.484099.

Papert, S. (1980). *Mindstorms: Children, computers, and powerful ideas*. New York: Basic Books.

Samra, H. E., Alice, S. L., Ben, S., & Mohammed, A. A. (2017). A cloud-based architecture for interactive e-training. *International Journal of Knowledge Society Research, 8*(1), 67–78. https://doi.org/10.4018/IJKSR.2017010104.

Tweddle, L. K. (2008). Reinventing papert's constructionism – Boosting young children's writing skills with e-learning designed for dyslexics. *The Electronic Journal of e-Learning, 6*(3), 227–234. ISSN: EISSN-1479-4403.

United Nations Development Programme. (2015). *Sustainable development goals*. Retrieved from: http://www.undp.org/content/undp/en/home/sustainable-development-goals.html. Accessed 26 June 2018.

Chapter 3
Towards a Definition of Educational Robotics: A Classification of Tools, Experiences and Assessments

David Scaradozzi, Laura Screpanti, and Lorenzo Cesaretti

Abstract Robotics in education (RiE) covers a variety of applications of robots to the world of teaching and learning. Despite all the benefits that robotics can bring to education, a clear definition of the purpose for introducing robotics in education is still missing. Authors aim at facing this issue proposing a classification of RiE experiences, stating the difference between RiE and educational robotics (ER). The need for this classification arises from the wide usage of ER to indicate a diverse range of activities using robots and from the lack of clarity when describing how ER impacts students' curricula. Moreover, a definition of ER can impact the definition of the policies on the integration of ER into formal and non-formal education; it can also provide a basis for further studies whose aim is to provide clear evidence on the benefits of ER activities; finally, it can enhance the replicability of ER activities. To better characterise ER, authors propose two more classifications: one for the robotic tools used in the ER activities and one for the evaluation of ER activities. Drawing upon the proposed classifications, authors point out some distinctive features of ER comparing them to literature. This general outline aims at creating a starting point to open a debate on the definition of ER.

David Scaradozzi, Laura Screpanti and Lorenzo Cesaretti contributed equally with all other contributors.

D. Scaradozzi (✉)
Dipartimento di Ingegneria dell'Informazione (DII), Università Politecnica delle Marche, Ancona, Italy

LSIS – umr CNRS 6168, Laboratoire des Sciences de l'Information et des Systèmes, Equipe I&M (ESIL), Case 925 - 163, Marseille, France
e-mail: d.scaradozzi@univpm.it

L. Screpanti
Dipartimento di Ingegneria dell'Informazione (DII), Università Politecnica delle Marche, Ancona, Italy

L. Cesaretti
Dipartimento di Ingegneria dell'Informazione (DII), Università Politecnica delle Marche, Ancona, Italy

TALENT srl, Osimo, Ancona, Italy

L. Daniela (ed.), *Smart Learning with Educational Robotics*,
https://doi.org/10.1007/978-3-030-19913-5_3

Keywords Educational robotics · Curricular robotics · Robotics in education · Primary school · Secondary school · STEM · Evaluation · Assessment

The present chapter will analyse the scientific literature reporting experiences in the field of educational robotics (ER). This analysis aims to provide a broad classification of experiences reporting the use of a robot for education, a classification of the available robots used in the ER context and a classification of existing evaluation methods to carry out the assessment of the ER activities. Starting from the distinction between robotics in education (RiE) and ER, this chapter will contribute to the discussion of what ER means and consists of. On the other hand, the proposed classifications aim to provide people working in the field of ER with a reference, by stating clearly what robotics can do for education and by providing a benchmark against which one can compare the activities carried out in the educational context. This comparison could improve teachers' and educators' understanding of how to bring robotics into the classroom. Moreover, all stakeholders could rethink existing experiences and work together to improve and replicate them.

Previous literature in the field of ER searched through databases to answer specific questions like "What topics are taught through robotics in schools?" (Alimisis 2013; Benitti 2012; Jung and Won 2018; Mubin et al. 2013), "What kind of skills does an ER activity develop?" (Jung and Won 2018; Miller and Nourbakhsh 2016), "How is student learning evaluated?" (Alimisis 2013; Benitti 2012; Jung and Won 2018; Miller and Nourbakhsh 2016; Toh et al. 2016), "What kind of robotic tools are employed in an ER activity?" (Alimisis 2013; Miller and Nourbakhsh 2016; Mubin et al. 2013), "Which pedagogical theories are supporting the implementation of ER activities?" (Jung and Won 2018; Mubin et al. 2013) and "Is robotics an effective tool for teaching and developing skills?" (Benitti 2012; Jung and Won 2018; Toh et al. 2016). Unfortunately, there is little agreement on what the essential features of ER are. This means that even if researchers are trying to answer the same questions, they are working on different sets of examples taken from the literature. For example, Benitti (2012) and Toh et al. (2016) excluded from their analysis those papers reporting activities using robotics as a subject in primary and secondary education. On the contrary, Jung and Won (2018) reviewed existing literature describing trends in two areas: robotics to teach robotics itself and robotics to teach other subjects. Moreover, Jung and Won (2018) analysed literature in robotics education using robotics kits for young children, excluding social robots, whereas Mubin et al. (2013) included them. Differences in carrying out activities and in researching on this field affect the results and their comparison.

Authors will provide in each section a classification for an aspect that characterises an ER activity. First of all, Section 1 states the difference between RiE and ER and provides a general classification of RiE and ER activities. Section 2 presents an overview of the robotic tools that are used to carry out activities and a classification of these tools based on four main features. Section 3 discusses a classification for the evaluation of ER activities and proposes the authors' first steps and considerations into a novel real-time technique for the assessment of the ER activities. Results from the proposed classifications can be found in the Appendix section.

Section 1: A Classification of Experiences Carried Out in Education Using Robots

Even if some literature uses "robotics in education" and "educational robotics" as synonyms (Benitti 2012; Eguchi 2017), authors believe that a distinction should be made between the two labels. Robotics in education (RiE) is a broader term referring to what robotics can do for people in education. For example, it can help impaired students to overcome limitations or it can help teachers to gain attention or to deliver content to pupils. Educational robotics (ER) refers to a specific field, which is the intersection of different kinds of expertise like robotics, pedagogy and psychology. ER builds on the work of Seymour Papert, Lev Vygotsky and Jean Piaget (Ackermann 2001; Mevarech and Kramarski 1993; Papert 1980; Vygotsky 1968) to bring not just robotics in education, but to create meaningful experiences on robotics since an early age (Scaradozzi et al. in press, 2015). ER is made of robots allowing a construction/deconstruction and programming activity, teachers/experts facilitating the activity and methodologies enabling students to explore the subject, the environment, the content of the activity and their personal skills and knowledge. These key elements of ER make it an integrated approach to STEM (Brophy et al. 2008) and an interdisciplinary and transdisciplinary subject (Eguchi 2014). Authors identified four different features to describe a RiE experience or project: the learning environment, the impact on students' school curriculum, the integration of the robotic tool in the activity and the way evaluation is carried out. Regarding how the robotic tool is integrated into the activity, we can distinguish ER as a subset of RiE (Fig. 3.1).

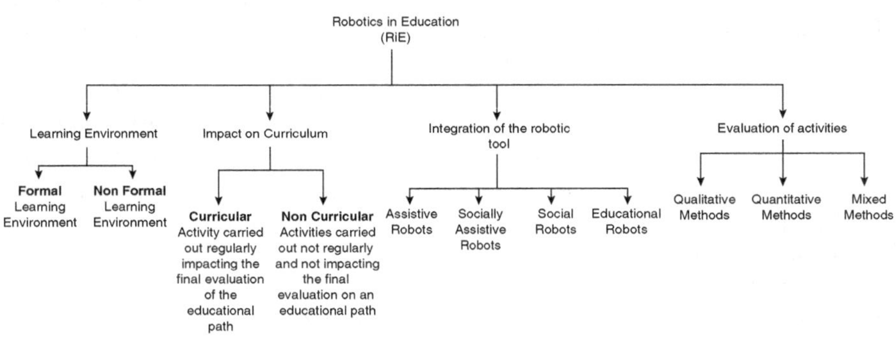

Fig. 3.1 The proposed classification for robotics in education (RiE)

In the next subsections, these four categories are described. Table 3.1, reported in the Appendix section, shows some examples of experiences using robotics in education and analyses them through the four main categories proposed by the authors.

Learning Environment: Formal or Non-formal Projects

Students can learn in a variety of settings (e.g. at school, at home, in an outdoor environment). Each setting is characterised by the physical location, learning context and cultures. Usually, each setting holds specific rules and ethos to define relationships, behaviours and learning activities. It's the authors' opinion that it is important to specify in a RiE activity whether the learning environment is formal or non-formal. Formal education is usually delivered by trained teachers in publicly recognised organisations providing structured activities and evaluation. Non-formal education can be a complement to formal education, but it may be apart from the pathway of the national education system, consisting in a shorter activity. Usually, non-formal activities lead to no qualification, but they can have recognition when they complete competences otherwise neglected. Formal environment is where formal education usually takes place (e.g. schools) and non-formal environment is where non-formal education usually happens (e.g. private houses, company's headquarters, museums).

Teaching methodologies, spaces, furniture and many other variables influence the outcome of a RiE or an ER activity, but they are out of scope in this part of the classification, which intends to make a distinction at a broader level.

School Curriculum Impact: Curricular or Non-curricular Projects

The way activities are integrated in education strongly impacts their design and their expected outcomes. Activities carefully designed to fit the curriculum needs, carried out regularly in the classroom to support students' learning of a concept and whose evaluation is recognised in the final evaluation of the school on students, are curricular activities. Seldom activities organised to better support the teaching of particular concepts, both inside and outside the classroom, and that lead to no final formal recognition are non-curricular activities. There may be activities performed at school (formal learning environment) that do not account for the final evaluation of the student (non-curricular activity). On the other hand, there may be an activity performed outside the classroom environment (non-formal learning environment) that is recognised into the final evaluation of the student provided by the school (curricular activity).

Integration of Robotic Tools

Robotic tools that are used into the activities should be distinguished according to the purpose they serve in the educational context. First, they can reduce the impairments for students with physical disabilities. These tools are usually medical devices that help people in their activities of daily living and they compensate for the lost function. These kinds of robots are assistive robots, and they are not intentionally produced to meet the need of education, but to meet the needs of impaired people.

Second, some robots can help people with a social impairment (e.g. autistic spectrum disorder). This kind of robots can be defined as socially assistive robots, because they are capable of assisting users through social rather than physical interaction (Matarić and Scassellati 2016). Socially assistive robots "attempt to provide the appropriate emotional, cognitive, and social cues to encourage development, learning, or therapy for an individual" (Matarić and Scassellati 2016, p. 1974).

Third, some robots can be companions to students' learning or to teachers while teaching (Belpaeme et al. 2018). These robots are called social robots, because they are designed to interact with people in a natural, interpersonal manner to accomplish a variety of tasks, including learning (Breazeal et al. 2016).

Fourth, robots can be a tool to study robotics and STEAM subjects and to develop transversal skills. ER projects use this kind of robots. Generally, they are presented to students as disassembled kits to give the possibility to create meaningful interdisciplinary pathways, letting students be free to build original artefacts. To build an artefact with fully functioning actuators and sensors, students need to master the fundamental concepts about robotics. Only when these concepts are reworked and absorbed by students that they can feel confident in reusing that kind of knowledge in another context. So, one of the main features of ER is the basic understanding of robotics fundamentals.

Evaluation: Qualitative, Quantitative or Mixed Methods

Evaluation in RiE activities could be carried out by using a qualitative method, a quantitative method or a mixed-methods approach. Qualitative methods in education pertains to research and to everyday practice. Teachers and researchers can analyse essays, focus groups, scenarios, projects, case studies, artefacts, personal experiences, portfolios, role play or simulation and many other outputs of the activities. This is a deep and rich source of information on students' learning, but sometimes impractical in a crowded classroom and always vulnerable to personal biases or external influence. On the opposite, quantitative methods are easier to replicate and administer. They try to summarise with numbers the outcome of an activity. Common tools in quantitative methods are based on questionnaires, tests and rubrics. Anyway, experiments and empirical method should be applied to prove these methods are valid, reliable and generalisable. Moreover, a quantitative evaluation in education is often deemed as poor and reductive. Lately, researchers in education have been overcoming the historical distinction between qualitative and quantitative methods to exploit the beneficial aspects that both methods provide. Researchers have been proposing the mixed-methods approach as an appropriate research method to address problems in complex environments, like education. The choice of mixed-methods design is usually well motivated because it could imply a lot of work as it requires that both quantitative and qualitative data are collected. In the last years, some novel real-time techniques have been introduced to monitor students during their activities. Technology and artificial intelligence seem to be promising in providing feedback on students' learning and in integrating both qualitative and quantitative methods of assessment. Moreover, it could be deployed into classroom seamlessly and give response on the activity to support the assessment.

Section 2: A Classification of ER Tools

The way the robotic tool is integrated into the experience can make the difference between a general RiE experience or an ER activity, but even among the ER tools we can make a distinction. In fact, there are many robots and robotic kits available on the market, but not all of these products are meant to be "educational". Reviewing ER tools available on the market, authors included those robots or robotic kits that respected these two criteria: tools that were designed purposefully for education OR tools that have been used in educational contexts, whose activities were reported in a scientific paper. Table 3.2 reports the analysis of those tools according to four sub-categories:

1. *Age* (kindergarten/primary school/secondary school): The age group for which the kit is recommended; it could be a large range; indeed sometimes varying the educational activity is possible to use the kit with different age groups.
2. *Programming language* (text-based/block-based/unplugged): There are three different kinds of programming languages associated with the kits. The most commonly used are the block-based environments (scratch or similar), where the students can create software sequences using blocks, without writing code and the possibility of making syntactical errors (namely visual programming technique). Considering tools that are more suitable for secondary school students, the trend is to propose text-based programming language as an alternative to the block-based environments. The third option, the unplugged way to program a robot, is very common for the kindergarten tools. There is no need to use a screen (tablet or computer) to create the sequence. Students can design different behaviours for their robot using some physical blocks (or physical buttons).
3. *Assembly feature* ("ready-to-use" robot/"to-build" kit): Using some of the commercial kit, students have the possibility of building the robot, interacting with mechanical and electronic parts (wheels, gears, sensors, motors, etc.). Other solutions are "ready to use": opening the box pupils find an already assembled robot, so they can program only the behaviour of the system, without the chance of modifying the robot's aspect.

4. *Robot's environment* (earth/water/air): educational robots usually move on the floor, or on the school desk (earth robot); but in recent years, some companies and research institutes have developed also educational drones (air) and marine vehicles (water).

In addition to the commercial kits presented in Table 3.2, there are other tools purposefully designed by researchers to implement ER activities (Bellas et al. 2018; Ferrarelli et al. 2018; Junior et al. 2013; Naya et al. 2017).

Section 3: A Classification of the Assessment of ER Activities

Section 1 introduced a distinction on how evaluation is carried out based on the way observation is designed, carried out and presented, and resulting in three categories: qualitative methods, quantitative methods and mixed methods. This is not the only way to characterise evaluation and research methods. Considering the target of the evaluation, evaluation can focus on performance, attitude and behaviour. Performance measurement can be a test whose aim is to evaluate the knowledge acquired on the subject and/or the ability to use it to perform a task (Blikstein et al. 2017; Di Lieto et al. 2017; Screpanti et al. 2018b) or it can be based on neuropsychological measures (Di Lieto et al. 2017). Complex task evaluation can also be related to the development of skills, not only knowledge. Moreover, written tests more often reflect theoretical knowledge, while practical exercises or tests demonstrate applied skills. Attitudes and skills are more often measured through surveys and questionnaires (Atmatzidou and Demetriadis 2016; Cesaretti et al. 2017; Cross et al. 2017; Di Lieto et al. 2017; Goldman et al. 2004; Lindh and Holgersson 2007; Screpanti et al. 2018a; Weinberg et al. 2007), which are easy to administer and useful for triangulation. Measures of student's behaviours in ER activities can help the design of the learning environment as well as deepen understanding of how students learn (Kucuk and Sisman 2017).

Another distinctive feature of evaluation regards when to measure. Measurements (or evaluation of a student's state) can be performed before the activity, iteratively during the activity and after the activity. In addition to this, stating the purpose of evaluation can help researchers and teachers to clarify how and when to perform such assessment. Summative assessment (or assessment *of* learning) is often related to the outcome of the activity and it is often regarded as the post-activity evaluation which relates to benchmarks. Formative assessment (or assessment *for* learning) is often a kind of evaluation taking place before the activity, but it can also be iterative, occurring periodically throughout the ER activity. The purpose of formative assessment is to adjust teaching and learning activities to improve student's attainment. More recently, the field of assessment *as* learning brought the idea that formative assessment, feedback and metacognition should go together (Dann 2014; Hattie and Timperley 2007).

At the end of an ER activity, it would be interesting to investigate the process that led to the resolution of a specific problem, or to the design of a software sequence. During an ER activity, students experiment and modify their sequence of instructions or robot's hardware structure, to obtain a specific behaviour. They usually work in team in a continuous process of software and/or hardware improvement, as specified by the TMI model (Martinez and Stager 2013). It would be very interesting for an educator to have the chance to observe and analyse this process, but it is not realistic to have one teacher per group that keeps track of the students' development inside the classroom. New experimentations in constructionist research laid the way into new possibilities of insights into the students' learning processes. Evaluation can be performed using the "offline" or "online" method. The offline methods are those assessments gathering information one or more times during the activity and then usually processed later by a human evaluator. The online methods are those assessments "continuously" gathering information on students' activity (e.g. camera recording students' behaviour, sensors collecting physiological parameters, log system recording students' interactions) aiming at providing an analysis of the student's learning while the student is still exploring the activity. Online methods are usually automated and rely on educational data mining (EDM) and learning analytics (LA). The first applications of these technologies tried to extrapolate information from data gathered from structured online learning environments (Baker et al. 2004; Beck and Woolf 2000; Berland et al. 2014; Merceron and Yacef 2004): in this type of condition, it was easier to deduce relations and recognise patterns in the data. Recent studies (Asif et al. 2017; Ornelas and Ordonez 2017) tried to predict students' success using machine learning algorithms on data gathered from closed environments. Blikstein et al. (2014) collected the code snapshots of computer programs to investigate and identify possible states that model students' learning process and trajectories in open-ended constructionist activities. Berland et al. (2013), extending the previous work by Turkle and Papert (1992), registered students' programming actions and used clustering to study different pathways of novice programmers. This led to the identification of three general patterns: tinkering, exploring and refining. To evaluate different aspects of constructionist activities, other works relied on external sensors (cameras, microphones, physiologic sensors) and automated techniques, like text analysis, speech analysis and handwriting analysis (Blikstein and Worsley 2016). A key for future developments and experimentations will probably be connected to the availability and cost of implementation of such technological solutions for classroom assessment. External sensors may be more expensive, whereas embedded software solutions and machine learning algorithms could be effective and reliable in extracting evidence of students' learning process and helping teachers to provide personalised feedback to students. Anyway, as stated by Berland et al. (2014), EDM and LA in constructionist environment aim at generating complementary data to assist teachers' deep qualitative analysis with quantitative methods.

A first experimentation that used data mining in the field of ER was conducted by Jormanainen and Sutinen (2012). They adopted the Lego Mindstorms RCX and collected data from students' activities with the main functions of a new graphical programming environment that they designed. They created an open monitoring environment (OME) for the teachers involved, obtaining promising results with decision trees algorithm (J48 implementation) for classifying students' progress in the ER setting. But probably there were some weaknesses in this experimentation: the kit chosen for the study was anachronistic, indeed in 2012 the new model of Lego Mindstorms (the NXT version) had been on the market since 2006; only 12 students and 4 teachers from primary school were involved, a very low number of participants to validate the method; a new graphic programming environment was developed, but it was without a block-based approach, maybe not so friendly for primary school pupils.

First Steps Towards Educational Data Mining with Lego Mindstorms EV3

The first steps in the application of educational data mining to Lego Mindstorms EV3 were made by the authors in an Italian upper secondary school, Liceo Volta Fellini in Riccione (a formal learning environment) during an alternating school-work course (a non-curricular activity). Thanks to a software development it was possible to track all the sequences of blocks made by the students using the Lego Mindstorms EV3 software environment. Three classes were involved in the project. Participants were divided into teams of 3–4 students who worked together to design software or hardware solutions to a set of tasks. The first challenge faced by the learners, after the robot's construction, was *programming the robot so that it covers a given distance (1 m), trying to be as precise as possible*. Solving the task, students had to consider a few constraints:

- Fifteen minutes to prepare the software solution and then the "final" competition between the teams.
- During the available time, the teams could test the solution as many times as they wanted.
- They could not use measuring instruments (set squares, rulers, etc.) to measure the distance covered by the robot on the floor during the test time; they had the possibility of using the instruments only to determine some robot's parameters (e.g. the radius of the wheel).

Some students realised that there were some cables with a known length inside the Lego Mindstorms box, and they were allowed to use them as a reference object for the trials.

This task was tricky because in the Lego software there are not blocks in which the designer can set a specific distance to cover. The trainer presented only one block for the challenge: the "move steering" function, where students can set three modes to control the motors ("on for seconds", "on for degrees" or "on for rotations") and the steering of the robot and the motors' power.

Students' teams mainly focused on the change in the last parameter: some groups calculated the wheel's circumference; other groups tried to measure the robot's speed (in order to calculate the number of seconds to set); other groups adopted a more practical and "trial and error" approach, for example, using the cable inside the box as a reference measurement. These different approaches to the solution to a given task seem to fit into the two different styles in problem-solving proposed by Turkle and Papert (1992). They suggested that students could achieve learning objectives while taking different pathways and strategies: the "bricoleur scientist" prefers "a negotiational approach and concrete forms of reasoning", while the "planner scientist" prefers "an abstract thinking and systematic planning".

Figure 3.2 shows an example of the log recorded by the modified Lego Mindstorms EV3 used during the challenge. It is interesting to take into consideration the rotations/seconds/degrees parameter set by the students during the trial time, and analyse the behaviour of three groups involved in the robotics course, which seem to have very definite features.

Figure 3.3 shows the sequence of the rotation parameter (the number of the rotations set for the motors) chosen by group 1: 9 tests were conducted by the team (the last one was the final competition), all of them with a rotation parameter very close to 5.78 rotations. In this case, planning seems to be the prevalent approach adopted by the group, probably with an initial mathematical calculus and then verification tests to check the robot's behaviour. This team obtained a 0.5 cm error from the desired measure.

Group 2 performed 8 tests (the last one was the final competition): the first one with a value equal to 1 rotation and the following with a value very close to 5.5 rotations (Fig. 3.4). Planning seems to be the prevalent approach adopted by the group. They probably did a first check of the robot's behaviour setting 1 rotation for the motors, then they inserted the value 5.5 in the rotation parameter. It is likely that they made a calculation (or a proportion) to reach the solution of the given task. This team obtained a 2 cm error from the desired measure.

```
MoveSteering, OnForRotations, Rotations = 5.7875, Speed = 100, Steering = 0, MotorPorts = 123;
STOP PROGRAM;

MoveSteering, OnForRotations, Rotations = 5.7875, Speed = 60, Steering = 0, MotorPorts = 123;
Sound, Play Tone, Frequency = 440, Volume = 78, Play Type = 0, Duration = 1;
STOP PROGRAM;

MoveSteering, OnForRotations, Rotations = 5.7875, Speed = 60, Steering = 0, MotorPorts = 123;
Sound, Play Tone, Frequency = 440, Volume = 78, Play Type = 0, Duration = 1;
STOP PROGRAM;

MoveSteering, OnForRotations, Rotations = 5.7875, Speed = 60, Steering = 0, MotorPorts = 123;
Sound, Play File, FileName = Motor stop, Volume = 83, Play Type = 0;
STOP PROGRAM;

Sound, Play File, FileName = Motor start, Volume = 100, Play Type = 0;
MoveSteering, OnForRotations, Rotations = 5.7875, Speed = 60, Steering = 0, MotorPorts = 123;
Sound, Play File, FileName = Motor stop, Volume = 100, Play Type = 0;
STOP PROGRAM;

Sound, Play File, FileName = Motor start, Volume = 100, Play Type = 0;
MoveSteering, OnForRotations, Rotations = 5.7875, Speed = 60, Steering = 0, MotorPorts = 123;
Sound, Play File, FileName = Motor stop, Volume = 100, Play Type = 0;
STOP PROGRAM;
```

Fig. 3.2 A log example, generated by the modified Lego Mindstorms EV3

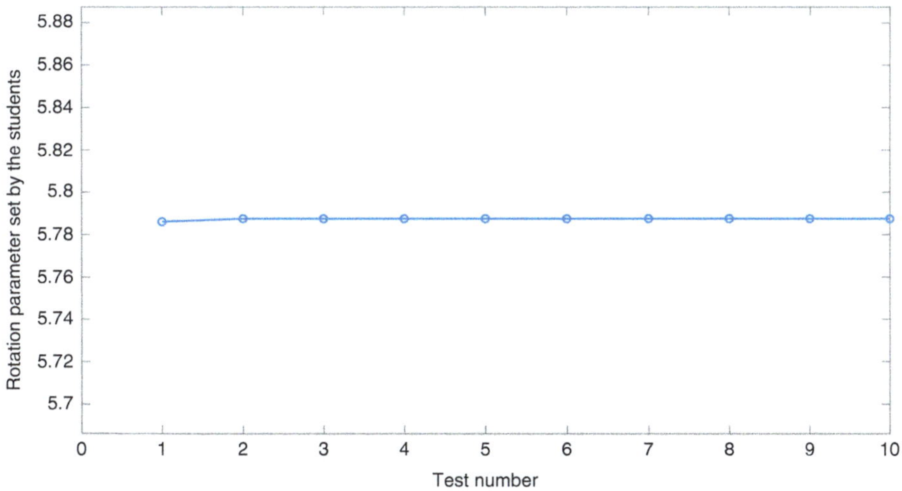

Fig. 3.3 Rotations graph for team 1

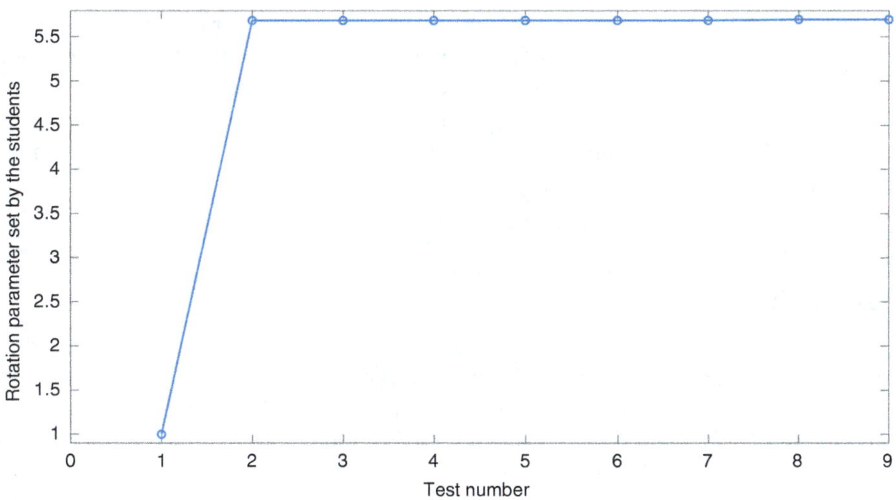

Fig. 3.4 Rotations graph for a team 2

Figure 3.5 shows the sequence of the values chosen by group 3 for the rotation parameter. They performed 15 tests (the last one was the final competition), and their strategy is represented by a broken line ranging from a minimum value of 1 rotation to a maximum value of 8. In this case, tinkering seems to be the prevalent approach adopted by the group, probably with a "trial and error" pathway more pronounced compared to the other teams. This team obtained a 1.5 cm error from the desired measure.

This preliminary analysis shows how such a tool can provide teachers with complementary information on students learning. Furthermore, such an automated tool

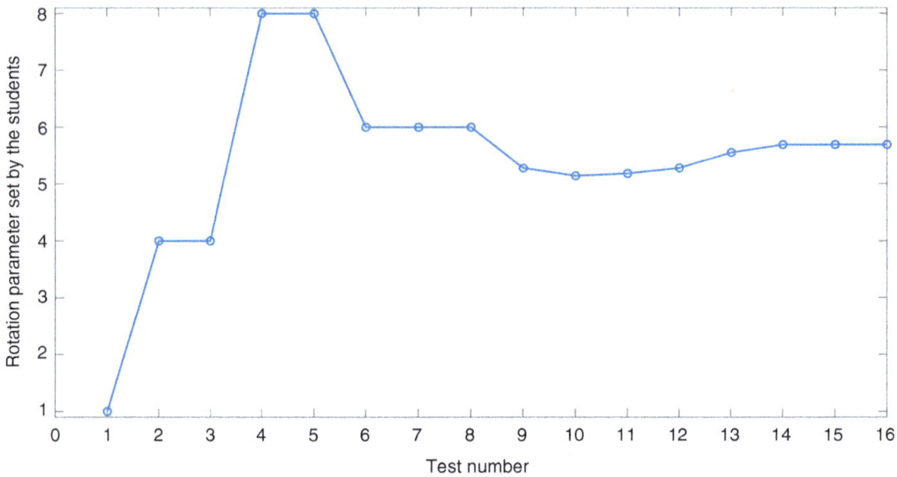

Fig. 3.5 Rotations graph for team 3

assessing the progress of the activity from each group (as an online method of evaluation) can provide feedback to the teacher, thus allowing a real-time evaluation. Experts are cooperating to identify meaningful indexes for students' performance and style of learning. More complex tasks, and therefore logs, are under analysis, to unravel the knot of different skills and knowledge applied in an open-ended environment. Moreover, different machine learning algorithms are compared to extrapolate knowledge from the raw data.

Discussions and Conclusion

Results from authors' classification of RiE experiences are shown in Table 3.1. Information on learning environment or on school curriculum impact is often missing (the word "Unknown" in the table means that authors didn't find these specifications). This can be related to the scope of some activities within the RiE field, namely, social robotics, socially assistive robotics and assistive robotics, where studies are mainly focused on interaction or physical or cognitive rehabilitation, not on education. But even in the ER subfield, it is hard to retrieve information on school's curriculum impact. Information about the impact of an ER research project on school curriculum is fundamental to the process of integrating ER at school and for the design of activities because it influences the learning outcomes and their evaluation. Moreover, clear consideration of the curriculum impact could make it easier for teachers and educators to replicate the project in other schools or institutions, spreading the academic results into the daily educational practice.

It is also important for ER designers to consider the appropriate tools, analysing the four features proposed in Section 2: age group, programming language, assembly feature and robot's environment. Table 3.2 shows that market has a variety of robotic tools to choose from. For a deep understanding of the core concepts of robotics, authors suggest choosing kits defined as "to build", especially in primary school. This kind of kits lets students manipulate basic elements of a robot, design experimental mechanisms, design creative robots and create personal and meaningful "public entity", as proposed by Papert (1991). Furthermore, the simultaneous analysis of hardware and software during the design process is more challenging for students: if the robot doesn't work, students have to consider how they assembled the various parts of the robot as well as how they programmed it. This can be even more challenging when integrating an open control board (e.g. based on Arduino or Raspberry Pi) in the activity. On the one hand, it would offer teachers the chance to explain the relevance of the open source culture and the community. On the other hand, it provides students with a "white-box" tool, whose construction and reconstruction is enabled to a deeper level. Authors agree with Alimisis (2013) on the need of a transition to a "white-box" or "black-and-white" approach for constructionist environments. Teachers and educators can choose according to their learning objectives how to introduce robotics in their class to support teaching and to produce a positive impact on student's learning. Literature supports observations like "ER helps in developing twenty-first-century skills"(Eguchi 2014, 2015, 2016), "ER prevents ESL"(Daniela and Strods 2018; Daniela et al. 2017; Moro et al. 2018) and "ER is effective in conveying knowledge about subjects" (West et al. 2018), but often those studies are too limited to generalise. Several studies focus on qualitative methods that do not provide indexes or a numeric indication on how to evaluate student's performance. Moreover, there is no homogeneity in conducting such studies because they do not all align on the purpose of the study, and when they do, they do not use the same protocol to bring ER to student or measurement instrument (Castro et al. 2018). ER needs longitudinal studies to validate ER curricula, valid and reliable assessment instruments, trained and motivated educators and teachers, stakeholders' engagement to help ER methodologies and tools to enter the education system and impact the future citizens.

Table 3.3 shows some literature's studies and their description through the four features of evaluation. It reports that several constructs belonging to performance, behaviour and attitude are explored in relation to the ER experience. This evaluation has almost always the purpose of assessing the intended constructs and hardly ever the purpose of providing feedback to students. Moreover, qualitative and quantitative assessments are widely used, often in a mixed approach. It can be noted that the categories "what", "when" and "how" can belong to all RiE subfields, but "for what purpose" pertains specifically to those fields directly targeting learning. In fact, in socially assistive robotics and assistive robotics, the assessment is often focused on the evaluation of the improvements of the lost function following the intervention with robots (Bharatharaj et al. 2018; Cook et al. 2005; Holt et al. 2013; Mengoni et al. 2017; Tapus et al. 2012). In the RiE subfield of social robotics, studies are mainly focused on the interaction between the robot and the student or the teacher (Fridin 2014; Fridin and Belokopytov 2014).

In the ER context, online measurement is not used, but for Jormanainen and Sutinen (2012). This may be because the data mining approach is relatively new, and it has become robust only recently. Though, mainly unexplored, this research direction is an interesting challenge which may eventually lead to a system informing teachers or students on the ER activity. To reach this goal, data should be gathered through transparent, replicable and open experiments that could thus produce comparable results. Moreover, integrating teachers' qualitative evaluation, new technologies and techniques, like educational data mining and learning analytics, it will be possible to validate and examine in depth the real potential of ER. In a future scenario, teachers will be able to analyse minute by minute the progression of their students, and they will have available meaningful information about students' learning. In this scenario, students will also benefit from personalised feedbacks, with a real chance to develop their personal learning style.

The proposed classifications are in line with some aspects proposed by relevant literature but they lead to some considerations in relation to other aspects. Moro et al. (2018) stated that ER does not mean to teach a specific discipline like robotics, but rather a didactical approach to learning, based on constructivist and constructivism theories. Authors agree with the fact that ER is a didactical approach to learning but argue that this is not enough to describe ER. In fact, constructivism alone does not build the ER field. The didactical approach is a key element in ER education, but another essential element in ER is robotics. Students should develop the technical knowledge on the object they are using to grasp the meaning of the activity. This aspect is also highlighted by Angel-Fernandez and Vincze (2018). They proposed a definition of ER as a field of study at the intersection of three broad areas: education (all those disciplines aiming at studying and improving people's learning), robotics (all those disciplines aiming at studying and improving robots) and human-computer interface (aiming at improving user experience). This definition covers categories like robotics as a learning object (robots used to teach robotics), robotics as a learning tool (robots are tools to teach other subjects) and robots as learning aids (social robots). As previously stated, authors disagree with the inclusion of social robotics in the field of ER. Social robots focus on the interaction between robots and people in a natural, interpersonal manner, often to achieve positive outcomes (Breazeal et al. 2016). Thus, social robotics is a RiE subfield dealing with robots as companions to teachers or peers to students with the aim of engaging them in a learning activity. Although robots for ER, described in Section 2, do not focus on just the interaction between humans and robots to achieve an outcome, they are designed, built and programmed by students in the context of a constructionist environment.

This chapters presented a novel description of some basic features of an ER activity to provide a common ground for researchers and common knowledge for teachers and educators. Moreover, specifying the impact on school curriculum and the learning environment, authors intended to remark that ER can actually enter the school curriculum. Robotics should be a subject within school's hours, with its own lesson and evaluation plan or, at least, afternoon activities strictly connected to the school program. Whether a whole curriculum-based education or a regular activity inside another broader subject, ER should be part of school's curricular offer since an early stage.

Appendix

Table 3.1 Results from the classification proposed in Section 1

Paper	Learning environment	Impact on education	Integration of technology	Evaluation of activities
Akagi et al. (2015)	Formal	Curricular	Educational robotics	Qualitative
Bers et al. (2014)	Formal	Unknown	Educational robotics	Quantitative Offline
Bharatharaj et al. (2018)	Formal	Unknown	Socially assistive robotics	Quantitative Offline
Cannon et al. (2007)	Formal	Non-curricular	Educational robotics	Mixed Offline
Castro et al. (2018)	Formal	Unknown	Educational robotics	Quantitative Offline
Cesaretti et al. (2017)	Formal	Non-curricular	Educational robotics	Quantitative Offline
Chalmers (2018)	Formal	Unknown	Educational robotics	Mixed Offline
Chang et al. (2010)	Formal	Unknown	Social robotics	Qualitative
Chen (2019)	Non-formal	Unknown	Educational robotics	Mixed Offline
Cook et al. (2005)	Formal	Unknown	Assistive robotics	Mixed Offline
Costantini et al. (2017)	Non-formal	Non-curricular	Educational robotics	Unknown
Cross et al. (2015)	Formal	Unknown	Educational robotics	Mixed Offline
Cross et al. (2017)	Formal	Unknown	Educational robotics	Mixed Offline
Daniela et al. (2017)	Formal	Unknown	Educational robotics	Mixed Offline
Denicolai et al. (2018)	Formal	Unknown	Educational robotics	Unknown
Di Lieto et al. (2017)	Formal	Non-curricular	Educational robotics	Quantitative Offline
Eguchi (2015)	Formal	Curricular	Educational robotics	Mixed Offline
Eguchi (2016)	Non-formal	Non-curricular	Educational robotics	Quantitative Offline
Ferrarelli et al. (2018)	Formal	Non-curricular	Educational robotics	Mixed Offline
Frangou et al. (2008)	Formal	Unknown	Educational robotics	Qualitative
Fridin (2014)	Formal	Non-curricular	Social robotics	Quantitative Online

(continued)

Table 3.1 (continued)

Paper	Learning environment	Impact on education	Integration of technology	Evaluation of activities
Fridin and Belokopytov (2014)	Formal	Non-curricular	Social robotics	Quantitative Offline
Goldman et al. (2004)	Non-formal	Non-curricular	Educational robotics	Quantitative Offline
Horn et al. (2008)	Non-formal	Non-curricular	Educational robotics	Mixed Online
Holt et al. (2013)	Formal	Unknown	Assistive robotics	Mixed Online
Iacobelli (2010)	Formal	Curricular/ non-curricular	Educational robotics	Mixed Offline
Iacobelli and Spano (2011)	Formal	Unknown	Educational robotics	Unknown
Jeon et al. (2016)	Formal	Non-curricular	Educational robotics	Unknown
Jormanainen and Sutinen (2012)	Non-formal	Non-curricular	Educational robotics	Quantitative Online
Junior et al. (2013)	Unknown	Unknown	Educational robotics	Quantitative Offline
Kandlhofer and Steinbauer (2016)	Formal	Non-curricular	Educational robotics	Quantitative Offline
Kim et al. (2015)	Formal	Unknown	Educational robotics	Mixed Offline
Kory Westlund et al. (2016)	Formal	Unknown	Social robotics	Mixed Offline
Kucuk and Sisman (2017)	Formal	Non-curricular	Educational robotics	Quantitative Online
Lindh and Holgersson (2007)	Formal	Unknown	Educational robotics	Mixed Offline
Lins et al. (2018)	Formal	Unknown	Assistive robotics	Quantitative Online
Mengoni et al. (2017)	Formal	Unknown	Socially assistive robotics	Mixed Offline
Micotti et al. (2017)	Formal	Unknown	Educational robotics	Unknown
Montero and Jormanainen (2016)	Unknown	Unknown	Educational robotics	Unknown
Oreggia et al. (2016)	Formal	Non-curricular	Educational robotics	Quantitative Offline
Ospennikova et al. (2015)	Formal	Unknown	Educational robotics	Unknown
Ozgur et al. (2018)	Non-formal	Unknown	Assistive robotics	Quantitative Online
Palsbo and Hood-Szivek (2012)	Unknown	Unknown	Assistive robotics	Quantitative Offline

Table 3.1 (continued)

Paper	Learning environment	Impact on education	Integration of technology	Evaluation of activities
Polishuk et al. (2012)	Non-formal	Non-curricular	Educational robotics	Unknown
Polishuk and Verner (2017)	Non-formal	Non-curricular	Educational robotics	Qualitative
Rusk et al. (2008)	Non-formal	Non-curricular	Educational robotics	Unknown
Ryu et al. (2013)	Unknown	Unknown	Assistive robotics	Unknown
Sahin et al. (2014)	Formal	Non-curricular	Educational robotics	Qualitative
Scaradozzi et al. (2018)	Formal	Unknown	Educational robotics	Quantitative Offline
Scaradozzi et al. (2015)	Formal	Curricular	Educational robotics	Quantitative Offline
Scaradozzi et al. (2016)	Formal	Non-curricular	Educational robotics	Quantitative Offline
Screpanti et al. (2018a)	Formal	Non-curricular	Educational robotics	Quantitative Offline
Screpanti et al. (2018b)	Formal	Non-curricular	Educational robotics	Quantitative Offline
Sullivan (2008)	Formal	Non-curricular	Educational robotics	Mixed Offline Online
Tapus et al. (2012)	Unknown	Unknown	Socially assistive robotics	Quantitative Online
Tocháček et al. (2016)	Formal	Non-curricular	Educational robotics	Quantitative Offline
Vitale et al. (2016)	Formal	Curricular	Educational robotics	Quantitative Offline
Weinberg et al. (2007)	Formal	Non-curricular	Educational robotics	Quantitative Offline
West et al. (2018)	Non-formal	Non-curricular	Educational robotics	Unknown

Table 3.2 Results from the classification proposed in Section 2

Kits name	Age	Assembly feature	Programming language	Robot's environment	Link
Bee-Bot/Blue-Bot	Kindergarten/ primary	Ready to use	Unplugged	Earth	https://www.bee-bot.us/
Clementoni Doc	Kindergarten/ primary	Ready to use	Unplugged	Earth	https://www.clementoni.com/ it/11112-doc-robottino-educativo-parlante/
Cubetto	Kindergarten/ primary	Ready to use	Unplugged/block based	Earth	https://www.primotoys.com/it/
Matatalab	Kindergarten/ primary	Ready to use	Unplugged	Earth	https://www.matatalab.com/
Dash & Dot	Kindergarten/ primary	Ready to use	Block based	Earth	https://www.makewonder.com/dash/
Clementoni Mind Designer	Primary	Ready to use	Unplugged/block based	Earth	https://www.clementoni.com/ it/12087-mind-designer-robot-educativo-intelligente/
Ozobot	Primary	Ready to use	Unplugged/block based	Earth	https://ozobot.com/
Codey Rocky	Primary	Ready to use	Block based/text based	Earth	https://www.makeblock.com/steam-kits/codey-rocky
Sam Labs kits	Primary/lower secondary	To build	Block based	Earth	https://samlabs.com/
Little Bits kits	Primary/lower secondary	To build	Unplugged/block based	Earth	https://littlebits.com/
Sphero	Primary/lower secondary	Ready to use	Block based/text based	Earth	https://www.sphero.com/
Thymio	Primary/lower secondary	Ready to use	Block based	Earth	https://www.thymio.org/
mBot kits	Primary/secondary	To build	Block based/text based	Earth	https://www.makeblock.com/steam-kits/mbot

airBlock	Primary/secondary	To build	Block based/text based	Air/water	https://www.makeblock.com/steam-kits/airblock
Lego Wedo	Primary	To build	Block based	Earth	https://education.lego.com/en-gb/product/wedo-2
Clementoni RoboMaker Pro	Primary/lower secondary	To build	Block based	Earth	https://www.clementoni.com/it/13992-robomaker-pro/
Jimu Robot	Primary/secondary	To build	Block based	Earth	https://ubtrobot.com/collections/jimu-robots
Lego Mindstorms EV3	Primary/secondary	To build	Block based/text based	Earth	https://education.lego.com/en-gb/product/mindstorms-ev3
Fischertechnik – ROBO TX Explorer	Secondary	To build	Block based	Earth	https://www.fischertechnik.biz/Robotics/fischertechnik-robo-tx-explorer-comp.html
BYOR	Secondary	To build	Block based/text based	Earth	https://byor.scuoladirobotica.it/it/homebyor.html
Pi2Go	Secondary	To build	Block based/text based	Earth	https://4tronix.co.uk/blog/?p=452
Nao	Secondary	Ready to use	Block based/text based	Earth	https://www.softbankrobotics.com/emea/en/nao
Alpha 1 Pro	Secondary	Ready to use	Block based	Earth	https://ubtrobot.com/pages/alpha
Parrot Mambo	Secondary	Ready to use	Block based/text based	Air	https://www.parrot.com/it/droni/parrot-mambo-fpv
Openrov 2.8	Secondary	To build	Text based	Water	https://store.openrov.com/collections/diy-rovs-part/products/openrov-v2-8-kit
SeaMATE AngelFish	Secondary	To build	–	Water	https://seamate.org/collections/rov-kits/products/3-seamate-angelfish-control-box-kit
SeaMATE PufferFish	Secondary	To build	–	Water	https://seamate.org/collections/rov-kits/products/seamate-pufferfish-control-box-kit-rev-7
Kuka Youbot	Upper secondary	Ready to use	Text based	Earth	http://www.youbot-store.com
Dobot Magician	Secondary	Ready to use	Block based	Earth	http://www.dobot.it

Table 3.3 Results from the classification of evaluation proposed in Section 3

Paper	What	When	To what purpose	How
Akagi et al. (2015)	Performance: Robotics	Post	Assessment of learning	Qualitative
Bers et al. (2014)	Performance: Computational thinking skills	During	Assessment of learning	Quantitative Offline (rubric, Likert scale)
Bharatharaj et al. (2018)	Behaviour	During	Unknown	Quantitative Offline (questionnaires) Qualitative (observation)
Cannon et al. (2007)	Attitude: STEM	Pre, during, post	Assessment of learning	Quantitative Offline (surveys, rankings) Qualitative (students' comments)
Castro et al. (2018)	Performance: Robotics	Pre, post	Assessment of learning	Quantitative Offline (questionnaires)
Cesaretti et al. (2017)	Attitude: STEM, teamwork	Post	Assessment of learning	Quantitative Offline (questionnaires)
Chalmers (2018)	Performance: Computational thinking skills	Pre, during, post	Assessment of learning	Quantitative Offline (questionnaires) Qualitative (interviews)
Chang et al. (2010)	Performance: Second language	Pre, during	Assessment of learning	Qualitative (video recording)
Chen (2019)	Performance: Problem-solving skill (self-assessment)	Post	Assessment of learning	Mixed Offline (interviews and data analysis)
Cook et al. (2005)	Behaviour	During, post	Unknown	Quantitative Offline (video recording and data analysis) Qualitative (interviews)
Costantini et al. (2017)	Unknown	Unknown	Unknown	Unknown
Cross et al. (2015)	Performance: Technological fluency, robotics Attitude: Teamwork	Pre, post	Assessment of learning	Quantitative Offline (questionnaires) Qualitative (interviews, observations)

Table 3.3 (continued)

Paper	What	When	To what purpose	How
Cross et al. (2017)	Performance: Technological fluency, robotics Attitude: Teamwork	Pre, post	Assessment of learning	Quantitative Offline (questionnaires) Qualitative (interviews, observations)
Daniela et al. (2017)	Attitude: School	Pre, during, post	Assessment of learning	Quantitative Offline (questionnaires) Qualitative (observations)
Denicolai et al. (2018)	Unknown	Unknown	Unknown	Unknown
Di Lieto et al. (2017)	Performance: Visuo-spatial working memory inhibition skills, attention, robotics	Pre, post	Assessment of learning	Quantitative offline (questionnaires)
Eguchi (2015)	Performance: Subjects Attitude: Twenty-first-century skills	Post	Assessment of learning	Mixed Offline (final essay analysed using text coding with quasi-grounded Theory)
Eguchi (2016)	Attitude: STEM, twenty-first-century skills	Post	Assessment of learning	Quantitative Offline (questionnaires)
Ferrarelli et al. (2018)	Performance: Physics Attitude: Technology, teamwork	Pre, during, post	Assessment of learning	Quantitative Offline (questionnaires) Qualitative (observations)
Frangou et al. (2008)	Performance: Subjects	Post	Assessment of learning	Qualitative
Fridin (2014)	Performance: Robot-children interactions	During	Assessment of learning	Quantitative Online (video recording with robot's camera)
Fridin and Belokopytov (2014)	Attitude: Technology	Pre	Unknown	Quantitative Offline (questionnaires)
Goldman et al. (2004)	Performance: Physics, math	Pre, post	Assessment of learning	Quantitative (questionnaires)

(continued)

Table 3.3 (continued)

Paper	What	When	To what purpose	How
Horn et al. (2008)	Behaviour	During	Unknown	Quantitative Online (computers log, evaluators log) Qualitative (observation)
Holt et al. (2013)	Behaviour	During	Unknown	Quantitative Online (system error logs and callouts) Qualitative (observation)
Iacobelli (2010)	Performance: Robotics, twenty-first-century skills	Post	Assessment of learning	Quantitative Offline (rubric) Qualitative (observation)
Iacobelli and Spano (2011)	Unknown	Unknown	Unknown	Unknown
Jeon et al. (2016)	Unknown	Unknown	Unknown	Unknown
Jormanainen and Sutinen (2012)	Performance: Robotics	During	Assessment of learning	Quantitative Online (data logging of students programming)
Junior et al. (2013)	Attitude: Robotics	Post	Assessment of learning	Quantitative Offline (questionnaires)
Kandlhofer and Steinbauer (2016)	Performance: Technical skills Attitude: Science, social and soft skills	Pre, post	Assessment of learning	Quantitative Offline (questionnaires)
Kim et al. (2015)	Performance: STEM teaching and learning Attitude: Engagement in robotics and STEM Activities	Pre, during, post	Assessment of learning	Quantitative Offline (questionnaires) Qualitative (observations)
Kory Westlund et al. (2016)	Attitude: Teachers' perception of social robots in their classrooms	Pre, post	Unknown	Quantitative Offline (questionnaires) Qualitative (interviews)
Kucuk and Sisman (2017)	Behaviour: Student-teacher interaction	During	Unknown	Quantitative Online (video recording)

Table 3.3 (continued)

Paper	What	When	To what purpose	How
Lindh and Holgersson (2007)	Performance: Math, problem-solving	Pre, during, post	Assessment of learning	Quantitative Offline (tests in mathematics and problem-solving) Qualitative (observation, interview and inquiry)
Lins et al. (2018)	Unknown	During	Unknown	Quantitative Online (brain wave sensor)
Mengoni et al. (2017)	Unknown	Pre, post	Unknown	Quantitative Offline (questionnaires) Qualitative (observation)
Micotti et al. (2017)	Unknown	Unknown	Unknown	Unknown
Montero and Jormanainen (2016)	Unknown	Unknown	Unknown	Unknown
Oreggia et al. (2016)	Attitude: Computer engineering	Pre, post	Assessment of learning	Quantitative Offline (questionnaires)
Ospennikova et al. (2015)	Performance: Physics	Unknown	Assessment of learning	Unknown
Ozgur et al. (2018)	Behaviour	During	Unknown	Quantitative Online (log of the motion and game data)
Palsbo and Hood-Szivek (2012)	Performance: Hand motor function	Pre, post	Unknown	Quantitative Offline (specific test for the motor function)
Polishuk et al. (2012)	Unknown	Unknown	Unknown	Unknown
Polishuk and Verner (2017)	Performance: Systems thinking skills	Post	Assessment of learning	Qualitative (rubrics)
Rusk et al. (2008)	Unknown	Unknown	Unknown	Unknown
Ryu et al. (2013)	Unknown	Unknown	Unknown	Unknown
Sahin et al. (2014)	Attitude: STEM, twenty-first-century skills	During	Assessment of learning	Qualitative (observation, interviews)

(continued)

Table 3.3 (continued)

Paper	What	When	To what purpose	How
Scaradozzi et al. (2018)	Performance: Robotics, coding, tinkering	Pre, post	Assessment of learning	Quantitative Offline (questionnaires)
Scaradozzi et al. (2015)	Performance: Robotics	Post	Assessment of learning	Quantitative Offline (final grade)
Scaradozzi et al. (2016)	Performance: Robotics, science Attitude: STEM, teamwork	Post	Assessment of learning	Quantitative Offline (questionnaires)
Screpanti et al. (2018a)	Performance: Robotics, science Attitude: STEM, twenty-first-century skills	Post	Assessment of learning	Quantitative Offline (questionnaires, crossword puzzle)
Screpanti et al. (2018b)	Performance: Robotics Attitude: STEM, twenty-first-century skills	Post	Assessment of learning	Quantitative Offline (surveys, rankings)
Sullivan 2008	Performance: Problem-solving skill, science process skill, technology literacy, systems understanding	Pre, during, post	Assessment of learning	Quantitative Offline (questionnaires) Online (descriptive written logs of student activity created through multiple viewing of the videotapes.)
Tapus et al. (2012)	Behaviour	During	Unknown	Quantitative Online (video recording)
Tocháček et al. (2016)	Performance: Technology literacy	Post	Assessment of learning	Quantitative Offline (not clearly described)
Vitale et al. (2016)	Performance: Robotics, teamwork	During	Assessment of learning	Quantitative Offline (rubric)
Weinberg et al. (2007)	Attitude: STEM	Pre, post	Assessment of learning	Quantitative Offline (questionnaires)
West et al. (2018)	Unknown	Unknown	Unknown	Unknown

References

Ackermann, E. (2001). Piaget's constructivism, Papert's constructionism: What's the difference. *Future of Learning Group Publication, 5*(3), 438.

Akagi, T., Fujimoto, S., Kuno, H., Araki, K., Yamada, S., & Dohta, S. (2015). Systematic educational program for robotics and mechatronics engineering in OUS using robot competition. *Procedia Computer Science, 76*, 2–8.

Alimisis, D. (2013). Educational robotics: Open questions and new challenges. *Themes in Science and Technology Education, 6*(1), 63–71.

Angel-Fernandez, J. M., & Vincze, M. (2018). Towards a formal definition of educational robotics. In P. Zech & J. Piater (Eds.), *Proceedings of the austrian robotics workshop 2018* (Conference series). Innsbruck: Innsbruck University Press. https://doi.org/10.15203/3187-22-1.

Asif, R., Merceron, A., Ali, S. A., & Haider, N. G. (2017). Analyzing undergraduate students' performance using educational data mining. *Computers & Education, 113*, 177–194.

Atmatzidou, S., & Demetriadis, S. (2016). Advancing students' computational thinking skills through educational robotics: A study on age and gender relevant differences. *Robotics and Autonomous Systems, 75*, 661–670.

Baker, R.S., Corbett, A.T., & Koedinger, K. R. (2004). Detecting student misuse of intelligent tutoring systems. In *Proceedings of the 7th international conference on intelligent tutoring systems* (pp. 531–540).

Beck, J. E., & Woolf, B. P. (2000). High-level student modeling with machine learning. In G. Gauthier, C. Frasson, & K. VanLehn (Eds.), *Intelligent tutoring systems* (Lecture notes in computer science) (pp. 584–593). Berlin, Heidelberg: Springer.

Bellas, F., Naya, M., Varela, G., Llamas, L., Prieto, A., Becerra, J. C., Bautista, M., Fain, A., & Duro, R. (2018). The Robobo project: Bringing educational robotics closer to real-world applications. In W. Lepuschitz, M. Merdan, G. Koppensteiner, R. Balogh, & D. Obdržálek (Eds.), *Robotics in education*. RiE 2017. Advances in intelligent systems and computing (Vol. 630). Cham: Springer.

Belpaeme, T., Kennedy, J., Ramachandran, A., Scassellati, B., & Tanaka, F. (2018). Social robots for education: A review. *Science robotics, 3*(21), eaat5954.

Benitti, F. B. V. (2012). Exploring the educational potential of robotics in schools: A systematic review. *Computers & Education, 58*(3), 978–988.

Berland, M., Martin, T., Benton, T., Petrick Smith, C., & Davis, D. (2013). Using learning analytics to understand the learning pathways of novice programmers. *Journal of the Learning Sciences, 22*(4), 564–599.

Berland, M., Baker, R. S., & Blikstein, P. (2014). Educational data mining and learning analytics: Applications to constructionist research. *Technology, Knowledge and Learning, 19*(1–2), 205–220.

Bers, M. U., Flannery, L., Kazakoff, E. R., & Sullivan, A. (2014). Computational thinking and tinkering: Exploration of an early childhood robotics curriculum. *Computers & Education, 72*, 145–157.

Bharatharaj, J., Huang, L., Krägeloh, C., Elara, M. R., & Al-Jumaily, A. (2018). Social engagement of children with autism spectrum disorder in interaction with a parrot-inspired therapeutic robot. *Procedia Computer Science, 133*, 368–376.

Blikstein, P., & Worsley, M. (2016). Multimodal learning analytics and education data mining: Using computational technologies to measure complex learning tasks. *Journal of Learning Analytics, 3*(2), 220–238.

Blikstein, P., Worsley, M., Piech, C., Sahami, M., Cooper, S., & Koller, D. (2014). Programming pluralism: Using learning analytics to detect patterns in the learning of computer programming. *Journal of the Learning Sciences, 23*(4), 561–599.

Blikstein, P., Kabayadondo, Z., Martin, A., & Fields, D. (2017). An assessment instrument of technological literacies in makerspaces and FabLabs. *Journal of Engineering Education, 106*(1), 149–175.

Breazeal, C., Dautenhahn, K., & Kanda, T. (2016). Social robotics. In B. Siciliano & O. Khatib (Eds.), *Springer handbook of robotics* (pp. 1935–1972). Cham: Springer.
Brophy, S., Klein, S., Portsmore, M., & Rogers, C. (2008). Advancing engineering education in P-12 classrooms. *Journal of Engineering Education, 97*(3), 369–387.
Cannon, K. R., Panciera, K. A., & Papanikolopoulos, N. P. (2007). Second annual robotics summer camp for underrepresented students. *ACM SIGCSE Bulletin, 39*(3), 14–18. ACM.
Castro, E., Cecchi, F., Valente, M., Buselli, E., Salvini, P., & Dario, P. (2018). Can educational robotics introduce young children to robotics and how can we measure it? *Journal of Computer Assisted Learning, 34*(6), 970–977.
Cesaretti, L., Storti, M., Mazzieri, E., Screpanti, L., Paesani, A., Principi, P., & Scaradozzi, D. (2017). An innovative approach to school-work turnover programme with educational robotics. *Mondo Digitale, 16*(72), 2017–2015.
Chalmers, C. (2018). Robotics and computational thinking in primary school. *International Journal of Child-Computer Interaction, 17*, 93–100.
Chang, C. W., Lee, J. H., Chao, P. Y., Wang, C. Y., & Chen, G. D. (2010). Exploring the possibility of using humanoid robots as instructional tools for teaching a second language in primary school. *Journal of Educational Technology & Society, 13*(2), 13–24.
Chen, X. (2019). How does participation in FIRST LEGO league robotics competition impact children's problem-solving process? In W. Lepuschitz, M. Merdan, G. Koppensteiner, R. Balogh, & D. Obdržálek (Eds.), *Robotics in education*. RiE 2018. Advances in intelligent systems and computing (Vol. 829, pp. 162–167). Cham: Springer.
Cook, A. M., Bentz, B., Harbottle, N., Lynch, C., & Miller, B. (2005). School-based use of a robotic arm system by children with disabilities. *IEEE Transactions on Neural Systems and Rehabilitation Engineering, 13*(4), 452–460.
Costantini, R., Laura, L., Mazza, L., & Santilli, R. (2017). STEAM – un nuovo framework didattico per l'Alternanza Scuola Lavoro: Coding, robotica e design nel Milano Luiss Hub (STEAM – a new educational framework for the alternating school-work programme: Coding, robotics and design at the Milano Luiss Hub). In *Proceedings of DIDAMATICA 2017*.
Cross, J. L., Hamner, E., Bartley, C., & Nourbakhsh, I. (2015). Arts & Bots: application and outcomes of a secondary school robotics program. In *Frontiers in Education Conference (FIE), 2015 IEEE* (pp. 1–9). IEEE.
Cross, J. L., Hamner, E., Zito, L., & Nourbakhsh, I. (2017). Student outcomes from the evaluation of a transdisciplinary middle school robotics program. In *Proceedings of Frontiers in Education Conference (FIE)* (pp. 1–9). IEEE.
Daniela, L., & Strods, R. (2018). Robot as agent in reducing risks of early school leaving. In L. Daniela (Ed.), *Innovations, technologies and research in education* (pp. 140–158). Newcastle upon Tyne: Cambridge Scholars Publishing. ISBN (10): 1-5275-0622-3.
Daniela, L., Strods, R., & Alimisis, D. (2017). Analysis of Robotics-based Learning Interventions for Preventing School Failure and Early School Leaving in Gender Context. In *Proceedings of the 9th International Conference on Education and New Learning Technologies (Edulearn17)* (pp. 810–818). 3–5 July, 2017, Barcelona, Spain. ISBN 9788469737774. ISSN 2340-1117.
Dann, R. (2014). Assessment as learning: Blurring the boundaries of assessment and learning for theory, policy and practice. *Assessment in Education: Principles, Policy & Practice, 21*(2), 149–166.
Denicolai, L., Grimaldi, R., & Palmieri, S. (2018). Robotica e linguaggio audiovisivo: Quando le tecnologie si parlano (robotics and audio-visual language: When different technologies communicate with each other). In *Proceedings of DIDAMATICA 2018*. Cesena, Italy: AICA.
Di Lieto, M. C., Inguaggiato, E., Castro, E., Cecchi, F., Cioni, G., Dell'Omo, M., Laschi, C., Pecini, C., Santerini, G., Sgandurra, G., & Dario, P. (2017). Educational robotics intervention on executive functions in preschool children: A pilot study. *Computers in Human Behavior, 71*, 16–23.

Eguchi, A. (2014). Robotics as a learning tool for educational transformation. In *Proceeding of 4th International Workshop Teaching Robotics, Teaching with Robotics & 5th International Conference Robotics in Education* (pp. 27–34). Padova (Italy).

Eguchi, A. (2015). Educational robotics to promote 21 st century skills and technological understanding among underprivileged undergraduate students. In *Integrated STEM Education Conference (ISEC), 2015 IEEE* (pp. 76–82). IEEE.

Eguchi, A. (2016). RoboCupJunior for promoting STEM education, 21st century skills, and technological advancement through robotics competition. *Robotics and Autonomous Systems, 75*, 692–699.

Eguchi, A. (2017). Bringing robotics in classrooms. In M. Khine (Ed.), *Robotics in STEM education* (pp. 3–31). Cham: Springer.

Ferrarelli, P., Villa, W., Attolini, M., Cesareni, D., Micale, F., Sansone, N., Pantaleone, L. C., & Iocchi, L. (2018). Improving students' concepts about Newtonian mechanics using Mobile robots. In *International conference on robotics and education (RiE) 2017* (pp. 113–124). Cham: Springer.

Frangou, S., Papanikolaou, K., Aravecchia, L., Montel, L., Ionita, S., Arlegui, J., Pina, A., Menegatti, E., Moro, M., Fava, N., & Monfalcon, S. (2008). Representative examples of implementing educational robotics in school based on the constructivist approach. In *Workshop proceedings of SIMPAR* (pp. 54–65).

Fridin, M. (2014). Storytelling by a kindergarten social assistive robot: A tool for constructive learning in preschool education. *Computers & Education, 70*, 53–64.

Fridin, M., & Belokopytov, M. (2014). Acceptance of socially assistive humanoid robot by preschool and elementary school teachers. *Computers in Human Behavior, 33*, 23–31.

Goldman, R., Eguchi, A., & Sklar, E. (2004). Using educational robotics to engage inner-city students with technology. In *Proceedings of the 6th international conference on learning sciences* (pp. 214–221). International Society of the Learning Sciences.

Hattie, J., & Timperley, H. (2007). The power of feedback. *Review of Educational Research, 77*(1), 81–112.

Holt, R., Weightman, A., Gallagher, J., Preston, N., Levesley, M., Mon-Williams, M., & Bhakta, B. (2013). A system in the wild: Deploying a two player arm rehabilitation system for children with cerebral palsy in a school environment. *Journal of Usability Studies, 8*(4), 111–126.

Horn, M. S., Solovey, E. T., & Jacob, R. J. (2008). Tangible programming and informal science learning: Making TUIs work for museums. In *Proceedings of the 7th international conference on interaction design and children* (pp. 194–201). ACM.

Iacobelli, C. (2010). I robot a scuola: l'esperienza insegna (Robots at school: experience teaches). Retrieved from http://www.robocupjr.it/roboscuola/2010_atti/cesareiacobelli_x.pdf

Iacobelli, C. and Spano A. (2011). Competizioni di robotica: Nuovi percorsi per la didattica (robotics competitions: New pathways for teaching and learning). In *Proceedings of DIDAMATICA 2011*. Turin, Italy. AICA.

Jeon, M., FakhrHosseini, M., Barnes, J., Duford, Z., Zhang, R., Ryan, J., & Vasey, E. (2016). Making live theatre with multiple robots as actors: Bringing robots to rural schools to promote STEAM education for underserved students. In *The 11th ACM/IEEE international conference on human robot interaction* (pp. 445–446). IEEE Press.

Jormanainen, I., & Sutinen, E. (2012). Using data mining to support teacher's intervention in a robotics class. In *Digital Game and Intelligent Toy Enhanced Learning (DIGITEL), 2012* IEEE Fourth International Conference on (pp. 39–46). IEEE.

Jung, S. E., & Won, E. S. (2018). Systematic review of research trends in robotics education for young children. *Sustainability, 10*(4), 905.

Junior, L. A., Neto, O. T., Hernandez, M. F., Martins, P. S., Roger, L. L., & Guerra, F. A. (2013). A low-cost and simple arduino-based educational robotics kit. *Cyber Journals: Multidisciplinary Journals in Science and Technology, Journal of Selected Areas in Robotics and Control (JSRC), December edition, 3*(12), 1–7.

Kandlhofer, M., & Steinbauer, G. (2016). Evaluating the impact of educational robotics on pupils' technical-and social-skills and science related attitudes. *Robotics and Autonomous Systems, 75*, 679–685.

Kim, C., Kim, D., Yuan, J., Hill, R. B., Doshi, P., & Thai, C. N. (2015). Robotics to promote elementary education pre-service teachers' STEM engagement, learning, and teaching. *Computers & Education, 91*, 14–31.

Kory Westlund, J., Gordon, G., Spaulding, S., Lee, J. J., Plummer, L., Martinez, M., Das, M., & Breazeal, C. (2016). Lessons from teachers on performing HRI studies with young children in schools. In the *11th ACM/IEEE International Conference on Human Robot Interaction* (pp. 383–390). IEEE Press.

Kucuk, S., & Sisman, B. (2017). Behavioral patterns of elementary students and teachers in one-to-one robotics instruction. *Computers & Education, 111*, 31–43.

Lindh, J¸ Holgersson, T. (2007). Does lego training stimulate pupils ability to solve logical problems? Computers & Education, 49 (4), 1097–1111.

Lins, A. A., de Oliveira, J. M., Rodrigues, J. J., & de Albuquerque, V. H. C. (2018). Robot-assisted therapy for rehabilitation of children with cerebral palsy-a complementary and alternative approach. *Computers in Human Behavior*. https://doi.org/10.1016/j.chb.2018.05.012.

Martinez, S. L., & Stager, G. (2013). *Invent to learn: Making, tinkering, and engineering in the classroom*. Torrance, CA: Constructing modern knowledge press.

Matarić, M. J., & Scassellati, B. (2016). Socially assistive robotics. In B. Siciliano & O. Khatib (Eds.), *Springer handbook of robotics* (pp. 1973–1994). Cham: Springer Handbooks. Springer.

Mengoni, S. E., Irvine, K., Thakur, D., Barton, G., Dautenhahn, K., Guldberg, K., Robins, B., Wellsted, D., & Sharma, S. (2017). Feasibility study of a randomised controlled trial to investigate the effectiveness of using a humanoid robot to improve the social skills of children with autism spectrum disorder (Kaspar RCT): A study protocol. *BMJ Open, 7*(6), e017376.

Merceron, A., & Yacef, K. (2004). Mining student data captured from a web-based tutoring tool: Initial exploration and results. *Journal of Interactive Learning Research, 15*(4), 319–346.

Mevarech, Z. R., & Kramarski, B. (1993). Vygotsky and Papert: Social-cognitive interactions within logo environments. *British Journal of Educational Psychology, 63*, 96–109. https://doi.org/10.1111/j.2044-8279.1993.tb01044.x.

Micotti, F., Fiocchi, E. and Giurato, M. (2017). Introdurre la robotica nelle scuole secondarie, un approccio progettuale (Introducing Robotics into secondary schools, a planning approach). In *Proceedings of DIDAMATICA 2017.*

Miller, D. P., & Nourbakhsh, I. (2016). Robotics for education. In *Springer handbook of robotics* (pp. 2115–2134). Cham: Springer.

Montero, C. S., & Jormanainen, I. (2016). Theater meets robot–toward inclusive STEAM education. In *International conference EduRobotics 2016* (pp. 34–40). Cham: Springer.

Moro, M., Agatolio, F., & Menegatti, E. (2018). The RoboESL project: Development, evaluation and outcomes regarding the proposed robotic enhanced curricula. *International Journal of Smart Education and Urban Society (IJSEUS), 9*(1), 48–60.

Mubin, O., Stevens, C. J., Shahid, S., Al Mahmud, A., & Dong, J. J. (2013). A review of the applicability of robots in education. *Journal of Technology in Education and Learning, 1*(1–7), 13.

Naya, M., Varela, G., Llamas, L., Bautista, M., Becerra, J. C., Bellas, F., Abraham Prieto, Alavaro Deibe & Duro, R. J. (2017). A versatile robotic platform for educational interaction. In *Intelligent Data Acquisition and Advanced Computing Systems: Technology and Applications (IDAACS), 2017 9th IEEE International Conference on* (Vol. 1, pp. 138–144). IEEE.

Oreggia, M., Chiorri, C., Pozzi, F., & Tacchella, A. (2016, July). Introducing Computer Engineering Curriculum to Upper Secondary Students: An Evaluation of Experiences Based on Educational Robotics. In *2016 IEEE 16th International Conference on Advanced Learning Technologies (ICALT)* (pp. 293–294). IEEE.

Ornelas, F., & Ordonez, C. (2017). Predicting student success: A Naïve Bayesian application to community college data. *Technology, Knowledge and Learning, 22*(3), 299–315.

Ospennikova, E., Ershov, M., & Iljin, I. (2015). Educational robotics as an innovative educational technology. *Procedia-Social and Behavioral Sciences, 214*, 18–26.

Ozgur, A. G., Wessel, M. J., Johal, W., Sharma, K., Özgür, A., Vuadens, P., Francesco Mondada, Friedhelm Christoph Hummel & Dillenbourg, P. (2018). Iterative design of an upper limb rehabilitation game with tangible robots. In *ACM/IEEE International Conference on Human-Robot Interaction (HRI)* (p. 187).

Palsbo, S. E., & Hood-Szivek, P. (2012). Effect of robotic-assisted three-dimensional repetitive motion to improve hand motor function and control in children with handwriting deficits: A nonrandomized phase 2 device trial. *American Journal of Occupational Therapy, 66*(6), 682–690.

Papert, S. (1980). *Mindstorms: Children, computers, and powerful ideas*. New York: Basic Books, Inc.

Papert, S. (1991). Situating constructionism. In S. Papert & I. Harel (Eds.), *Constructionism* (pp. 1–11). Norwood, NJ: Ablex Publishing.

Polishuk, A., & Verner, I. (2017). Student-robot interactions in museum workshops: Learning activities and outcomes. In M. Merdan, W. Lepuschitz, G. Koppensteiner, & R. Balogh (Eds.), *Robotics in education. Advances in intelligent systems and computing* (Vol. 457, pp. 233–244). Cham: Springer.

Polishuk, A., Verner, I., Klein, Y., Inbar, E., Mir, R., & Wertheim, I. (2012). The challenge of robotics education in science museums. *International Journal of Robots, Education and Art (IJREA), 2*(1), 30–37.

Rusk, N., Resnick, M., Berg, R., & Pezalla-Granlund, M. (2008). New pathways into robotics: Strategies for broadening participation. *Journal of Science Education and Technology, 17*(1), 59–69.

Ryu, G. J., Kang, J. B., Kim, C. G., & Song, B. S. (2013). Development of a robot remote support system for student with health impairment. In *Proceedings of the 7th International Convention on Rehabilitation Engineering and Assistive Technology* (p. 24). Singapore Therapeutic, Assistive & Rehabilitative Technologies (START) Centre.

Sahin, A., Ayar, M. C., & Adiguzel, T. (2014). STEM related after-school program activities and associated outcomes on student learning. *Educational Sciences: Theory and Practice, 14*(1), 309–322.

Scaradozzi, D., Cesaretti, L., Screpanti, L., Costa, D., Zingaretti, S., & Valzano, M. (in press). Innovative tools for teaching marine robotics, iot and control strategies since the primary school. In: Daniela, L. (Ed.), *Smart learning with educational robotics – using robots to scaffold learning outcomes*, Springer. ISBN 978-3-030-19912-8.

Scaradozzi, D., Sorbi, L., Pedale, A., Valzano, M., & Vergine, C. (2015). Teaching robotics at the primary school: An innovative approach. *Procedia-Social and Behavioral Sciences, 174*, 3838–3846.

Scaradozzi, D., Screpanti, L., Cesaretti, L., Mazzieri, E., Storti, M., Brandoni, M., & Longhi, A. (2016). Rethink Loreto: We build our smart city!" A stem education experience for introducing smart city concept with the educational robotics. In *9th annual international conference of education, research and innovation (ICERI 2016)*, Seville, Spain (pp. 750–758).

Scaradozzi, D., Screpanti, L., Cesaretti, L., Storti, M., & Mazzieri, E. (2018). Implementation and assessment methodologies of teachers' training courses for STEM activities. *Technology, Knowledge and Learning*, 1–21.

Screpanti, L., Cesaretti, L., Storti, M., Mazzieri, E., & Longhi, A. (2018a). Advancing K12 education through Educational Robotics to shape the citizens of the future. In *Proceedings of DIDAMATICA 2018*. AICA.

Screpanti, L., Cesaretti, L., Marchetti, L., Baione, A., Natalucci, I. N., & Scaradozzi, D. (2018b, July). An educational robotics activity to promote gender equality in STEM education. In *Procedings of the eighteenth International Conference on Information, Communication Technologies in Education (ICICTE 2018)* (pp. 336–346). Chania, Crete, Greece.

Sullivan, F. R. (2008). Robotics and science literacy: Thinking skills, science process skills and systems understanding. *Journal of Research in Science Teaching: The Official Journal of the National Association for Research in Science Teaching, 45*(3), 373–394.
Tapus, A., Peca, A., Aly, A., Pop, C., Jisa, L., Pintea, S., Rusu, A., & David, D. O. (2012). Children with autism social engagement in interaction with Nao, an imitative robot: A series of single case experiments. *Interaction Studies, 13*(3), 315–347.
Tocháček, D., Lapeš, J., & Fuglík, V. (2016). Developing technological knowledge and programming skills of secondary schools students through the educational robotics projects. *Procedia-Social and Behavioral Sciences, 217*, 377–381.
Toh, E., Poh, L., Causo, A., Tzuo, P. W., Chen, I., & Yeo, S. H. (2016). A review on the use of robots in education and young children. *Journal of Educational Technology & Society, 19*(2).
Turkle, S., & Papert, S. (1992). Epistemological pluralism and the revaluation of the concrete. *Journal of Mathematical Behavior, 11*(1), 3–33.
Vitale, G., Bonarini, A., Matteucci, M., & Bascetta, L. (2016). Toward vocational robotics: An experience in post-secondary school education and job training through robotics. *IEEE Robotics & Automation Magazine, 23*(4), 73–81.
Vygotsky, L. S. (1968). *Thought and language*. Cambridge, MA: M.I.T. Press.
Weinberg, J. B., Pettibone, J. C., Thomas, S. L., Stephen, M. L., & Stein, C. (2007). The impact of robot projects on girls' attitudes toward science and engineering. In *Workshop on research in robots for education* (Vol. 3, pp. 1–5).
West, J., Vadiee, N., Sutherland, E., Kaye, B., & Baker, K. (2018). Making STEM accessible and effective through NASA robotics programs. *Tribal College Journal of American Indian Higher Education, 29*(4).

Chapter 4
Introducing Maker Movement in Educational Robotics: Beyond Prefabricated Robots and "Black Boxes"

Dimitris Alimisis, Rene Alimisi, Dimitrios Loukatos, and Emmanouil Zoulias

Abstract Nowadays, several studies assure that digital fabrication and making technologies, if coupled with proper learning methodologies such as constructivism and constructionism, can provide learning experiences that promote young people's creativity, critical thinking, teamwork and problem-solving skills – the essential skills necessary in the workplace of the twenty-first century. Robotic technologies combined with digital fabrication and DIY electronics emerge as unique making tools that can create a learning environment attracting and keeping learners interested and motivated with hands-on, fun learning activities. The "maker movement" is seen as an inspiring and creative way for youth to deal with our world and able to develop technological interest and competences. However, in the field of educational robotics, the focus is often on prefabricated robots and ready-made code to program behaviours for the robots. This way, robotics is conceived as "black box" for children who are invited to play or interact with a robot without understanding "what's inside" and how it works. The project eCraft2Learn researched, designed, piloted and evaluated an ecosystem intended to introduce digital fabrication and maker movement in formal and informal education, to make robots transparent for children and finally to help them make their own robotic artefacts.

Keywords Educational robotics · Maker movement · Constructionism · eCraft2Learn project

D. Alimisis (✉) · R. Alimisi · E. Zoulias
EDUMOTIVA - European Lab for Educational Technology, Athens, Greece
e-mail: alimisis@edumotiva.eu

D. Loukatos
EDUMOTIVA - European Lab for Educational Technology, Athens, Greece

Agricultural University of Athens - Department of Natural Resources Management & Agricultural Engineering, Athens, Greece

L. Daniela (ed.), *Smart Learning with Educational Robotics*,
https://doi.org/10.1007/978-3-030-19913-5_4

Introduction

While robotic technologies are radically changing the way people work in industry, finance, services, media and commerce, the introduction of robotics in education has emerged in recent years as a challenge for education systems. Engaging youth in robotics education even from early school years helps their familiarisation with robotic technologies and promotes a first understanding of the way robots are created and operate to serve people in their everyday life (Castro et al. 2018; Scaradozzi et al. 2015).

Robotic technologies if coupled with proper learning methodologies such as suggested by constructivism (Piaget 1974) and constructionism (Papert and Harel 1991) can provide learning experiences that promote young people's creative thinking, teamwork and problem-solving skills (Alimisis 2014) – the essential skills necessary in the workplace of the twenty-first century (Fullan and Langworthy 2013). Studies report a potential impact on learners, both in subject areas (physics, electronics, mathematics, engineering, computer science and more) and on personal development including cognitive, metacognitive and social skills (Alimisis 2013; Alimisis et al. 2017; Moro et al. 2018).

However, the progress in the robotics education arena is rather slow. Robotic technologies in education are often used in a way reinforcing old methods of teaching (Alimisis 2013), which is ineffective in a society and labour market demanding creativity, entrepreneurship, critical thinking, collaboration skills, computational fluency and so forth (Fullan and Langworthy 2013).

This chapter introduces the connection of educational robotics with the maker movement, exemplifies this connection with the case of an exemplary project presenting both its pedagogical and technological solution, reports evaluation results and lessons learnt from the project pilots with students and finally concludes that the future of educational robotics should be envisioned in close connection with the maker movement.

Connecting Robotics Education with the Maker Movement

While education in robotics is so important and fruitful, current curricula and educational resources in school education are very often developed according to a widespread misconception that robotics is "hard" science and suitable only for gifted children or science- and technology-oriented students. This misconception is often coupled with gender-biased views that robotics subjects are only for boys and poses real obstacles to the adoption of robotics in education (Alimisis 2013).

Furthermore, lack of robotic equipment in schools and limited resources often cause additional difficulties. The commercial robotics kits are usually too expensive for schools or cheap and unreliable (Darrah 2018). In addition to this, most of the robotics kits available in the market come with ready-made robots, inherent lock-in mechanisms, closed hardware and/or software or cookbook-like recipes for step-by-step assembly of one or few predefined models treating users rather as consumers than creative makers and learners. Consequently, robots are still conceived as "black boxes" by children who are invited to play or interact with a robot without understanding "what's inside" and how it works (Alimisis and Kynigos 2009; Alimisis 2013).

We have observed in several robotics courses using this approach that, not surprisingly, children's interest, after the first assembly and some trials, is diminishing and learners don't advance beyond some mere knowledge acquisition. Moreover, this approach of the ready-made robots is not compatible with the principles of the maker movement (Blikstein 2013; Schon et al. 2014), which has recently emerged in education with the great promise to democratise access to opportunities for learning by making, for skills development and more importantly for fostering positive attitudes and openness to making culture for the future generations of citizens.

In this chapter, we argue for moving away from elitist misconceptions to the recognition that fluency with robotic technologies is no longer just a vocational skill, but it is knowledge and skills valuable for every citizen (Alimisis 2013). This, in turn, implies that we need to provide learners a hands-on open-source robotics learning environment that will be inexpensive, reliable and able to support learners to make their own robotic creations from scratch. According to this approach, learners are invited to act as makers and developers of their own projects instead of being passive consumers of ready-made robots. For this purpose, the educational robotics community needs to explore the potential of the making culture, the emerging digital fabrication and do-it-yourself (DIY) philosophy and connect with the broader community of makers.

The maker movement has its roots in Papert' s constructionism (Papert and Harel 1991) and can offer a vision for a robotics education that will enable learners to make their own robotic artefacts using the twenty-first-century technologies. The incorporation of the maker movement in robotics education implies a paradigm shift in robotics curricula from step-by-step-guided tasks and predefined robots towards open projects and making practices, from "black box" and silo products to the "white box" paradigm where learners become "makers" of their own transparent robotic artefacts (Alimisis 2013).

Inspired from the maker movement, the eCraft2Learn project (eCraft2Learn 2018) researched, designed, piloted and validated an ecosystem aimed to introduce digital fabrication and making technologies in education. The constructivist "learning by making" methodology is strongly related to the digital fabrication and "do-it-yourself" (DIY) philosophy (Schon et al. 2014) and is the driving force behind the eCraft2Learn pedagogy. Hence, the project suggests that the twenty-first-century learning ecosystems should be designed in a way that can actively engage students with learning tasks, hands-on activities and learning experiences that promote young people's creativity, critical thinking, teamwork and problem-solving.

To exemplify this paradigm in concrete terms, we present in the next sections the eCraft2Learn project: the pedagogical methodology, the technological solution, the pilots with children, indicative projects, evaluation results and lessons learnt.

The eCraft2Learn Project in Action

The Context

This section describes the small-scale pilots with learners that were carried out in Athens during the project implementation period (Autumn 2017–Spring 2018). The small-scale pilots were conducted in two rounds in formal and informal education settings. Prior to deploying the pilots, the involved teachers participated in capacity-building workshops that aimed at enabling them to use the eCraft2Learn tools and the corresponding pedagogical model (Alimisis et al. 2018).

The pilots in formal education settings were carried out in a secondary vocational school where the main challenge was how to integrate the eCraft2Learn lab in the school curriculum (Asimakopoulos et al. 2018). This chapter focuses on the informal eCraft2Learn lab, which was established in the Technopolis City of Athens, a hub of cultural and educational events and a focal point in the cultural identity of Athens. Technopolis is a former gas factory that was restored to an industrial park. A wide variety of cultural and educational events are held in Technopolis every year: music, dance, theatre and performing arts, plastic and applied arts, educational programs for children and youth, entrepreneurship and temporary exhibitions, attracting over 600,000 people annually. It is centrally located in Athens and well accessible. All these made it an ideal place for accommodating the eCraft2Learn initiative. The lab was set up in the first floor of a former industrial building, the Gas Water Tower. The old machinery remained in the place, creating an inspiring scenery for making with strong conceptual symbolism.

An open invitation to the local school community was distributed. No student was excluded from the selection process; the only limitation was the room space. The first pilot round was conducted with 24 students from 13 to 17 years old. The same number of students participated in the second pilot round. The parents were informed and agreed signing a consent form that their children's activities are observed and analysed. Table 4.1 summarises the key information regarding the two pilot rounds in Athens.

The Learning Methodology

Embedded in a constructionist pedagogical model, the learning methodology is aimed to encourage teachers and students to work together and explore the fun and challenges of the making process. The methodology proposes five stages highly

Table 4.1 Key information on the two pilot rounds

	First pilot round	Second pilot round
Total participants	24 students (13–17 years old)	24 students (13–17 years old)
Number of teachers	6–8 per class	4–5 per class
Teamwork approach	Work in groups of 3–4	Work in groups of 3–4
Time	3 hours every Saturday	4 hours every day in 1 week
Total duration	November 2017–February 2018 Totally 30 hours	June 2018 Totally 20 hours

interlinked: ideation, planning, creation, programming and sharing (Alimisis et al. 2018). The main pillars of the pedagogical model are presented shortly in the following lines.

Project-Based Learning

The eCraft2Learn learning methodology focuses on the project-based learning, a model for classroom activities, that shifts away from the traditional classroom practices of short, isolated, teacher-centred lessons. The methodology encourages learners' engagement in a real-life scenario that requires taking an action for making or using a robot in a creative way, planning and designing their own robotic projects, making and programming their own robotic artefacts, testing and reflecting on their solutions and finally sharing their experiences with the community. Students are encouraged and supported to devise their own heuristic approach to a solution which offers much more space for creativity and involvement in creative design for learners compared to closed problem-solving.

Teamwork

Following the pedagogical ideas underpinning the eCraft2Learn methodology, the teamwork is highly encouraged. The students, early from the beginning, are invited to form groups of 3–4. As the sessions are going by, the students can move to support other groups as well, to exchange tips and to allocate roles. In some groups, the students may be equally involved in the project tasks, but in most cases, there is a role rotation. For example, some students may be more involved into programming, others more into electrical circuit making while some others are taking care of the handcrafting tasks or 3D modelling. The reasons for this role allocation is usually related to time constraints and personal interests.

During the first session, focus is placed on familiarising the students with the eCraft2Learn tools, technologies and resources. Some groups need more time for familiarisation than others, but the whole familiarisation process is integrated into the making process and occurred through the practical engagement in projects for computer-supported artefact constructions. It is worth mentioning that as the workshops are progressing, the students are expected to become more confident in using the available tools and more eager in trying out different ideas. As the participant teachers have pointed out, initially the students are a bit reticent but, as soon as they become familiar with the available tools and technologies, are noticed to take initiatives for new projects and extended implementations.

Ice-Breaking and Setting the Rules in the Informal Pilot Site

The first session started with ice-breaking activities, the setting of the ground rules and the elaboration of the process which the students will go through. Given that the students and the teachers in the informal site did not know already one another, part of the first session was dedicated to ice-breaking activities. These activities were selected in advance by the teachers with the aim to activate the necessary mechanisms for the "group-development process" and the establishment of a positive and warm atmosphere.

In the context of the ice-breaking activities, the students were encouraged to form a circle and introduce themselves and to talk about their hobbies and interests; through playful techniques, they were also invited to have short one-to-one conversations. It is noteworthy that the teachers took also part in the ice-breaking activities and the discussions. The formation of the groups was partly based on these discussions and partly random. These discussions were also seen as important steps towards team bonding and good relationship establishment.

During the first session, focus was also placed (at group level) on creating a set of rules that would reflect the accepted behaviours in the group and in the lab, for both teachers and students. The teachers considered that the best way to create a set of rules was to decide on them as a group. It was important to ensure that the appropriate rules have been set for establishing a positive atmosphere for peer learning, smooth project deployment and ideas and experiences sharing. The discussion focused on the following topics: group/classroom behaviour, the importance of sharing and ways to support it, the use of the equipment, storage/uploading/downloading of files and lab safety rules.

The discussion about lab safety rules was revisited as the sessions were progressing. The ice-breaking activities and the setting of the rules were followed by the exploration of the lab equipment at group level. Supported by their teachers, the students set up their working stations and did their very first steps into electrical circuit making using the Tinkercad circuits simulator (Tinkercad 2018).

Implementing the eCraft2Learn Methodology

The ideation stage was considered a challenging process, and the teachers, especially in the beginning, supported it a lot. The first projects that the students involved into were proposed by the teachers, who exploited the list of the indicative scenarios introduced during their teacher training. Easy-to-start-with projects were selected with the aim to smoothly familiarise the students with the available tools and the kind of artefacts that can be created. As the sessions were progressing, the teachers were reducing the level of support on this matter encouraging free choice in project selection.

More precisely, the students were asked about any possible idea that they would like to implement soon. It is noteworthy that through their diaries, they were also encouraged to periodically document their ideas for new projects. Their responses on this matter were not very enlightening in the beginning. However, as they became more familiar with tools and technologies, they started expressing interest in working on specific or thematic projects. In December, being in Christmas mood, some teams were noticed to give a Christmas touch to their artefacts and discuss the implementation of Christmas-related artefacts. The review of the students' diaries brought also additional interesting ideas into focus: many students expressed an interest in creating a moving robotic artefact that could be controlled by them. Some of their ideas were vague enough while some others more specific. For example, they were referring to robots that move and change colours, solar cars, vehicles with many sensors, cars that move around and follow commands and more. Building upon this interest, the teachers supported a relevant project for DIY automobiles, providing students with the freedom to personalise their automobiles, to add specific behaviours and functionalities and to give them the form that they liked.

During the second pilot round, most of the projects came directly by the students. It is noteworthy that one group (the members of which had met one another during the first pilot round) mentioned that they arranged a meeting in a cafe before the beginning of the second round to get together and to discuss some ideas that had emerged for a new computer-supported artefact construction. This was another encouraging sign that the eCraft2Learn practices had entered learners' daily life in a meaningful way.

The generation of ideas was also important during failures; failures were part of the making process (i.e. failed prints, artefacts that did not operate properly), and often the students were invited to share their ideas regarding possible solutions for overcoming the emerging problems.

The teachers discreetly observed and supported this process; in some cases, teachers' intervention was more dynamic by providing useful explanations (i.e. in making circuitry more transparent, increasing students' understanding of electronics) to help students move forward. Frequently, the teachers were encouraging the team members to bring these ideas in the plenary session for the benefit of the whole group. Sharing existing ideas, plans for implementation, problem-solving practices and thoughts in group and in the plenary session was seen as a process that can significantly boost the generation of ideas for new computer-supported artefact constructions.

Bringing Ideas in the Plenary Sessions

There was also encouragement towards analysing ideas, breaking down complex activities into subtasks, keeping notes about science, technology, engineering, arts and math (STEAM) concepts related to their project, listing the materials that will be needed, sketching the structure of the construction and visualising the key processes. This was the stage of planning that in many cases was embedded in the ideation process, revisited and creatively re-approached by the groups during the creation of the artefacts and the programming phase. In a way, these practices show how interlinked the stages of the eCraft2Learn methodology are. Most of the teams did paper-based plans, while few others were based on oral agreements at team level.

As the sessions were progressing, the students were engaging more naturally in the creative production of different artefacts based on their interests and at their own pace. Different projects were on at the same time, different challenges were calling for solutions and lots of hands-on making was on that was inspiring students to dig deeper and to extend their ideas. Moved by the fun of making, many students were noticed to stay longer in the eCraft2Learn lab than initially planned.

Teamwork, Role Allocation and Challenges

Role allocation was also noticed in some teams; some students were in charge of the electrical circuit making, others more into programming and some others more involved into 3D modelling and handcrafting. The role allocation happened at team level and was not enforced by the teachers. However, there were teams where all the members were involved in all the parts of the development of the computer-supported artefact, supporting one another. The teachers intervened only in few cases where one member of the team was inactive. They were mainly trying to understand the reasons behind the inactivity and to create a situation where, through the interaction with the other team members, a role for him/her would emerge.

For example, in one of the teams, there was a young boy rather introvert that was absorbed by his smartphone. The teacher/coach of the team told him that it would be very useful to record the artefact construction process using his smartphone as this would allow the sharing of the work online ensuring greater visibility. The student took happily the challenge and started observing what was going on but (initially) only through his smartphone video-recording the process of the construction. Smoothly, he was taken over by the making spirit and was noticed to participate more, to express ideas for alternative solutions and become active member of the team.

A Closer Look into the Aspect of "Sharing"

The sharing of the making processes and projects with others was considered of great significance. The teachers encouraged all the teams to share the current status of their work in the end of almost each session and to talk about the processes that they went through and their future plans.

In addition, the teams were encouraged to showcase their work in the school community and the wider public. In this light, the students presented their projects in the Athens Science Festival 2018 and interacted with people of all ages and from varying scientific backgrounds as well as with other groups of students that participated in the festival either as exhibitors or visitors.

The students and the teachers were also noticed to record their work using their smartphones or cameras. At a later stage, some of this material was uploaded by them in their social media accounts. Although not practiced by all the teams, some of them, after encouragement by their teachers, were seen to upload their 3D models from the Tinkercad (2018) environment to the Thingiverse (2018) Community.

The Role of the Teachers

The description above revealed already many interesting aspects of the role of the teachers. Given the different ages in the students' group, their contribution on the formation of the teams early in the beginning and their remedial actions were also of great significance. The teachers were supporters of the learning process, co-makers and boosters of the collaborative work, the discussion and the sharing at team level and beyond.

Most of them adopted smoothly these roles, dealing with their duties and concerns at their own pace. Some teachers were quite concerned about their self-image in the class. Their concerns revolved around the question: "What if we do not manage to support the students? What if we cannot answer their questions". As long as they started seeing the eCraft2Learn lab as a making environment and themselves as co-makers, co-designers and facilitators of the learning process, their stress smoothly eliminated allowing them to stand by the students as coaches.

The teachers supported significantly the generation of ideas prompting for relevant group discussions and existing project ideas extension. In addition, they boosted a lot the "can-do" attitude, sharing their enthusiasm with the students and creating an atmosphere conducive to learning.

The Technological Solution

Several innovative technologies have been combined in order to design and implement the eCraft2Learn ecosystem to support the pilot activities with teachers and students. This ecosystem consists of several hardware and software components, glued together using a specially designed web platform called unified user interface (UUI) (2018) and having several RPi3 (2018) units to play a dominant role in this ecosystem.

The Hardware Components

The hardware core of the eCraft2Learn ecosystem includes (Fig. 4.1):

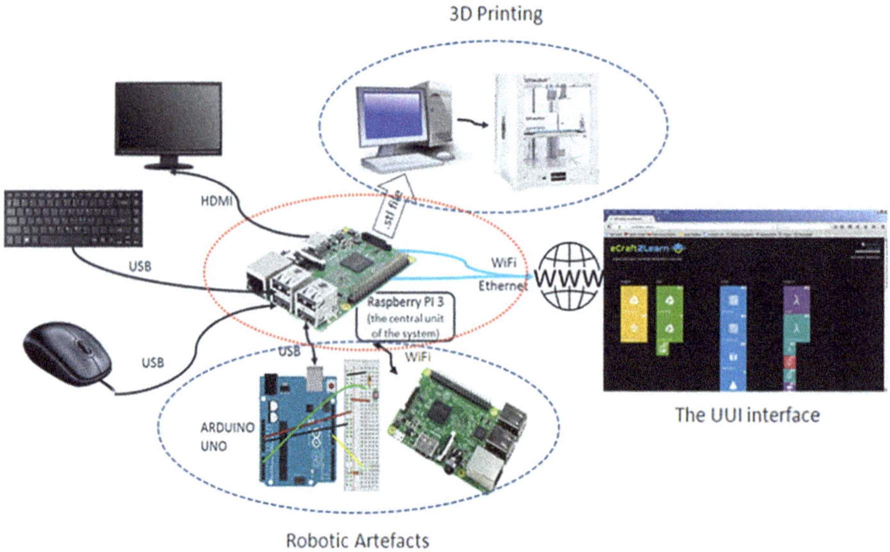

Fig. 4.1 Overview of the eCraft2Learn ecosystem

- A set of Raspberry Pi3 (RPi3) units serving as experimental development computers (workstation units) for the students. These units are equipped with TFT screens and keyboard-mouse sets. The adoption of this solution tackles the problems related to the vast software and hardware diversity characterising a typical school lab and reduces the cost and the overall energy consumption.
- A second set of Raspberry Pi3 units that can be used for a potential implementation in the designed artefacts (the more complex ones).
- One or more Arduino Uno (2018) boards connected with a variety of electronic components and/or the Raspberry Pi3 units. The Arduino boards are the core module inside most of the designed artefacts.
- Various DIY electronic components (e.g. photoresistors, potentiometers, servomotors, LEDs) that are used in conjunction with the RPi3 and the Arduino boards to realise the designed artefacts.
- Various DIY-modified parts brought from home during a recycling process, like broken toys, plastic bottles, pieces of paperboard, computer fans, speakers, etc.
- A Raspberry Pi unit (or a laptop) to act as 3D printing server and potential file and/or web server. This unit runs the appropriate 3D printing software.
- A 3D printer used for preparing customised physical components to be integrated in the creation-programming-sharing workflow.
- Some extra networking equipment to facilitate the interconnection of the pilot sites and provide further access to the Internet and project's resources, like Wi-Fi routers, access points and switches.
- The necessary power supply equipment like power banks and small solar panels.

The selection of all the hardware components is intended to minimise the cost, the size and the power consumption and maximise the reusability of materials and electronics.

The Software Components

The most important parts of the eCraft2Learn soft ecosystem include:

- Visual programming tools like the Snap! and its modified version Snap4Arduino (2018) that allows RPi3-Arduino interoperation, custom visual block creation and easily exportable/importable.xml code and cooperates with many web-based tools for useful data exchange.
- The native Arduino software (integrated development environment, compiler and library). Although a visual, block-based environment is considered more suitable for rapid prototyping and in terms of usability for children, familiarisation with the Arduino software (and the wiring language) is useful since this is the basis of Arduino programming and in many cases a prerequisite software package for many visual programming tools.
- The Ardublock (2018) tool is another tool that provides visual programming functionality and stand-alone code generation. If the creation of minimalistic autonomous artefacts is a priority, this tool is a necessary asset. The Ardublock software is offering one of the most mature and stable set of blocks for visual programming.
- The very flexible MIT App Inventor (2018) environment. Indeed, as the vast majority of today's students are very familiar with smartphones and tablets, the MIT App Inventor tool allows for rapid mobile programming even by inexperienced users.
- Simple necessary code blocks in the Python language offering server functionality (e.g. general-purpose inputs/outputs specific or web-specific) on RPi3 units and thus facilitating the interaction with the artefacts.
- 3D modelling software allowing the design of models to be printed like the FreeCAD (2018) or the easy to use by the beginners and web-based Tinkercad. To complete the printing tasks, 3D printing (slicing) software is necessary as well, like the Cura (2018) package.
- Generic-purpose software packages allowing actions like file sharing (WinSCP or FileZilla), remote access (PuTTY, VNC), web support and so on.

The software tools of the eCraft2Learn ecosystem have been selected to run on the RPi3 environment, to have reduced need of installation or update of software elements; to have a user-friendly interface; to be pedagogically meaningful, easy to integrate with the external hardware, to be open source and free (or at least of low cost); and finally to be capable to support advanced extensions based on artificial intelligence (AI) cloud services. This basic set of software tools may be extended with more tools dedicated to more advanced users.

The RPi3 units are playing a principal role, as they are hosting the software tools (or methods to access them) that students and teachers need, in their microSD cards. Indeed, all the necessary software components are hosted (or accessed) via the microSD card of the RPi3 units. This microSD card can be seen as one of the most critical core components of the eCraft2Learn ecosystem.

The Unified User Interface (UUI)

The UUI (2018) is the main hub for using the eCraft2Learn ecosystem. Through the UUI, it is possible to access all the tools (and the relevant instructions) that are available in the eCraft2Learn environment. The tools are represented by tiles grouped in several categories based on the stages of the eCraft2Learn pedagogical model. The educational extension of the UUI also provides services for collecting learning analytics. Resources like code or data can be retrieved using the cloud support.

Indicative Projects Realised During the Pilots

The Lighthouse Project

The lighthouse project was one of the first projects that was implemented by the students. The main task of the students was to look online for information about the functionality of the lighthouses, and to make their own functional lighthouse that blinks only at dark (Fig. 4.2).

This project was proposed by the teachers as it was considered as an ideal start towards exploring the available eCraft2Learn tools and technologies and a first simple step towards becoming familiar with visual programming and electrical

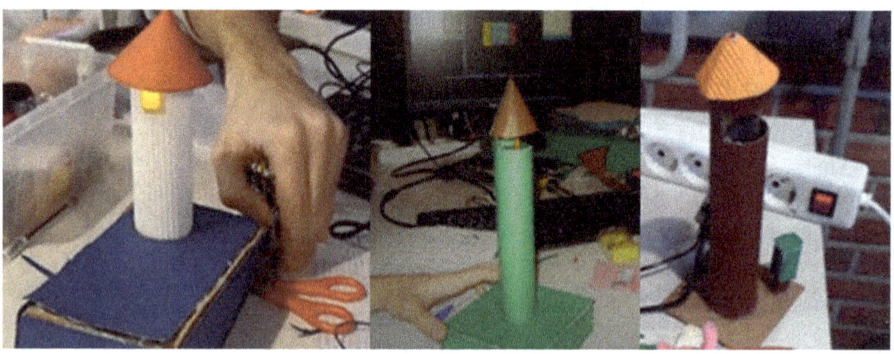

Fig. 4.2 Different lighthouses created by the students

circuit making. A worksheet was given to the students to support their engagement in the eCraft2Learn process.

The scenario of this project offered students with opportunities to express their creative skills and be involved into handcrafting. Some teams were noticed to make drawings and mock-ups inspired by structures of lighthouse buildings that they found online, while some others were making more abstractive or creative designs. The project offered also opportunities for discussion, supported by the teachers, which mainly revolved around the following topics: continuity and change of lighthouses over time, technological and scientific developments over long periods and maritime history.

***Time allocated*:** Most of the groups completed the project within 3 hours.

***Hardware and materials that were used*:** Cardboards, recycled materials and many different types of paper for making the structure, wooden sticks, wires and LEDs, photoresistors, resistors, Arduino Uno boards.

***Technical details and software used*:** Students used Ardublock or Snap4Arduino visual programming environments to program the Arduino board, so an LED is blinked at specific *on/off* pattern, at a first stage, and, at a second stage, to eliminate this activity only at dark. The basic idea was to periodically poll the photoresistor (available via the A0 input of the Arduino) and compare these readings with an experimentally defined threshold value. Whenever the readings were below that threshold, the lighthouse LED was set to *high* and after that to *low*, for several milliseconds, according to a characteristic blinking pattern.

The Sunflower Project

This project revolved around the phenomenon of phototropism: the orientation of plants according to the location of light. The students were encouraged to discuss this phenomenon in teams and to search for information online about phototropism in sunflowers. They were then encouraged to use the available tools, technologies and resources in order to simulate this phototropic response when the light stimulus is present.

A worksheet was given to the students to support their engagement in the eCraft2Learn process for this specific project. Some groups were noticed to make plans on the paper (Fig. 4.3) and to keep notes related to their project (Fig. 4.4).

This project was chosen due to two main reasons:

- It introduces students into the parallel use of photoresistor sensors and angle servos, allowing the implementation of more complex functionalities from a STEAM perspective. After becoming smoothly familiar with more complex combinations of the eCraft2Learn tools and technologies, it was considered more likely to pinpoint possible topics of interest and to take initiatives in working on their own project ideas.

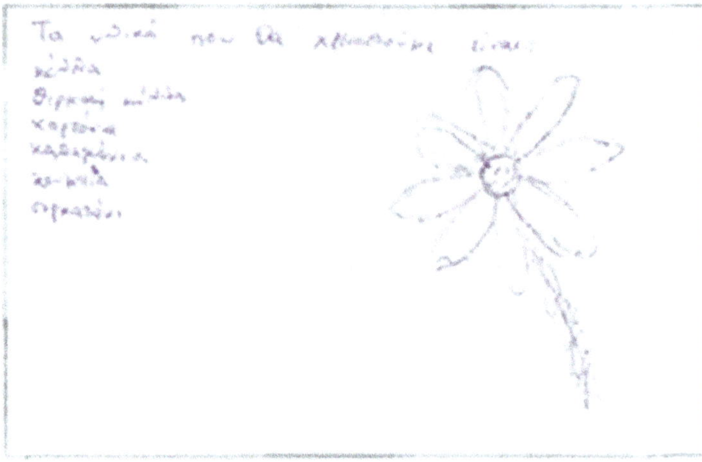

Fig. 4.3 Students make a list of the materials that will be needed for the implementation of the project

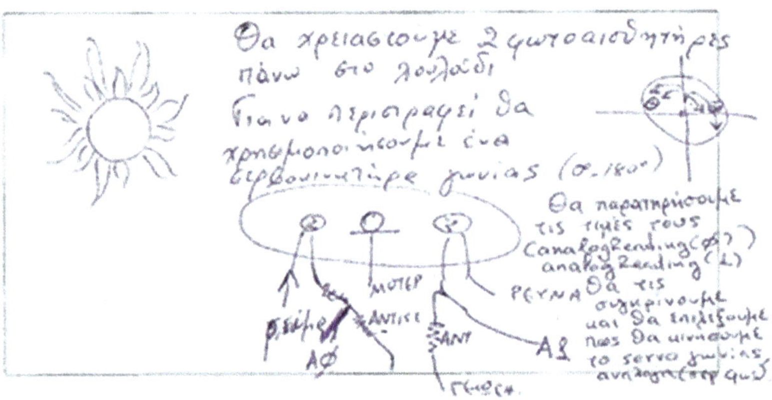

Fig. 4.4 Students break down the problem in the main subtasks

- It provides excellent opportunities for opening discussion about the phenomenon of phototropism and environmental factors that might cause a plant to move or face a different direction. Thereby, it is ideally suited in the flexible zone of the school curriculum for cross-thematic activities.

***Time allocated*:** Most of the teams completed the project within 4–5 hours.

***Hardware and materials that were used*:** Cardboards, recycled materials and many different types of paper for making the flower, wooden sticks, metal wire, buttons, knitting kit, photoresistors, resistors, wires, breadboards, angle servomotors and Arduino Uno board (Fig. 4.5)

Fig. 4.5 Different implementations under the *sunflower* topic

***Technical details and software used*:** Students, via the UUI, were able to have access to information related to the Snap4Arduino or the Ardublock visual programming tools, so as to proceed with this project. More specifically, the sunflower project can be considered as an advanced version of the lighthouse project because students are now challenged to experiment with two light sensors (photoresistor-based ones) and give motion to a flower artefact to make this turn towards the direction with the highest light reading. For this purpose, the light sensor outputs were directed to the A0 and A1 inputs of the Arduino board, respectively. The code on Arduino periodically polled these two values, compared them to each other and ordered an angle servomotor to turn towards the direction exhibiting the strongest reading (Fig. 4.6). The servomotor had to be connected to a PWM-capable Arduino digital PIN.

Evaluation: Did the Methodology Worked in Practice?

The eCraft2Learn learning ecosystem was designed to actively engage students with the technical environment and the 5-stages learning methodology. To evaluate the extent of the students' engagement and the implementation of the proposed paradigm in real settings, the method of the direct observation in the class was adopted. One of the authors undertook the role of the observer monitoring and recording field notes in each session using a written observation form which was filled in during the sessions. The aim of the observation tool was to evaluate how the teams worked in each of the five stages of the eCraft2Learn educational model as well as the accomplishment of a final product by the students. The observer had to respond to open and closed questions (categorical), together with any additional comments (Table 4.2).

The results from the observation grid related to the 5-stage methodology are presented in Table 4.3.

First, the observers found that it was difficult to observe always what exactly was going on in each team as each team progressed at a different rate. Second, the progress

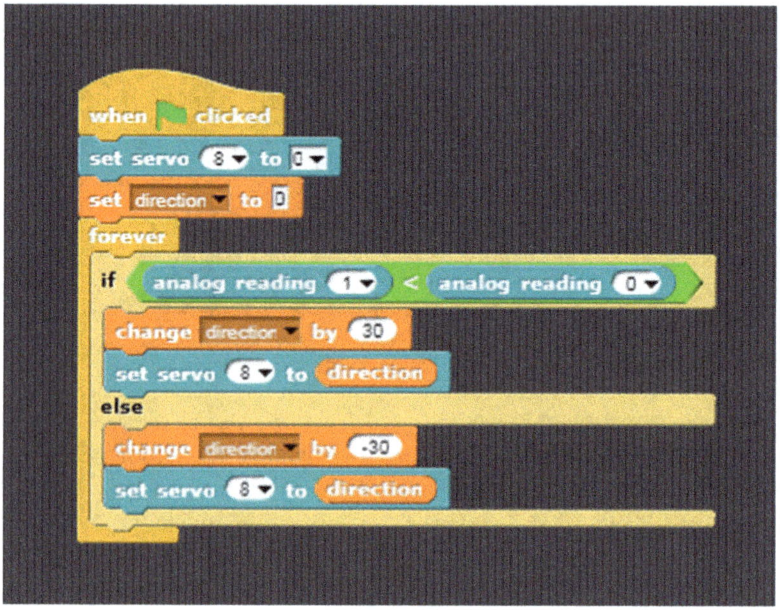

Fig. 4.6 Indicative code for making the sunflower to direct towards the light

Table 4.2 The observation grid (shortened version)

The 5 stages	Did the students go through this stage today?	Comments/interesting quotes
IDEATION	Yes	
	Yes, but to a limited extent	
	No	
PLANNING	Yes	
	Yes, but to a limited extent	
	No	
CREATION	Yes	
	Yes, but to a limited extent	
	No	
PROGRAMMING	Yes	
	Yes, but to a limited extent	
	No	
SHARING	Yes	
	Yes, but to a limited extent	
	No	
Project/artefact accomplished?	Yes	
	In some extent	
	No	

Educational resources used by the teachers/ (i.e. videos, web links, printouts)…
Needs emerged (if any)…
Overall workshop impressions…

Table 4.3 Data from the observation grid filled in during the pilots

Stage	First round	Second round	Total	Percentage of subtotal
IDEATION	teams	teams	teams	
Yes	31	7	38	54.29%
Yes, but to a limited extent	14	5	19	27.14%
No	10	3	13	18.57%
Subtotal			70	
PLANNING	teams	teams	teams	
Yes	28	7	35	48.61%
Yes, but to a limited extent	19	6	25	34.72%
No	8	4	12	16.67%
Subtotal			72	
CREATION	teams	teams	teams	
Yes	49	11	60	81.08%
Yes, but to a limited extent	2	5	7	9.46%
No	4	3	7	9.46%
Subtotal			74	
PROGRAM	teams	teams	teams	
Yes	50	8	58	80.56%
Yes, but to a limited extent	2	3	5	6.94%
No	0	9	9	12.50%
Subtotal			72	
SHARE	teams	teams	teams	
Yes	31	3	34	53.97%
Yes, but to a limited extent	10	2	12	19.05%
No	5	12	17	26.98%
Subtotal			63	
Project/artefact completion	teams	teams	teams	
Yes	42	1	43	57.33%
Yes, but to a limited extent	13	13	26	34.67%
No	0	6	6	8.00%
Subtotal			75	

throughout the stages was not linear. The teams very often returned to a previous stage or jumped to a later one without finalising the previous stages. This non-linear process was well anticipated and even encouraged by the research team and the teachers. We found out that learning by making is a dynamic process, which takes often unpredictable paths and makes difficult both the observation and the coaching task. In addition to this, the number of the participants was varying from session to session, and this resulted in different number of teams observed in each session. Finally, during the first round, we recorded observations for 8 sessions from 10 totally carried out (the first session was introductory and the final one was devoted to students' demonstrations) with around 7 teams per session and during the second round observations for 4 sessions from a total of 5 sessions (the final one was again devoted to students' demonstrations) with around 5 teams per session.

The analysis of the observation sheets revealed several interesting findings. In general, it was found that in each session, more than a half of the teams did *Ideation* (only 18.57% "No" responses), *Planning* (only 16.67% "No" responses), *Creation* (81.08% "Yes" responses) and *Programming* (80.56% "Yes" responses). A significant result of the analysis concerns *Sharing*. It was found that 53.97% did this stage, 19.05% did it to some extent and a worrying 26.98% did not share at all.

Finally, regarding the last question about "Project/artefact completion", in all sessions, the majority accomplished this goal fully (57.33%) or to a limited extent (34.67%). The 6 observations of not accomplished projects (8.00%) in only the second session does not reflect reality and there is an explanation. These six cases camefrom the very first two sessions of the second round, which was about an open project for students to develop where it was reasonable to complete *Ideation* and possibly *Planning* but impossible to complete a project or artefact.

Regarding the question about *educational resources used by the teachers*, the observers noted several times the following with different frequencies:

YouTube videos, online images, online information, paper-based sketches, worksheets, printouts, paper, pairs of scissors, glue, silicon, LEDs, resistors, Arduino, Raspberry Pi, servomotors, sound sensors, motion sensors, handcraft, speakers, wires, pictures of a chip, distance sensor, Snap4Arduino, Ardublock, Tinkercad, Cura, AI blocks, microphone, web search, unified user interface

This list confirms the big variety of resources and materials that the teachers provided for students' projects. They included not only technical resources (e.g. Arduino and Raspberry Pi boards, electronics and more) but also online resources (YouTube videos, online images, online information) intended to help students use effectively or understand tools and techniques necessary for their projects. The resources included printed materials (e.g. paper-based sketches, worksheets, printouts) as well. Crafting tools like pairs of scissors, glue, silicon and so on are also reported, providing evidence for the "traditional" crafting work made by hand or by using only simple tools that took place in parallel with the assembly of electronic circuits and computing.

As for the overall workshop impressions, a positive working atmosphere was observed. The teams were working on their projects usually with high interest and without serious problems. The teachers considered the stage of sharing very important, and they tried hard to provide time for this. The students were also enjoying this stage and spent some time on deciding together how to present their work and talking about their project.

We have also observed that for every piece of equipment provided during the session, the teams assumed it was for immediate use and rushed to test it. When the students had difficulties in using a new tool, they were inventive enough to select other tools more convenient for them and their work. In addition, cross-team collaboration was observed taking place quite often among the teachers of different teams as well as among students of different teams, with no intervention from the research team. In some cases, students were complaining for technical problems: their Raspberry Pi board had crashed or the Snap4Arduino or Arduino programming environment was very slow, the microphones did not work and so on. While the research team usually found solutions in these cases, the delays were quite annoying for students and teachers. Finally, during their projects, the students and the teachers were observed many times taking pictures or video of their creations to share online which has promoted the sharing stage.

Lessons Learnt from Pedagogical Perspective

During the pilots, we had the opportunity to test and review the applicability of the eCraft2Learn hands-on, open-source learning environment intended for STEAM education with a special focus on educational robotics and making practices. In contrast with the commercially available robots that are often expensive and not easy to modify for specific educational purposes, the eCraft2learn learning ecosystem is of low cost and at the same time was found efficient, flexible and reliable.

The eCraft2Learn platform allowed students to design 3D-printable parts for their robots, to be creative using open-source mechanics and electronics, to construct and modify iteratively their robotic artefacts, to program them with visual programming tools and finally to develop interesting and meaningful projects.

However, the emphasis was put not on the final product but on the making process. Students' teams worked at their own pace to develop their projects without any pressure to finish with a predefined product. The results of their work may be not "spectacular". They were rather simple projects but very inviting for novices, relevant for the children involved including those with low technical background or those coming from low status or immigrant profile families with poor skills. More importantly, the students could proudly claim in the end ownership of their projects. This became obvious during their presentations in the Athens Science Festival 2018 where the children were provided the opportunity to demonstrate their projects to the public.

However, the simplicity of the student's projects does not mean their work was "easy fun". We have observed learners working with passion to manage some difficult tasks, going sometimes beyond their comfort zone. The process was not without difficulties, frustration and some failures, especially in the cases when we wanted to provide only minimal or none facilitation. In these cases, we have seen some students (mainly those with less familiarisation and lower technical background) to feel frustrated, eventually to withdraw from the technical work or to leave the "hard work" (mostly circuits assembly or programming) for the high-achieving students and to "shelter" in the crafting corner to work with more familiar cardstock-based tasks.

Though trial-and-error efforts are welcome in the constructivist approach, and some frustration is inevitable when we engage learners in open projects, we have found, in good line with other researchers (Blikstein and Worsley 2016), that novices coming into a robotics lab need a considerable amount of support and facilitation before they can start making their own projects. Learners, especially the novices, should be carefully introduced to the lab activities and not to be exposed to excessive levels of frustration (Blikstein and Worsley 2016).

Furthermore, the technologies we put in students' hands must be appropriate for their age, knowledge and skills level. We have seen learners (and teachers as well) struggling to use platforms such as Arduino or DIY electronics sets that come with too many technical functions and services unnecessary for their learning tasks and in too small dimensions and sizes, convenient for skilled professionals or hobbyists but not for young learners. We have seen, for instance, one of our inventive teachers to bring and use with her students a magnifying lens over their Arduino board and electronic circuits trying to identify pins, wires and other small-size components.

These findings are in line with the relevant literature (Blikstein and Worsley 2016; Alimisis et al. 2017; Cesaretti et al. 2017) arguing that technologies like robotics or electronics kits intended for education should not come directly from platforms intended for professionals or hobbyists; they should be specifically designed for education, compatible with learners' skills, needs and interests. The challenge for platforms intended for professionals or hobbyists with a high learning potential, like for instance the popular Arduino boards, is how to transform or adapt before bringing them in the class to make convenient for learners to use in meaningful ways that will serve the learning purposes. We conceive this transformation as an abstraction that removes or keeps out of sight the unnecessary components while highlights the useful parts in ways that make them clear and easy to use for learners. Moreover, preparatory classes and well-designed materials should familiarise learners with basic skills and knowledge in using electronics and boards before exposing them to open projects.

Similar concerns have emerged from the pilots regarding the programming tools. We need simple, visual, block-based environments like Scratch, Snap4Arduino, Ardublock and so on, specifically designed for children instead of professional or general-purpose programming languages.

Collaboration within the teams was observed to be good enough in most cases and was shown as a crucial factor for successful project work. Most of the children met others for first time in the lab; however, by the end of the pilots, they had formed social groups and friendships surviving long after the pilots. Parents have evidenced in oral communications that their kids became more social and self-esteemed during the pilots and especially thanks to their participation in the Athens Science Festival where they had proudly presented their projects. Kids from low-status and immigrant profile were integrated well in their teams and in the pilots in general which, according to their mentors' or parents' evidence, helped them boost their self-esteem, social inclusion and interest for receiving more training in robotics and STEAM technologies. However, social interaction within teams was not without problems. In some cases, a newcomer had upset the team until they find a common language and a new balance. In other cases, a high achiever in electronics or in programming had marginalised the others in the team in passive roles so that corrective actions should have been taken.

Conclusions and Future Plans

In the eCraft2Learn pilots in Athens, we have seen the students planning their robotics projects in teams, creating their own robotic artefacts collaboratively, programming behaviours for their robots and finally sharing their projects with the whole class and the community. We have seen a lot of hands-on activity in a real making atmosphere with a lot of "trial and error", shifts in the roles of the students and good collaboration. The evaluation, still in progress, has provided so far evidence for good acceptance of the eCraft2Learn ecosystem by teachers and students, positive attitudes and curiosity to learn more about robots. Interestingly, we have been informed by teachers or parents that children from immigrant families or from low social status families living in Athens suburbs have continued their robotic projects at home or in educational centres after the end of the pilots.

In conclusion, this exemplary project has provided promising results and support to our claim that the future of educational robotics should be envisioned in close connection with the maker movement. Well-designed educational actions in robotics and with robotics that incorporate the making culture and technologies can promote knowledge and skills relevant not only to robotics but to a wide spectrum of disciplines, understanding and acceptance of robotics by the young generations, and finally can contribute to the development of a future robotics society.

In the short future, we plan new training courses for teachers and learning activities for children to explore further the learning potential of the eCraft2Learn ecosystem and to disseminate its benefits to more children. Our plans include also the communication of the ecosystem at European level and the establishment of European summer schools in Athens that will receive in the eCraft2Learn lab teachers/student-teachers and secondary school students from all the European countries. The establishment of a master course specialised in the connection of educational robotics with maker movement might offer a continuation of studies at academic level for those of the teachers who will wish to deepen more in the topic after their participation in the summer schools.

Acknowledgement This research was supported by the eCraft2Learn project funded by the European Union's Horizon 2020 Research and Innovation Action under Grant Agreement No. 731345.

Disclaimer This communication reflects the views only of the authors, and the European Commission cannot be held responsible for any use which may be made of the information contained therein.

References

Alimisis, D., & Kynigos, C. (2009). Constructionism and robotics in education. In D. Alimisis (Ed.), *Teacher education on robotics enhanced constructivist pedagogical methods* (pp. 11–26). Athens: School of Pedagogical and Technological Education.

Alimisis, D. (2013). Educational robotics: Open questions and new challenges. *Themes in Science and Technology Education, 6*(1), 63–71.

Alimisis, D. (2014), Educational Robotics in Teacher Education: an Innovative Tool for Promoting Quality Education, In L. Daniela, I. Lūka, L. Rutka and I. Žogla (Eds) *"Teacher of the 21st Century: Quality Education for Quality Teaching"*, Cambridge Scholars Publishing, Newcastle upon Tyne, UK, p.14-27.

Alimisis, D., Moro, M., & Menegatti, E. (Eds.). (2017). Educational robotics in the makers era. In *Advances in intelligent systems and computing book series* (Vol. 560). Springer Nature, Switzerland.

Alimisis, R., Loukatos, D., Zoulias, E., & Alimisis, D. (2018). Introducing the making culture in teacher education: The eCraft2Learn project. In M. Moro, D. Alimisis, & L. Iocchi (Eds.), *Educational robotics in the context of the maker movement*. Springer Nature, Switzerland (Pre-Print).

Ardublock. (2018). http://blog.ardublock.com. Accessed 21 October 2018.

Arduino Uno. (2018). https://www.arduino.cc/. Accessed 21 October 2018.

Asimakopoulos, K., Zoulias, E., Saranteas, I., Makaris, I. Sandali, G., & Spiliou, T. (2018). The eCraft2Learn project in 1st EPAL KORYDALLOS: Integrating a FabLab in the school curriculum. *Proceedings of the 11th Pan-Hellenic and international conference ICT in Education* (Accepted for publication).

Blikstein, P. (2013). Digital fabrication and 'making' in education: The democratization of invention. In J. Walter-Herrmann & C. Büching (Eds.), *FabLabs: Of machines, makers and inventors*. Bielefeld: Transcript Publishers.

Blikstein, P., & Worsley, M. (2016). Children are not hackers: Building a culture of powerful ideas, deep learning, and equity in the maker movement. In: *Makeology: Makerspaces as learning environments* (Kindle ed., Kindle Locations 56–59, Vol. 1). Taylor and Francis. New York, Accessed 30 November 2018.

Castro, E., Cecchi, F., Valente, M., Buselli, E., Salvini, P., & Dario, P. (2018). Can Educational Robotics introduce young children to robotics and how can we measure it? *Journal of Computer Assisted Learning, 34*, 970–977. https://doi.org/10.1111/jcal.12304.

Cesaretti, L., Storti, M., Mazzieri, E., Screpanti, L., Paesani, A., Principi, P., & Scaradozzi, D. (2017). An innovative approach to School-Work turnover programme with Educational Robotics. Mondo Digitale, 16(72), 2017-5.

Cura. (2018). https://ultimaker.com/en/products/cura-software. Accessed 21 October 2018.

Darrah, T. (2018). *Design and development of a low-cost open-source robotics education platform*. Proceedings of the 50th International Symposium on Robotics, Munich.

eCraft2Learn. (2018). https://project.ecraft2learn.eu. Accessed 21 October 2018.

FreeCAD. (2018). https://www.freecadweb.org/. Accessed 21 October 2018.

Fullan, M., & Langworthy, M. (2013). *Towards a new end: New pedagogies for deep learning*. http://www.newpedagogies.nl/images/towards_a_new_end.pdf. Accessed 21 October 2018.

MIT App Inventor. (2018). http://appinventor.mit.edu/explore. Accessed 21 October 2018.

Moro, M., Agatolio, F., & Menegatti, E. (2018). The development of robotic enhanced curricula for the RoboESL project: overall evaluation and expected outcomes. *International Journal of Smart Education and Urban Society, 9*(1), 48–60. https://doi.org/10.4018/IJSEUS.

Papert, S., & Harel, I. (1991). *Constructionism*. New York: Ablex Publishing Corporation.

Piaget, J. (1974). *To understand is to invent*. New York, NY: Basic Books.

Raspberry Pi. (2018). https://www.raspberrypi.org/. Accessed 21 October 2018.

Scaradozzi, D., Sorbi, L., Pedale, A., Valzano, M., & Vergine, C. (2015). Teaching robotics at the primary school: an innovative approach. *Procedia-Social and Behavioral Sciences, 174*, 3838–3846.

Schon, S., Ebner, M., & Kumar, S. (2014). The Maker Movement Implications from modern fabrication, new digital gadgets, and hacking for creative learning and teaching. In L. Canals & P.A.U. Education (Eds.), *eLearningPapers Special edition*, Barcelona, (pp. 86–100).

Snap4Arduino. (2018). http://snap4arduino.rocks/. Accessed 21 October 2018.

Thingiverse. (2018). https://www.thingiverse.com/. Accessed 21 October 2018.

Tinkercad. (2018). https://www.Tinkercad.com/. Accessed 21 October 2018.

UUI. (2018). https://ecraft2learn.github.io/uui/index.html. Accessed 21 October 2018.

Chapter 5
Modbloq. Design of a Modular Robot Made with 3D Printing for Educational Purposes

Pedro de Oro and Silvia Nuere

Abstract The main goal of this study is to design and build an open source-based prototype of a modular educational robot, called Modbloq. We will achieve it by subdividing the process into three phases.

First, a market survey is done comparing and studying tendencies and trends of similar projects and products. We analyze strengths and weaknesses of these data, once collected, to get better conclusions and help design a better product.

Later, we set the specifications of the modular robot and then design and develop it. Arduino is the coding language used to program the product.

Fused Deposition Modeling (FDM) 3D printing is used to make the prototype. This technology will make it easy to produce the robot in a classroom for educational purposes.

Keywords Educational Robot · Programming · Open Source · Modular Robot · Educational STEAM · Arduino · Creativity

Introduction

Nowadays, robotics or software programming are used as new and innovative ways of problem solving. Papert (1980) proposes a combination of both to promote logical thinking at primary levels of education. This kind of approach in school curriculum also entwines creativity. Other subjects of knowledge also result in a full development of STEAM (Science, Technology, Engineering, Arts and Mathematics) disciplines and a whole new branch of education focusing on innovation, communication and real world problem solving.

We propose the design and manufacturing of a modular prototype educational robot. The robot has a set of elements where functionalities are divided into units

P. de Oro · S. Nuere (✉)
ETS de Ingeniería y Diseño Industrial, Universidad Politécnica de Madrid, Madrid, Spain
e-mail: pedrodeoro@gmail.com; silvia.nuere@upm.es

L. Daniela (ed.), *Smart Learning with Educational Robotics*,
https://doi.org/10.1007/978-3-030-19913-5_5

that can work separately or combined. Students can quickly and easily arrange different shapes.

Students from 6 years old and above can develop and test new ideas to understand the basic ideas of programming, physics and mathematics. They can also improve their logic and computational skills as they build and program a robot. In a society growing more focused on the digital world, this educational goal helps educators to introduce technical skills and awareness in a different and entertaining way.

The aim is to share all the information in an open source web platform called *GitHub* (2018). This platform enables users not currently involved in developing the project to create alternative ones, vary existing content or contribute in many different ways. Thus, making a community where anybody can share their knowledge in a collaborative and altruist way.

The project is based on the Open Source idea (Perens 1999). This movement aims to spread knowledge easily and without barriers, by contributing to development of the technological human inheritance.

Therefore, the robot must perform these four features:

- Modular. Different combinations of the modules. The use has to be easy and fast, allowing a large number of combinations.
- Open Source. All the elements of the robot, including the plans, designs, code, and documentation, must be accessible for everyone. Any user should be able to make their own robot.
- 3D printing technology. It is fundamental to share the physical parts through the Internet.
- Programmable. The programming platform must be Open Source compatible with *Arduino* (2018).

State of the Art

Educational Robotics Nowadays

Recently, computer programming and robotics in the curriculum of non-university education has gradually begun to be generalized in many countries. In Spain, the approach to teaching technological education is focused on a humanistic vision.

Students must participate in technical processes. They have to solve problems and be actively involved in the development of projects. It is not only a matter of personal needs but also collective interests. Additionally, not only is there a clear interest in using robots in education but also for considering them as a tool to engage students and keep them in school (Alimisis 2013). Robots are being incorporated in our day-to-day life, and they can be introduced for the development and intellectual growth of children (Mubin et al. 2013). The study of Mubin (2013) and his team summarizes that robots provide a tangible and physical representation of learning outcomes. The use of robots has become a valuable aspect to be used in education.

There is a growing interest for teachers to incorporate these types of skills and tools into their daily work due to the important educational benefit. Principles proposed by Catlin and Blamires (2012 at the Educational Robotics Applications (ERA)) have been considered, and those that are a better fit were chosen. Students must participate and interact in educational activities. Robots must facilitate engagement, sustainable learning, and personalization.

The use of a robot is an experiential activity. As Eguchi (2017) says, educational robots are linked with the constructionist learning theories engaging hands-on exercises. Developing robotic projects improves the interest of students in learning, as Moro, Agatolio and Menegatti (2018) point out. The use of educational robots also increases their self-esteem. Students understand the world by building mental models from their experiences as a constructivist approach to education. They can assimilate or allow new experiences into their existing ideas (Catlin and Blamires 2012).

There are many studies around Educational Robotics Relevance in education. Some of them focus on robotics-based learning as a tool for preventing school failure and motivate students to improve their academic achievements as Daniela and Strods point out (2018). It also becomes a useful tool and a good solution to prevent early school leaving, as shown by Daniela, Strods, and Alimisis' research "The robotics activities were aimed at students' getting actively involved in programming, testing their knowledge and constructing new knowledge" (2017).

Information and Communication Technologies (ICT) is not only a technological tool of teaching–learning but has the faculty of awakening the interest and motivation of our students, especially those with a certain scientific–technological interest (Ocaña 2012). The educational value of its incorporation does not reside exclusively in contributing to favor the development of knowledge and skills within the STEAM disciplines. Educational robotics and programming provide students, among other skills and abilities, mental strategies to solve problems. Thus, the learning based on challenges and the developments of what has come to be called computational thinking involve a new pedagogical scenario in which the programming of small robots acquires full meaning. To help teachers in this task, there are different types of programmable robots on the market that use different programming languages. The research of Cesaretti, Storti, Mazzieri, Screpanti, Paesani, Principi, and Scardozzi (2017) presents an innovative approach to an alternate School–Work turnover programme based on Educational Robotics and on project-based learning dealing with different tools, such as Lego Mindstorms EV3 or the Arduino BYOR platform.

There are two ways of viewing the use of robots in schools symbolized by the acronym TRTWR, referring to the 3rd International Workshop "Teaching Robotics, Teaching with Robotics", organized by Moro and Alimisis in 2012. Any of these approaches can be selected with Modbloq Robot. Nowadays, teachers mostly use two types of educational robots; those that need to be assembled before playing with them and those that are already built. Within the first type of robots, a double classification can be made between those that have an open or flexible character and those whose construction is oriented to obtain a predefined robot model. Students must interact with robots, including with their hardware components and software, source code, and programming environments. Physical robot programming projects make it possible to achieve a level of significance and engagement from students. In the end, the study of robotics can be sufficiently rewarding, intellectually and emotionally, so they study with interest (Miller et al. 2008). An active Learning Model applied to robotics education, as Saygin, Yuen, Shipley, Wan, and Akopian point out (2012), keeps learners engaged, active, and motivated.

To facilitate the task of programming the robots, there are different programming languages based on a graphic interface. They are what has come to be called programming languages by blocks. The blocks are different elements related to variables, conditional, loop, and so on. To program you have to drag the corresponding blocks to the work area and stack them. There are two common programming languages *Scratch* (2013) and *Bitbloq* (2015).

Open Source World

Following the open design approach, we will build our robot in Open Source; this is a technology based on the idea of sharing knowledge freely and free of charge. Hundreds of people working on the same code will improve and develop the software faster. In addition, open software is linked to ethical and altruistic currents. Sharing knowledge without limits defends four freedoms:

- Freedom to use the program at your leisure.
- Freedom to study or modify the operations of the program.
- Freedom to distribute copies of the program.
- Freedom to improve the program and make improvements publicly.

Thanks to initiatives like these, great software has been born that is used today by millions of people, as for example *Linux* (2010). Although the Open Source stream was born within the programming, it has jumped to other fields, examples of which are initiatives, such as *Thingiverse* (2016), a repository of 3D objects, or repositories like *GitHub* (2018), in which projects of all kinds can be found, in the field of software or hardware.

3D Technologies

One of these new Open Source movements is the so-called *Rep-Rap* movement (2018). This movement began in 2005, when the mathematician and engineer Adrian Boywer, a professor at the University of Bath in the United Kingdom, began the project. The 3D printer was capable of self-replicating because it could print many of the items needed to make a new one. This was the beginning of the 3D homemade printers.

In Spain, this movement came from the hand of Juan González Gómez (2012), a Telecommunications Engineer from the UPM, under the name of Clone Wars

(2017). This movement has created a community of thousands of people who share this knowledge and open hardware, such as CNC machines and Cyclone PCB Factory (García-Saura 2013). You can also find laser cutters or even printable robots, such as *Printbots*, like Miniskybot (García-Saura & González-Gómez 2012). Mostly, it is necessary to assign a license to the project to ensure that it remains open and no one can take ownership of it. The most widespread license for this type of projects is the Creative Commons License (2018) created by the non-profit organization with the same name.

The most used way to share this knowledge is through Wiki (González-Gomez & Valero-Gómez 2008) or repositories platforms, such as Github (2018) or Thingiverse (2016), where everyone can freely upload their projects and share them with the community.

Owing to the open source nature of the design, it is essential the robot can be printable in 3D. A 3D printer allows the manufacturing of pieces with a multitude of materials, such as polymers, resins or ceramics. The previously designed object can be printed quickly and economically. Fusion Deposition Modeling (FDM), the technology for the creation of pieces, was chosen because it is the most widespread.

Market Study

Before the design stage starts, an analysis of the main characteristics of other similar products is recommended. A market study has been done analyzing some of the most relevant products for this project, educational robots with modularity. Some of them are listed below.

Lego Mindstorm (Denmark)

- Age from 10 years old and above.
- Price from €399.
- Lego environment.
- Programming and control App.
- Lego specific programming environment.
- Few varieties of functional blocks.
- Blocks are connected by mechanical fit.

Cubelets (USA)

- Age from 4 years old and above.
- Price Twelve Cubelets $329.95.
- Plug and play.
- Lego compatibility.
- App.
- Educational Curriculum with 40 h of class material.
- Cubes with a large variety of functionalities.

- Blocks are connected by mechanical fit and by magnets.

Little Bits (USA)

- Age from 8 years old and above.
- Price from $34.95 to $1999.
- Heavily focused on education with a lot of class material.
- Several documented projects.
- Cloudbit.
- It has a large community.
- Blocks are connected by magnets.

Moss (USA)

- Age from 8 years old and above.
- Prices from $199.95 to $599.95.
- App.
- It is programmable with Scratch and C.
- Educational Curriculum with 40 h of class material.
- Blocks are connected by magnets.

Tinkerbots (Germany)

- Age from 6 years old and above.
- Price €249.95.
- App.
- Mix blocks with electronic and construction blocks.
- Programmable Arduino.
- Blocks are connected by mechanical fit.

Therefore, there are two main ways of connecting blocks, by magnets and by mechanical fit.

Magnetic connection is used by *Little bits, Cubelets,* and *Moss*. After analyzing the connections with a sample, the conclusion is the magnetic connection is strong if the module is light (*Little bits*) but not strong enough if the module is heavy (*Cubelets*).

Mechanical connection is used by *Lego Mindstorm*, *Tinkerbots,* and *Cubelets*. This is the best way to connect as it has high durability and enables the creation of large constructions. The major disadvantage is the manufacturing process. The injection molding is the best way to ensure the high requirements in tolerance so everything connects properly. At the moment, the 3D printing FDM technology can not ensure this tolerance.

The final decision is to use a combination of mechanical and magnetic connection systems like Cubelets.

We consider these characteristics in this project:

- Locomotion blocks.
- Bluetooth connection so it can be use with an App.
- A wide variety of configuration possibilities, adding sensors and actuators.

Table 5.1 Market study price comparison

Name	Mindstorm	Cubelets	Little bits	Moss	Tinkerbots
Price	€399	$329.95	$199.95	$199.95	249.95€

To ensure the competitiveness of the robot, the cost should be below € 200, as shown in the market study price comparison (see Table 5.1). An average price has been used to compare robots with similar possibilities.

Specifications

First, we classify our robot following the EduRobot Taxonomy made by Catlin, Kandlhofer, and Holmquist (2018), as Build Bot: Maker Robots - Manufactured Parts with the tag of Expandable and Modular. It is a mobile robot from 3D printable parts. It uses open source mechanical and electronical parts that students can modify or expand. The specifications, such as cost, public, market study or production, will lay the foundations of the design.

This project is aimed at two groups, children and adolescents, both in the domestic and school environment, interested in robotics and educational centers. Usually, the interest is a technological product, robust, safe, ergonomic, flexible, open source, and multifunctional. They like to develop different projects, with fundamental technological aspects encouraging creativity. The design must allow creating a complete and affordable robot, not only for individual users but also for educational centers.

On the other hand, children are also going to use the robot. The special characteristics of the robot make it advisable for use with those 8 and older. For this group, the product has to be attractive, invite them to pick it up and play with it. This is achieved by using curved geometries, with simple shapes, such as spheres, complementary and garish colors, not binding to any gender.

From the market prices, a price range between € 200 and 400 can be established. For this study, it is possible to consider that our robot should be below € 200 to be competitive.

The specifications of form and aesthetics:

- The shape is intended to be simple, minimalist, and polyvalent.
- Abstract construction, because it allows more space for imagination.
- Accessible and easy to understand with the use of simple shapes and visible moving parts.
- Minimum personality to increase the possibility of personalization.

Among the different basic geometric bodies, the cube is the one chosen. Each edge of the blocks will be rounded to avoid damage when being manipulated. The blocks have to give a sensation of softness so students want to catch and use them.

Another important aspect to consider is the color. Color is more than an optical phenomenon and a technical medium (Heller 2004). That is why you should not choose stereotyped colors associated with gender. The robot has to produce a sense of balance, harmony, and calm sensations before facing any constructive and creative process. The studies carried out by Hallock (2003), in which he relates colors and age or gender among other parameters, have been considered.

Reading the results of the research of Hallock (2003) and Heller (2004), it is clear that the blue and green colors are preferred. Around 70% of potential users in the sample preferred blue and green. The preference for males is 57% blue and 14% green, and for females, the data are 35% blue and 14% green. Turquoise has been chosen because it is a mixture between the two.

Owing to the Open Source character of the project, it is essential for the hardware to be open. The ZUM Bloqs range of *bq* (2007) was chosen for the initial design and manufacturing of the prototype. This range of electronic components is open and compatible with *Arduino* (2018). In addition, the electronic components are conditioned and mounted on Printed Circuit Boards (PCBs) within their own standard; they have polarized connections using standard cables that facilitate the connection. In addition, all hardware is not limited to a single supplier or manufacturer.

Manufacturing by 3D FDM printing fixes the design specifications using the Open Source character of the robot (2016).

This issue involves a series of restrictions. The main specifications of the design are the following:

- It is essential that the design be printable by 3D FDM printers.
- It has to be robust and support the continued use of the user.
- The modules and their unions must allow the cables of the items to pass through them.
- The material has to be used in 3D FDM printing.

The material chosen for the mechanical parts of the robot must be a material used in 3D printing. It is not necessary that the chosen material is especially resistant to extreme temperature conditions. Of all the materials used in 3D printing by FDM technology, Polylactic Acid (PLA) (Matbase 2017) was chosen since it is a biodegradable polymer (Scott 2002). It minimizes the environmental impact and achieves a sustainable design, in line with the current trend, Eco-Design (Luttropp 2017).

Design

Designing a piece for 3D printing is not the same as designing it for manufacturing by injection molding or other kinds of manufacturing and has its own restrictions. 3D printing technology has specific manufacturing tolerances that must be considered when designing. It is difficult to give concrete values of tolerance for this type of manufacturing since each machine has its own tolerances. Most of them are homemade machines and easily go out of adjustment. To solve this problem, the tolerance values have been obtained from the printing of parts using different machines, to obtain, in this way, an average tolerance value.

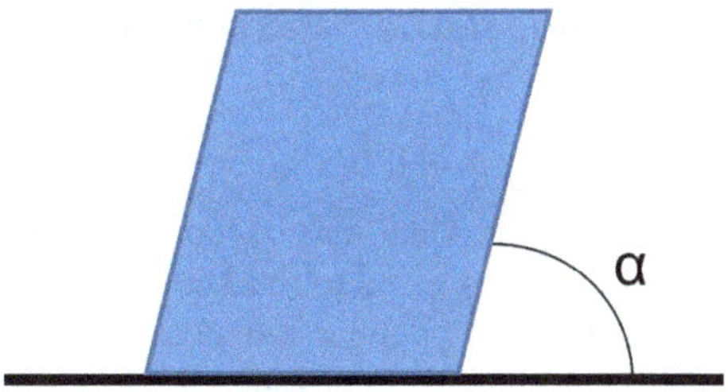

Fig. 5.1 Maximum inclination angle

- Hole-axis: −0.25 mm at the axis radius DIN ISO 2768 C.
- Pocket: −0.5 mm inside the box, on each side DIN ISO 2768 V.

In addition, each layer has to be supported by the previous layer; thus, it is necessary to fix a minimum inclination between the base and the piece's vertical face. For the same reason, it is necessary to avoid the corbels since these only can be printed by means of the use of supports. The angle α can be at most 30° with respect to the base (see Fig. 5.1).

Another restriction to the design is the thickness of the walls. Unlike injection molding where very small thicknesses can be achieved without affecting the integrity of the piece, in 3D printing it is necessary to give higher values of thickness to ensure good quality printing and durability.

As well as the considerations already mentioned, the following have been considered:

- The use of standardized items, such as screws, nuts or other elements.
- Reduce the variety of screws to the minimum necessary.
- Minimize the size of the robot as much as possible.

The main board is the control system of the robot to which the rest of the components and the power supply are connected. The program defined by the user is executed on the controller board. This program receives the data by the sensors and sends the control signals to the actuators according to the code written in the program. Specific modules have been designed to house each type of component, such as controller, sensors and actuators, servos and power.

All components are connected by cables to the controller board. To be able to connect the modules without mounting problems, a connection system has been designed. In this way the cables can pass through entire modules if necessary, until they reach the controller board.

Mechanical and Structural Design

Joining System

The most important part of modular design is the joining system. As mentioned in the section dedicated to market study, the solution chosen in the products analyzed was the use of magnets, the mechanical union or both.

One of the difficulties that the development of this project has presented, specifically in the design of the different pieces, derives from its Open Source character (2016). The chosen materials must be available to any user. Such is the case, for example, of the screws and nuts, which must be available in ironmongery stores.

The union system (see Fig. 5.2) consists of a male part that contains a neodymium magnet with strength of 3.2 kg. The female part is held with a screw clamped by a nut (see Fig. 5.2). The screw, being a ferromagnetic element, allows enough bond strength to be generated and to keep the pieces together through the magnetic field of the magnet.

The male connection piece has to meet four requirements (see Fig. 5.3):

- It must hold a magnet inside.
- It must allow the passage of the cables of the components.

Fig. 5.2 Explosion of the joining system

Fig. 5.3 Male union piece

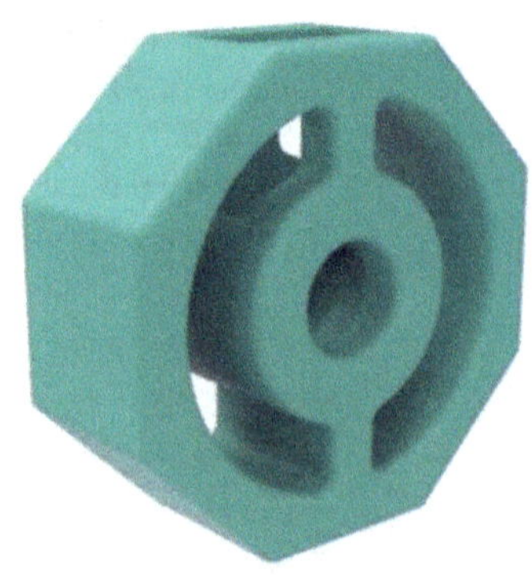

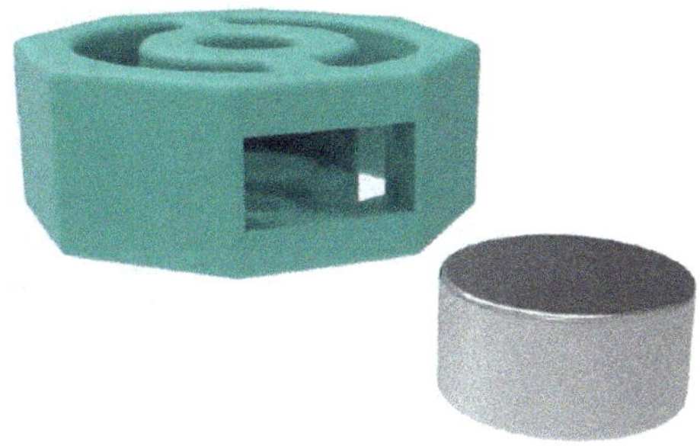

Fig. 5.4 Insertion of the magnet in the connection piece

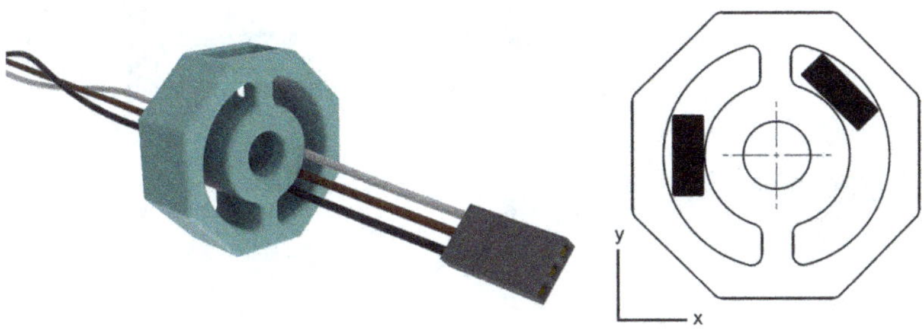

Fig. 5.5 Example of passage of the connector through the connecting piece

- It must have a shape that allows it to fit safely into the female module.
- It must be printable in 3D.

The solution achieved meets all these proposed requirements and also has a careful aesthetic design.

On the one hand, the piece houses a neodymium magnet of 10 mm diameter by 5 mm in height. The magnet is introduced under pressure by one of the faces of the piece (see Fig. 5.4). Once placed, the magnet is completely set in the piece.

To achieve this adjustment, the tolerance of the mortise described in the design considerations was reduced, at the beginning of the chapter, by −0.25 mm inside the box, for each side.

On the other hand, one of the design specifications is that the cables of the components can pass through the modules and their joints (see Fig. 5.5). The cables of the components have dimensions of 7.65 × 2.79 mm. To allow its passage through the pieces, an opening space was designed in order not to compromise its structural integrity.

The octagonal shape was chosen since it allows turns of 45°, enabling different combinations intended to be built with the robot (see Fig. 5.6). As already explained, for the piece to be printable it is required that there are no overhangs, so it was necessary to join the center of the piece with the contour through a column.

This design allows an angle of rotation between two modules of 180° in case a cable passes through both modules (see Fig. 5.7).

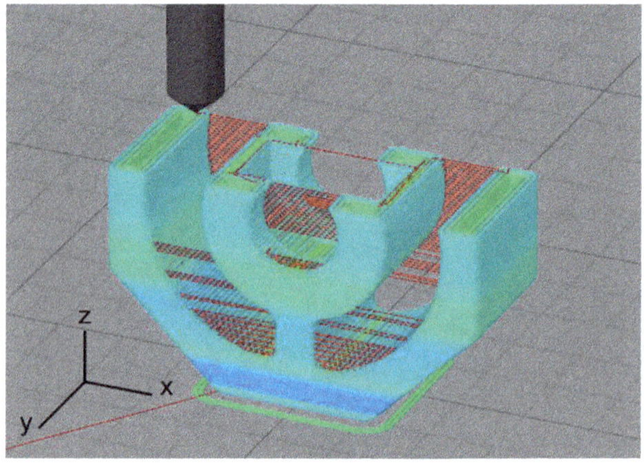

Fig. 5.6 Print 3D simulation (Cura 3D, Ultimaker 2017)

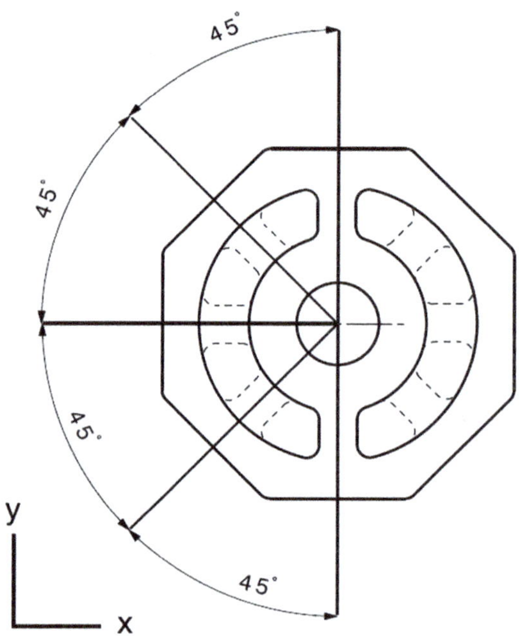

Fig. 5.7 Rotation diagram of the union piece

The female union piece also has to meet four requirements (see Fig. 5.8):

- It must have a ferromagnetic contact.
- It must allow the passage of the cables of the components.
- It must have a shape that allows it to fit safely into the female module.
- It must be printable in 3D.

The solution achieved meets all these requirements.

The central hole of 1.75 mm in diameter allows screwing a M3 × 5 mm hexagonal head DIN 912 screw and is secured on the inside with an M3 nut. In this way a ferromagnetic contact is provided (see Fig. 5.9).

Fig. 5.8 Female union piece

Fig. 5.9 Mounting diagram of the screw and nut

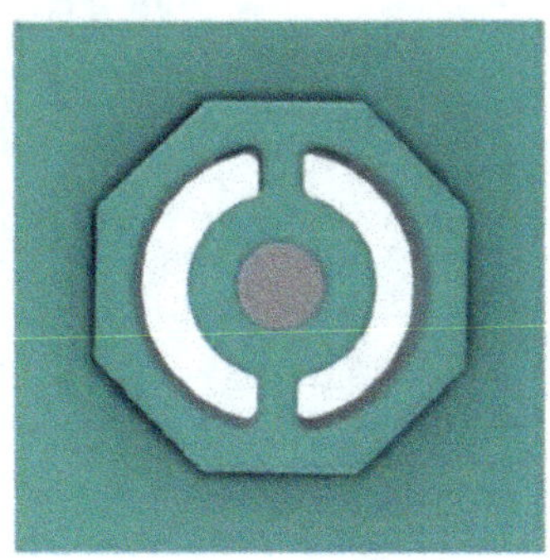

Fig. 5.10 Male union piece fitted in the female union piece

To allow the passage of the cables through the piece, the same shape was designed as for the male connection piece. When both pieces fit the wires can pass freely (see Fig. 5.10).

The piece is printable vertically with the inner column at a right angle to the printing base. This being the best alternative, since in other arrangements it would be necessary to use supports.

Modules

Once the bonding system is designed and defined, modules can be designed. As seen before, the geometric body chosen is the cube since it has a series of advantages:

- The inside space is optimized to house components.
- It allows the robot to develop in the 6 orthogonal directions of space.

In this way, organized growth is allowed in space and no empty spaces are left.

The size of the modules is conditioned by the larger parts that will be housed inside (control board, servos). It is also conditioned by the size of the female union piece. Therefore, the minimum unit size on which the modules will be built will be the size of the female union piece (see Fig. 5.11). In this way the controller board fits in a module of 2×2×1 basic units and the servo in one of 2×1×1 basic units.

The controller module must house the ZUM BT328 controller board (see Fig. 5.12). The best module size for these dimensions is 2×2×1 basic units. It is also compatible with the *Arduino* Uno Board and control boards from other manufacturers (see Fig. 5.13).

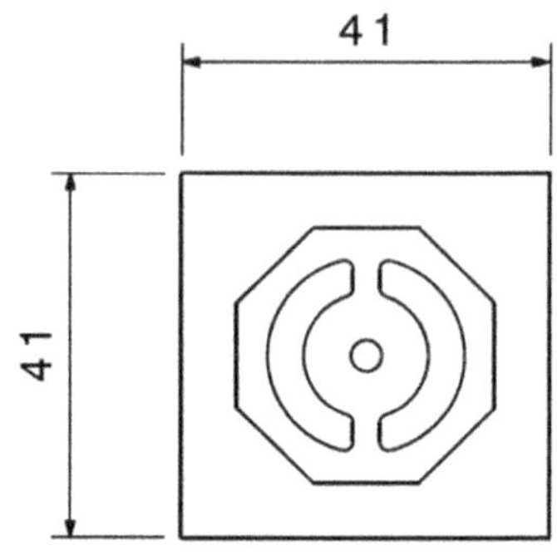

Fig. 5.11 Measurements of the size of the basic unit in millimeters

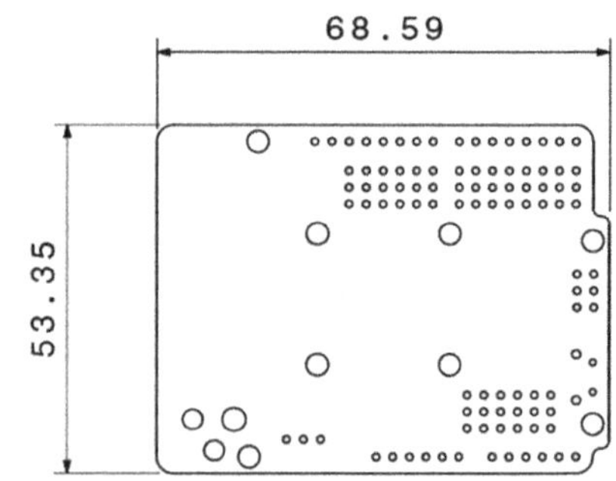

Fig. 5.12 Dimensions of the ZUM BT328 board in millimeters

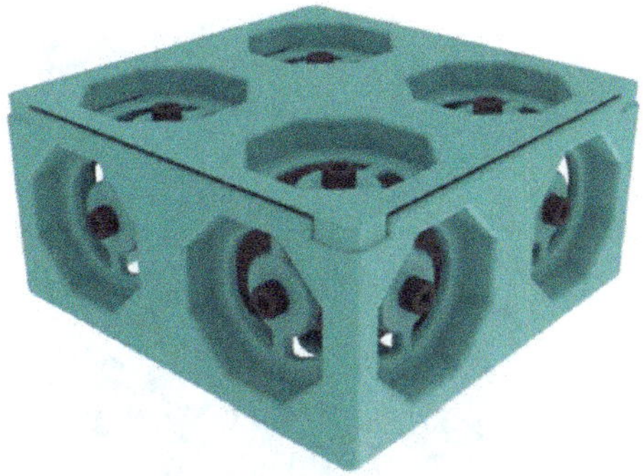

Fig. 5.13 Controller module

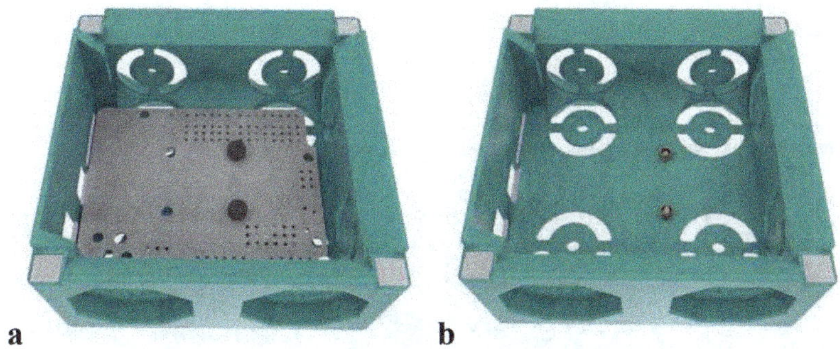

Fig. 5.14 (**a**, **b**) ZUM BT328 screwed to the module and inserts fixed in the module

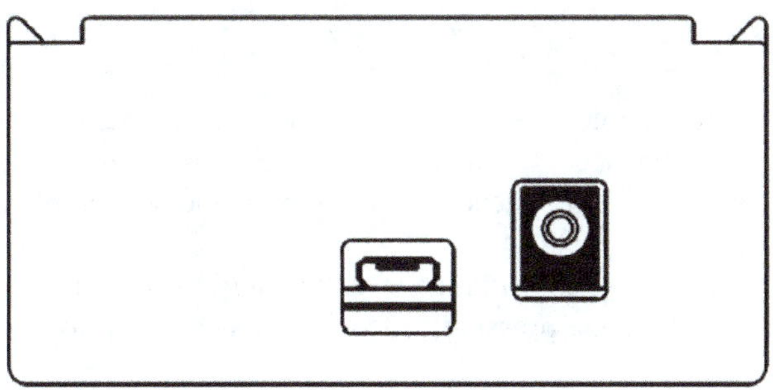

Fig. 5.15 Module port panel

Controller Module

The board is attached to the module by two M3 × 5 mm DIN 912 screws. The screws can be screwed into two inserts M3 40/TH030H55 or two M3 nuts. The module has been designed for both possibilities (see Fig. 5.14).

The ports of the controller board for power (DC Jack female) and programming (micro USB type B female) must always be accessible from the outside of the module. The connectors of the cables used by these ports are too large to be able to pass through the female connection piece. Therefore, one side of this module cannot hold module connections in order to give access to the ports (see Fig. 5.15).

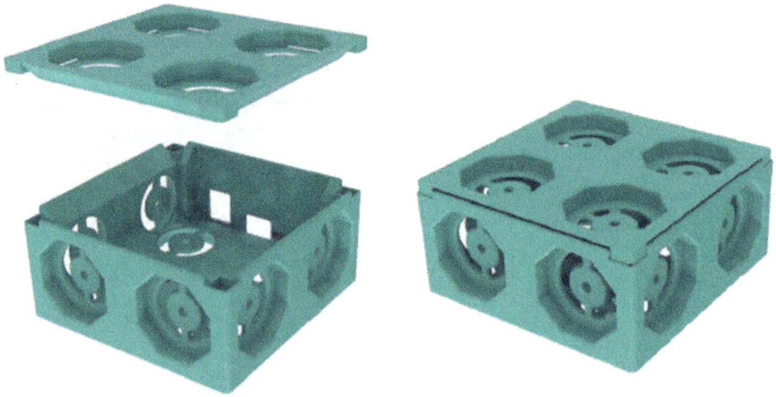

Fig. 5.16 Placement of the lid

Fig. 5.17 Details

To be able to easily connect the components on the controller board, you must have easy access to it. To solve this, a cover has been designed that is held by four neodymium magnets of 6×5×2 mm and a clamping force of 600 g each (see Fig. 5.16).

The magnet is placed in its housing using adhesive (see Fig. 5.17). The nut is housed under pressure, so it is not necessary to use cyanoacrylate adhesive to hold it.

Servo Module

The module that houses the servo motor must take into account a standard size (see Fig. 5.18a, b). The best module size for these dimensions is 2×1×1 basic units. The servo is fastened using 4 M3 × 5 DIN 912 screws. The screws are screwed into 4 M3 nuts embedded in the module.

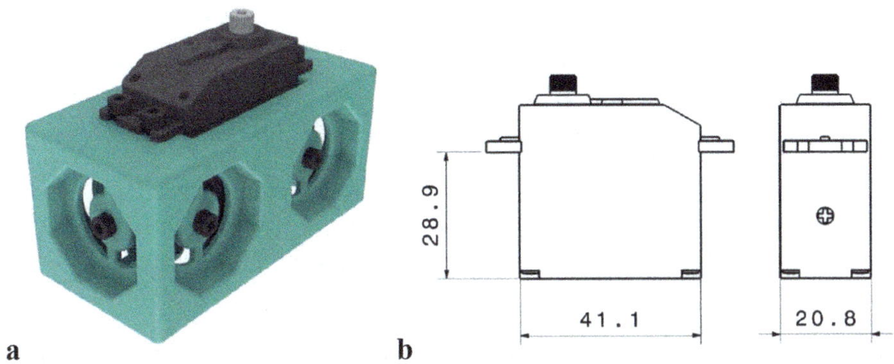

Fig. 5.18 (**a**, **b**) Servo module and servo SM-S4303R dimensions

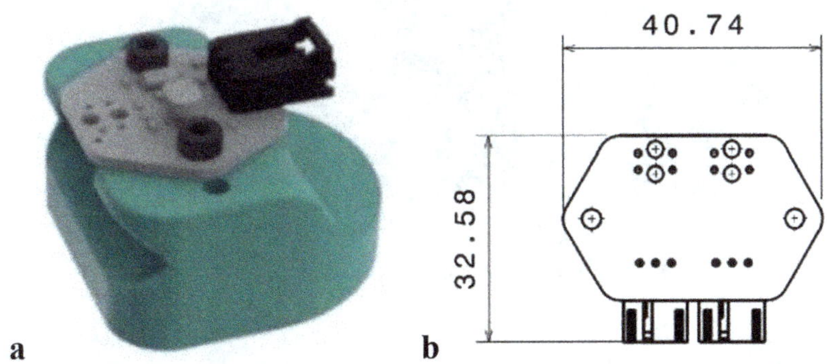

Fig. 5.19 (**a**, **b**) ZUM Bloq and ZUM Bloq dimensions in millimeters

ZumBloq Module

This module is designed to house the ZUM Bloq (2018), and the best module size for these dimensions is 1×1×1 basic units.

There are two sizes of ZUM Bloq (see Fig. 5.19a, b), each one with a different separation between the holes destined to the board's support. The design of the module was made compatible with both sizes by adding two holes in each side (see Fig. 5.20a, b).

L Module

The module in L is a structural module capable of joining or reinforcing the union of the modules with one another (see Fig. 5.21a). The design of this piece is characterized by having a rounded part, enabling rotating the piece to not interfere with the close module (see Fig. 5.21b).

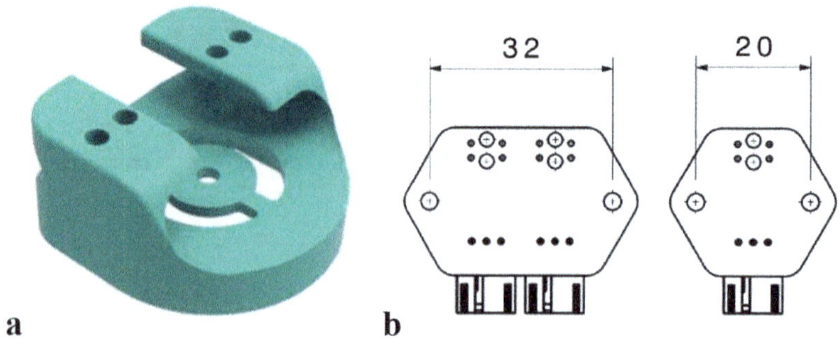

Fig. 5.20 (**a**, **b**) ZUM Kit module and dimensions between holes in both ZUM Bloq in millimeters

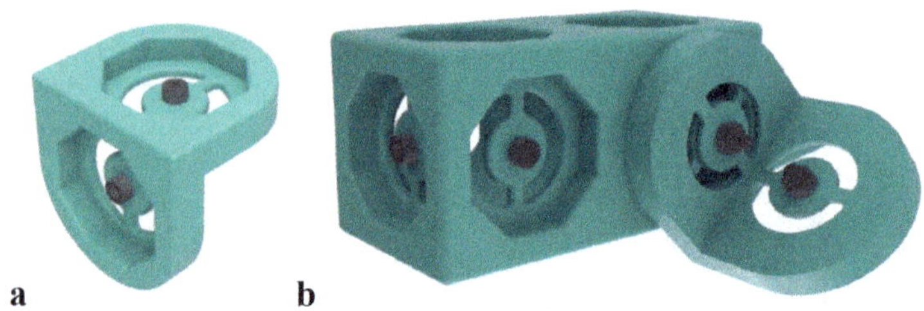

Fig. 5.21 (**a**, **b**) L module and turned position of the module

Power Module

This module should house a battery compartment for eight standard size AAA batteries (see Figs. 5.22 and 5.23). The best module size for these dimensions is 2×2×1 basic units.

The lid has been designed with ergonomic slits that facilitate its extraction. In addition, to remove the power cable, there is a hole with the size of the DC Jack connector of the battery compartment (see Fig. 5.24).

Wheels

The wheels must have a radius greater than the dimensions of the basic unit in order to make contact with the ground.

Some grooves in the edge of the wheel to house three O-rings were added. The O-rings ensure the correct friction with the ground, since the PLA does not have a large enough coefficient of friction, which causes the wheel to slide with the ground when turning (see Fig. 5.25a, b).

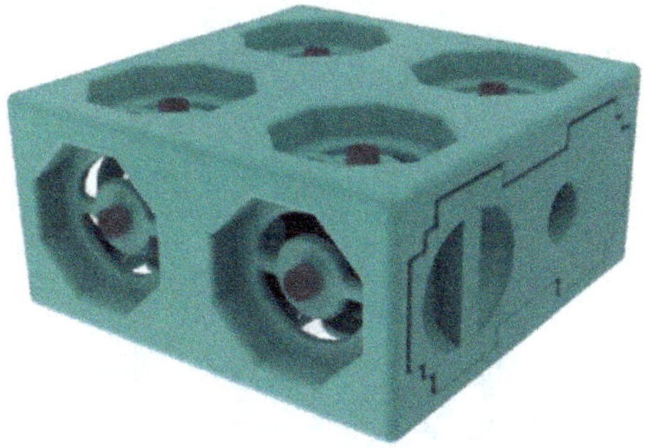

Fig. 5.22 Power module

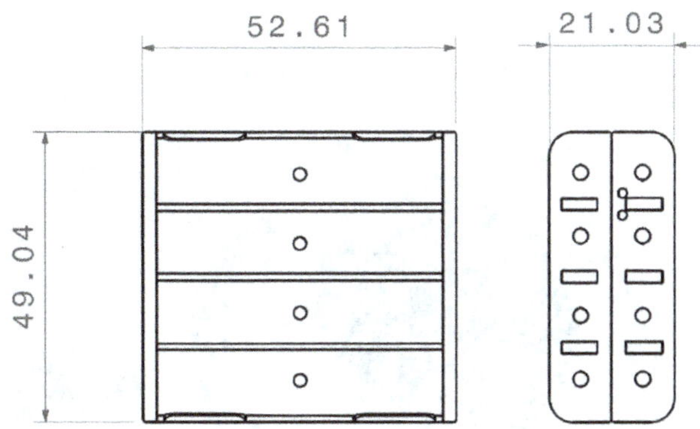

Fig. 5.23 Dimensions of the battery compartment in millimeters

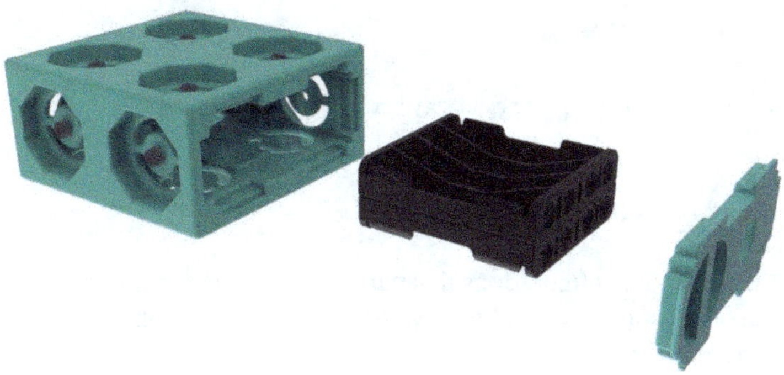

Fig. 5.24 Assembly of the power module

The connection with the servo is made by the head in the form of a cross, attached to the wheel by a specific screw. The screw and the head are included with the servo as accessories. To ensure the rotation of the head together with the wheel and that it does not slip, the head engages in a groove designed with its shape (see Fig. 5.26a, b). The shape of these heads is standard.

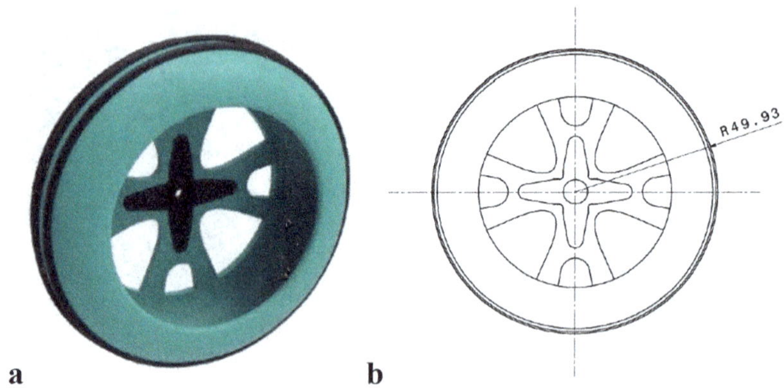

Fig. 5.25 (**a**, **b**) Wheel and wheel dimensions

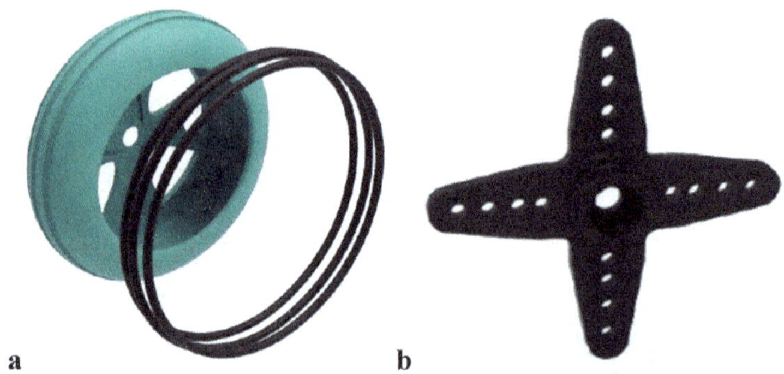

Fig. 5.26 (**a**, **b**) Assembly of O-rings and servo cross head

Possible Configurations and Activities

Some of the possible configurations that can be made with the different components we previously mentioned are addressed below. There are some ideas to promote creativity and ways of combining elements.

Teachers can start with the simplest one and continuing adding different features, for example, the capacity of the robot to follow a line.

It is not only a matter of building a robot but also a matter of making them work and therefore multiple choices can be found. We need to awake interest and curiosity and make students feel they can configure whatever they want to imagine.

Table 5.2 shows the average market price of the components needed to configure the robot (Figs. 5.27 and 5.28).

Table 5.2 Bill of materials and configurations cost

Description	Quantity	Unit average price	Total price
3D printed parts			**10.00 €**
Controller module	1	1.00 €	1.00 €
Servo module	4	0.75 €	3.00 €
ZumBloq module	2	0.25 €	0.50 €
L module	6	0.25 €	1.50 €
Power module	1	1.00 €	1.00 €
Wheel	4	0.75 €	3.00 €
Ironmongery			**32.28 €**
M3 Nut	84	0.02 €	1.65 €
M3 × 5 mm Screw. DIN 912	86	0.02 €	1.72 €
Inserts 40/TH03OH055	2	0.01 €	0.02 €
O-ring 88 mm diameter	6	0.87 €	5.22 €
Neodymium magnet 10 mm, height 5 mm	20	0.63 €	12.40 €
Neodymium magnet 5 mm × 6 mm × 2 mm	4	0.31 €	1.24 €
Neodymium magnet 5 mm × 6 mm × 2 mm	4	0.31 €	1.24 €
Electronic components			**91.00 €**
Arduino board or ZUM BT328	1	30.00 €	30.00 €
Continuous rotation servo	4	14.00 €	56.00 €
Infrared sensor	2	2.50 €	5.00 €
Power supply			**7.20 €**
AAA batteries	8	0.90 €	7.20 €
TOTAL			**140.48 €**

Four-Wheel Drive Mobile Robot

Fig. 5.27 Four-wheel drive mobile robot

Mobile Robot Follows-Lines Two-Wheel Drive

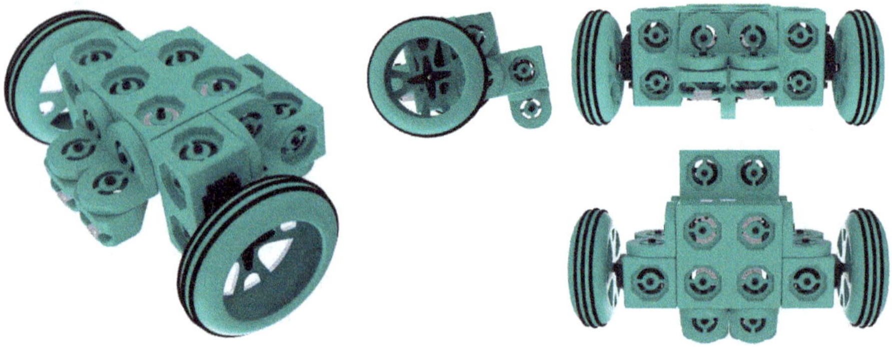

Fig. 5.28 Mobile robot follows-lines, two-wheel drive

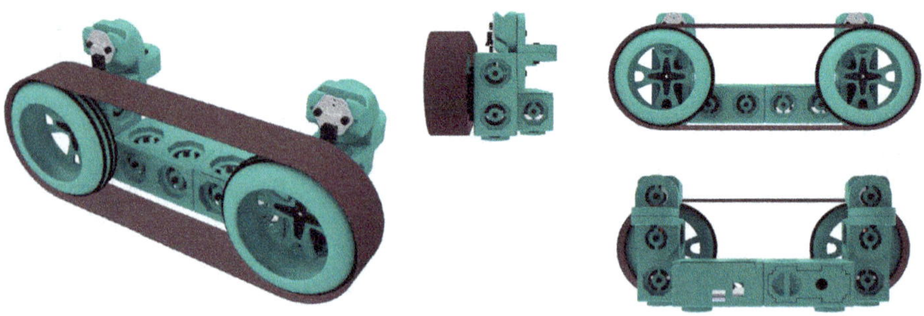

Fig. 5.29 Conveyor belt with infrared sensors at the end and start of the line

Conveyor Belt[1] with Infrared Sensors at the End and Start of the Line

Wheels are not only made for locomotive means, they can also work in other possible configurations. As teachers we need to make students think and go further than common thoughts (Fig. 5.29).

Traffic Light with Barrier

Occasionally, it will be necessary to include other elements, for example, the barrier made from cardboard or designed in 3D, that they can print themselves (Fig. 5.30).

[1] The conveyor can be made with cardboard or cloth.

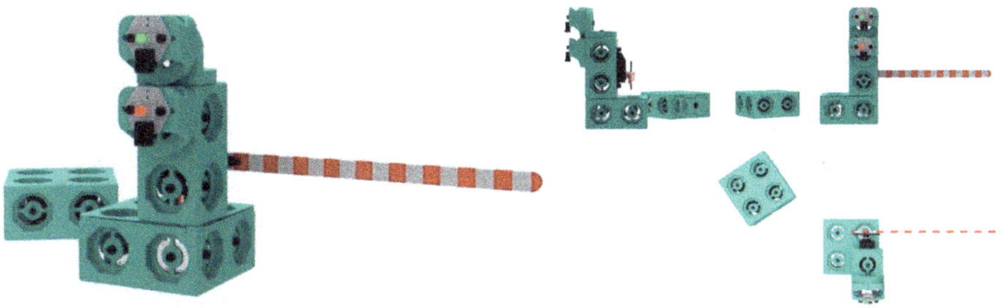

Fig. 5.30 Traffic light with barrier

Control System

As mentioned, the main control board is based on Arduino. The rest of the components, such as sensors and actuators, are connected to this main control board. The robot programs are stored and performed from the controller board.

Arduino is an electronic open hardware platform. It consists of a simple board with a micro controller that allows development of multiple prototypes and applications. They are powerful control boards and at the same time very economical. The microcontroller is programmed from the PC through its own Arduino software using its own programming language based on C++.

ZUM BT 328 (2017) is one of the versions of the original *Arduino* card, based on the ATMEGA328P-AU microcontroller. This control board was chosen instead of the boards officially distributed by *Arduino* since it presents significant improvements. It has additional strings of pins that duplicate the analogue and digital outputs and inputs. These added pin strings include ground and voltage signals, allowing a fast, safe, and simple connection of sensors and actuators.

Actuators

Servomotor

A Servo is a device composed of a motor, a power and control circuit, a potentiometer, and a gear train. The potentiometer takes the position of the rotor shaft. The control of the motors is carried out by pulse width modulation (PWM).

Spring RC SM-S4303R continuous rotation servos are used in this project, with a torque of 3.3/5.1 Kg.cm (4.8 V/6 V).

Buzzer

The buzzer is a component capable of emitting a continuous buzzing of the same tone. Its operation is simple; it is composed of an electromagnet and a steel sheet.

Sensors

Infrared sensor

The infrared sensor is typically used to measure distances or detect the passage of an object in front of it.

Electrical Power Supply

Battery holder

The power supply of the robot is supported by AAA batteries. This solution was chosen over others, such as using LiPo batteries, because it is the most accessible, thus contributing to the Open Source character of the project. The battery holder chosen is a standard model to house 8 AAA batteries in two floors. The battery can provide a voltage of 12 V; it is enough to power all the components of the Kit without losing effectiveness.

Prototype Manufacturing

The prototype manufacturing is carried out using an FDM printer, in this case *Hephestos 2* of *bq* (2018).

The first step is to export the 3D model in a format compatible with the slicing programs. Most programs support the .stl format and the .obj format. In this case, all the pieces were exported to the .stl format. The .stl format was created by the company 3D Systems (1998) and was designed for use in rapid prototyping machines. The abbreviations of STL refer to "Standard Triangle Language", which defines only the geometry of 3D objects excluding other typical properties of CAD files.

The slicing consists of generating a set of commands called .gcode the machine understands and uses for the manufacturing process (see Fig. 5.31). These commands contain the information of positions, speeds, and settings that the 3D printer will read and execute. Those commands are generated by cutting the object into slices or more commonly called layers. The code is generated in a similar way as that done for numerical control machines, such as lathes and milling machines. There are many free-to-use or proprietary software that develop this function, the best-known being *3D Cura* (2011), *Slic3r* (2017) or *Replicatorg* (2011). In this case, the *3D Cura* software was used with the following main settings:

- Layer height: 0.2 mm
- Thickness of the walls: 1.2 mm
- Thickness of the upper part
- Temperature: 210 °C

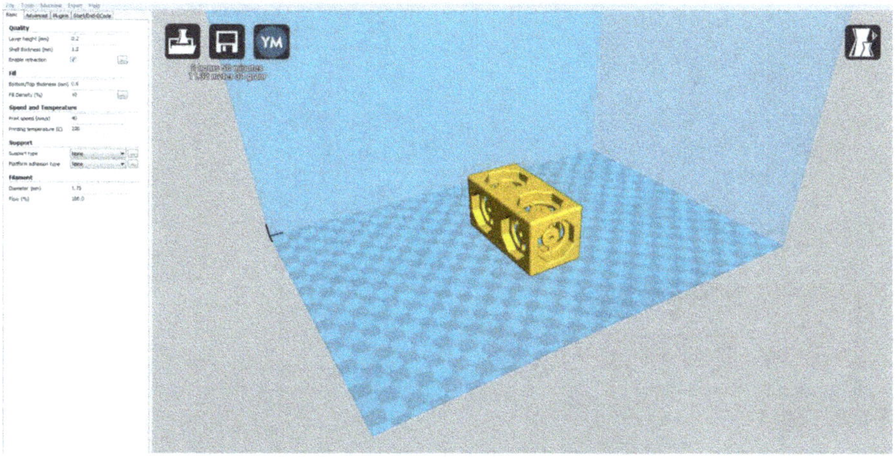

Fig. 5.31 Slicing of the pieces in *Cura 3D Cura 3D, Ultimaker, 2017*

Once sliced, the software shows us the route that will be made in each layer, and the .gcode file can be generated. The piece is then ready to be manufactured in the 3D printer.

Control Programming

Thanks to the use of *Arduino*-based electronics, it is possible to program the robot in a simple way. Two ways of programming are proposed, according to the knowledge of user programming. Programming by blocks using the *Bitbloq* tool, designed for use in school environments, and programming by code using *Arduino*, for advanced users.

Bitbloq

Bitbloq (2018) is an online block programming tool created by the company *bq*. It is a free tool that enables the programming of a large number of *Arduino*-based boards easily.

A block contains a section of code referring to the role developed by it. In this way it is not necessary for the user to learn the syntax of the programming language, you should only understand its logic to be able to program (see Fig. 5.32).

Blocks are classified by their function. You can find blocks with specific functions as variables, including logic or control ones.

Programming by blocks is therefore a powerful tool that greatly facilitates access to programming by users without knowledge of programming languages. It is perfect to start and learn the fundamentals of programming, increasing the speed of program creation.

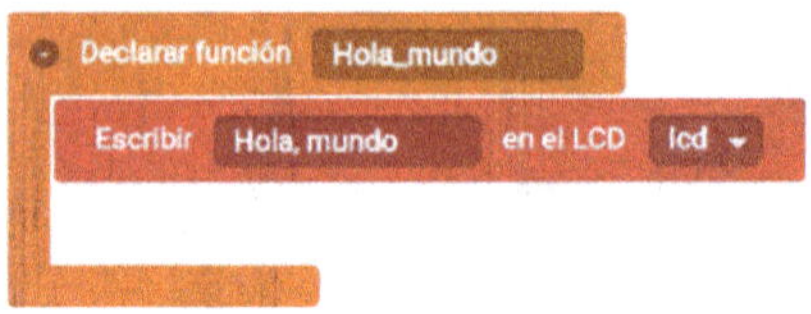


```
/***   Included libraries   ***/
#include <Wire.h>
#include <BitbloqLiquidCrystal.h>


/***   Global variables and function definition   ***/
LiquidCrystal lcd(0);

Big_Bang = ;



/***   Setup   ***/
void setup() {
    lcd.begin(16, 2);
    lcd.clear();

}


/***   Loop   ***/
void loop() {
    void Hola_mundo() {
        lcd.print("Hola, mundo");
    }
}
```

Fig. 5.32 Example of a block and the matching code (Bitbloq, bq, 2017)

Arduino

The most advanced users with previous programming knowledge can program the robot directly by writing the code. This allows greater freedom when programming. The programming language of *Arduino* is actually C++ adding a series of libraries with roles ready to control electronic units. Some of them are the position of a servo, the reading of a sensor or turning on an LED.

To be able to write and load the code in the robot, a compiler is necessary. To facilitate this task, *Arduino* offers its users a free compiler with all the tools to compile their code and load the programs on the boards without any problems (see Fig. 5.33).

Robot Cost

The cost of robot materials is broken down in Table 5.3. The total cost of the materials needed to build a robot with all the configurations shown in section "Possible Configurations and Activities" is €149.38.

Final Prototype

The final prototype has a configuration of two-wheel and follows-lines drive. It has the possibility of including a sensor to detect a line drawn on the floor and follow it. Students can see that with only two wheels the robot is stable and maintains balance (See Fig. 5.34).

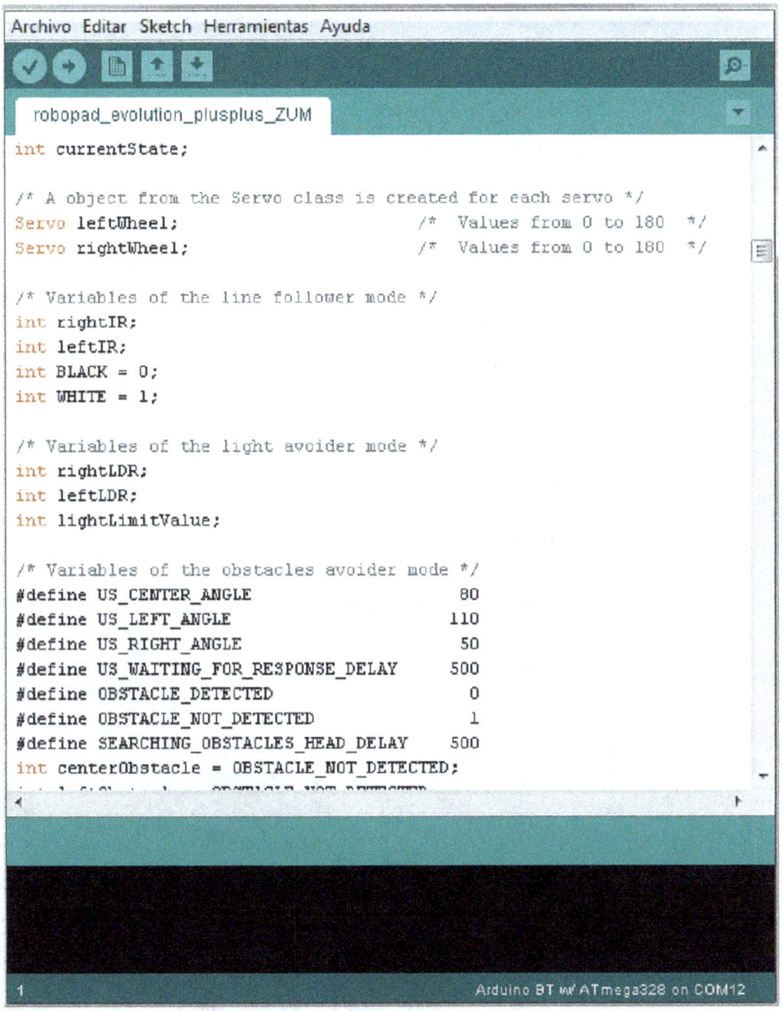

Fig. 5.33 *Arduino* Integrated Development Environment (IDE) (Arduino 2018)

As seen in Fig. 5.35, the robot is easy and fast to assemble. Perhaps the only difficulty you can find is the passage of the cables through the different modules, and depending on the age of the student, they might need some help.

Table 5.3 Robot cost

Code	Unit of measurement	Description	Quantity	Unit price	Total price
01.		**Chapter 1 – Software**			**0.00 €**
01.01	ud	Cura 3D	1	0.00 €	0.00 €
01.02	ud	Arduino IDE	1	0.00 €	0.00 €
02.		**Chapter 2 – Materials**			**32.28 €**
02.01	Kg	PLA	0.5	20.00 €	10.00 €
02.02	ud	M3 Nut	84	0.02 €	1.65 €
02.03	ud	M3 × 5 mm Screw. DIN 912	86	0.02 €	1.72 €
02.04	ud	Inserts 40/TH030H055	2	0.01 €	0.02 €
02.05	ud	O-ring 88mm diameter	6	0.87 €	5.22 €
02.06	ud	Neodymium magnet 10 mm. height 5 mm	20	0.63 €	12.40 €
02.07	ud	Neodymium magnet 5 mm × 6 mm × 2 mm	4	0.31 €	1.24 €
03.		**Chapter 3 – Electronic components**			**109.90 €**
03.01	ud	ZUM Kit bq	1	109.90 €	109.90 €
04.		**Chapter 4 – Power supply**			
04.01	ud	AAA batteries	8	0.90 €	7.20 €
		TOTAL			**149.38 €**

Fig. 5.34 Final prototype in the configuration Mobile robot follows-lines, two-wheel drive

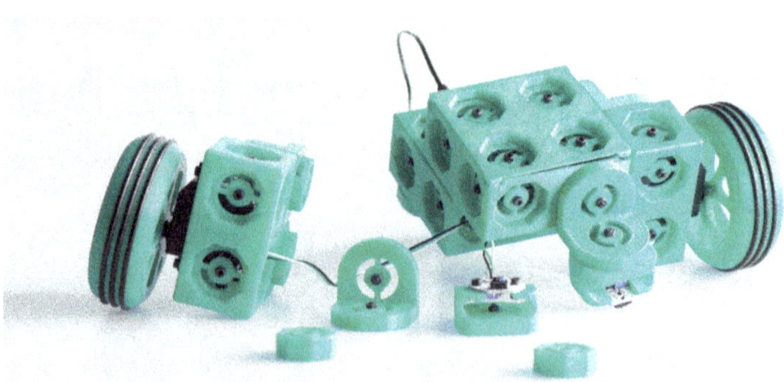

Fig. 5.35 Final prototype. Example of modularity

Conclusions

This project had two main goals:

- To design a prototype of an educational robot whose cost is below the average of the robots currently available on the market.
- To develop a replicable robot that any user with a 3D printer could reproduce with cheap and easily acquired electronic components. Ultimately, the intention was to create a robot whose cost would allow access to this technology to a greater number of users. Also, considering the school environment, we find that on many occasions, budget limits make it difficult to buy technological tools.

It has been possible to design a prototype with a price lower than the common commercial educational robots with easy-to-obtain materials.

On the other hand, one of the values of this product is its dual functionality. It can be used in a similar way to a robotics kit and it can be used to create robots equipped with movement. This characteristic, together with its special design, makes this material an open and multipurpose educational resource. It will enable its users to develop personal technology projects, as well as research and experimentation.

Another value lies in the possibility of programming the electronic components and, above all, the programming languages chosen with which this task can be carried out. In that sense, it is worth mentioning the possibility of using *Bitbloq*, a language based on graphic blocks. This option enables children to approach the world of programming, avoiding in these ages the difficulties of the syntax of written code. It favors developing computational thinking without the added difficulty of writing lines of code. This enables the child to "get rid of" that difficulty and focus more on creative aspects and elaborate strategies for solving problems.

The designed product, besides a domestic use or research tool, offers many educational possibilities in school environments. The physical characteristics of the design, as well as the programming language chosen, make it interesting for both Primary and Secondary educational purposes.

References

3D Systems. (1998). https://www.3dsystems.com/. Accessed 25 Apr 2017.

Alimisis, D. (2013). Educational robotics: Open questions and new challenges. *Themes in Science and Technology Education, 6*(1), 63–71.

Arduino. (2018) https://www.arduino.cc/

BQ. Bitbloq. (2015). http://bitbloq.bq.com/#/. Accessed 15 Jan 2018.

Catlin, D. & Blamires, M. (2012). The principles of educational robotic applications (ERA). A framework for understanding and developing educational robots and their activities. http://create.canterbury.ac.uk/10970/2/ERA%2520Paper%2520for%2520Constructionsim%25202010.pdf. Accessed 8 Jan 2019.

Catlin, D, Kandlhofer, M. and Holmquist, S. (2018). EduRobot taxonomy. Classification of educational robots. https://www.researchgate.net/project/EduRobot-Taxonomy. Accessed 8 Jan 2019.

Cesaretti, L., Storti, M., Mazzieri, E., Screpanti, L., Paesani, A., Principi, P., & Scaradozzi, D. (2017). An innovative approach to school-work turnover programme with educational robotics. *Mondo Digitale, 16*(72), 2017–2015.

Creative Commons. (2018). https://creativecommons.org/. Accessed 17 Feb 2018.

Cura 3D Ultimaker. (2011). https://ultimaker.com/en/products/cura-software. Accessed 25 Apr 2017.

Daniela, L., & Strods, R. (2018). Robot as agent in reducing risks of early school leaving. In L. Daniela (Ed.), *Innovations, technologies and research in education* (pp. 140–158). Newcastle upon Tyne: Cambridge Scholars Publishing.

Daniela, L., Strods, R., & Alimisis, D. (2017). Analysis of robotics-based learning interventions for preventing school failure and early school leaving in gender context. In *9th international conference on education and new learning technologies (Edulearn17)* (pp. 810–818). Barcelona: Conference Proceedings Barcelona.

Eguchi, A. (2017). Bringing robotics in classrooms. In M. Khine (Ed.), *Robotics in STEM education* (pp. 3–31). Cham: Springer.

García Saura, C. (2013). Cyclone PCB Factory – GitHub. https://github.com/CarlosGS/Cyclone-PCB-Factory. Accessed 17 Feb 2018.

García-Saura, C., & González-Gómez, J. (2012). *Low cost educational platform for robotics, using open-source 3d printers and open-source hardware. ICERI 2012 proceedings* (pp. 2699–2706).

Github. (2018). https://github.com/. Accessed 5 Feb 2018.

González-Gómez, J. (2012). http://www.iearobotics.com/blog/category/modular-robot/. Accessed 20 June 2017.

González-Gómez, J. (2017). Clone Wars – RepRap. http://www.reprap.org/wiki/Proyecto_Clone_Wars. Accessed 20 Nov 2017.

González-Gómez, J., Valero Gómez, A. (2008) IEARobotics. http://www.iearobotics.com/.%20. Accessed 20 Jun 2017.

Hallock, Joe. Colour assignment. (2003). http://www.joehallock.com/edu/COM498/preferences.html. Accessed 20 Nov 2017.

Heller, E. (2004). *Psicología del color. Cómo actúan los colores sobre los sentimientos y la razón. [Color psychology. How do colors act on feelings and reason]*. Barcelona: Gustavo Gili, S.A.

Hephestos 2bq. (2018). https://www.bq.com/es/hephestos-2. Accessed 20 Apr 2018.

Linux. (2010). https://www.linux.org/. Accessed 15 Nov 2017.

Luttropp, C. (2017). 10 Golden Rules in EcoDesign. http://lpmc.lv/uploads/media/10_golden-rules_in_eco-design.pdf. Accessed 20 Jan 2017.

Matbase. (2017) https://www.matbase.com/material-categories/natural-and-synthetic-polymers/thermoplastics/agro-based-polymers/material-properties-of-polyactic-acid-copolymer-pla-c.html#general. Accessed 20 Nov 2017.

Miller, D. P., Nourbakhsh, I. R., & Siegwart, R. (2008). Robots for education. In *Springer handbook of robotics. Chapter 55* (pp. 1283–1301).

Moro, M. & Alimisis, D. (2012) Proceedings of 3rd international workshop teaching robotics, Teaching with robotics. Integrating robotics in school curriculum. Riva del Garda (Trento, Italy). http://www.terecop.eu/TRTWR2012/proceedings.htm. Accessed 7 Jan 2019.

Moro, M., Agatolio, F., & Menegatti, E. (2018). The Robo ESL project: development, evaluation and outcomes regarding the proposed robotic enhanced curricula. *International Journal of Smart Education and Urban Society*.

Mubin, O., Stevens, C. J., Shahid, S., Al Mahmud, A., & Dong, J. J. (2013). A review of the applicability of robots in education. *Technology for Education and Learning., 1*, 1–7.

Ocaña Rebollo, G. (2012). Robótica como asignatura en enseñanza secundaria. Resultados de una experiencia educativa. [Robotics as a subject in secondary school. Results from a learning experience] Espiral. *Cuadernos del Profesorado, 5*(10), 56–64. http://ojs.ual.es/ojs/index.php/ESPIRAL/article/view/940. Accessed 23 May 2017.

Papert, S. (1980). *Mindstorms: Children, computers, and powerful ideas*. Basic Books, Inc.

Perens, B. (1999). The open source definition. In *Open sources: voices from the open source revolution* (Vol. 1, pp. 171–188).
Replicatorg. (2011). http://replicat.org/. Accessed 25 June 2017.
RepRap. (2018). http://www.reprap.org/wiki/RepRap. Accessed 15 Feb 2018.
Saygin, C., & Yuen, T. T., & Shipley, H. J., & Wan, H., &Akopian, D. (2012). *Design, development, and implementation of educational robotics activities for K-12 students*. Paper presented at 2012 ASEE Annual Conference & Exposition, San Antonio. https://peer.asee.org/21162
Scott, G. (2002). *Degradable polymers. Principles and applications. s.l.* Kluwer Academic Publishers. 1-4020-0790-6.
Scratch. (2013). https://scratch.mit.edu/. Accessed 24 Nov 2017.
Slic3r. (2017) http://slic3r.org/. Accessed 25 June 2017.
Thingiverse. (2016). Makerbot. https://www.thingiverse.com/. Accessed 20 Nov 2017.
ZUM Box. (2017). https://www.bq.com/es/zum-kit. Accessed 20 Jan 2018.

Chapter 6
Robotics in Primary School: A Realistic Mathematics Approach

Francisco Bellas, María Salgado, Teresa F. Blanco, and Richard J. Duro

Abstract Robots are technological tools of great interest in primary education for many reasons, but mainly for their compatibility with science, technology, engineering, and mathematics (STEM). However, it is very important to minimize the impact of the technical issues associated to robotics on the teachers, providing simple and functional tools that allow them to focus their attention in the creation of STEM content. To this end, this chapter presents a methodology, based on realistic mathematics, for the integration of educational robotics in primary schools. This methodology has been tested during one semester in the Sigüeiro Primary School (Spain) in the subject of mathematics, with students of different ages ranging from 7 to 11 years old. Two different educational robots, with different features, were used to highlight that the methodology is independent of the robotic platform used. Motivation surveys were administered to the students after the classes. Surveys reported highly successful results, which are discussed in the chapter.

Keywords Educational Robots · STEM · Realistic Mathematics · Primary Education · Robobo

Introduction

Educational robotics is a broad term typically associated with the use of real robots in pre-university education. In the last 10 years, the introduction of robots as didactical tools in primary and secondary schools has been very remarkable. The main reason behind this boom comes from the decrease of hardware cost, and from the development of programming environments adapted to younger students, mainly

---

F. Bellas (✉) · R. J. Duro
Integrated Group for Engineering Research, Universidade da Coruña, A Coruña, Spain
e-mail: francisco.bellas@udc.es; richard@udc.es

M. Salgado · T. F. Blanco
Applied Didactics, Universidade de Santiago, Santiago de Compostela, Spain
e-mail: maria.salgado@usc.es; teref.blanco@usc.es

L. Daniela (ed.), *Smart Learning with Educational Robotics*,
https://doi.org/10.1007/978-3-030-19913-5_6

based on blocks. Robots are used in classes as highly motivating platforms where students can learn programming, electronics, and basic mechanics.

But the fast development of educational robotics has led to different approaches toward the integration of robots in general education. Different countries, regions, or even single schools, have adopted their own didactical model that introduce robots in different subjects, without a formal analysis of the most convenient way to do it. As a consequence, nowadays one can find many educational robots in the market, all of them in use, with different technological features, target ages, and offering different learning options.

In primary schools, robots have been used mainly as platforms, alternative to classic computers, in which to run computer programs. At this educational level, the students acquire basic programming skills while they can observe the consequences of their programs in a real device that, typically, can move, thus increasing their motivation. Typical robots used in this age range are the Dash&Dot (https://www.makewonder.com/robots/dash/), LEGO WEDO (https://education.lego.com/en-us), Cubetto (https://www.primotoys.com), or Root (https://rootrobotics.com), which are simple and robust, and which can be programmed using a block-based language.

Here, a more formal perspective of educational robotics is presented. Robots are a very powerful tool to introduce the STEM methodology in primary schools. To this end, this chapter follows an approach where robots are introduced in the official curriculum of the mathematics subject, in a progressive way from the first grades and in particular topics. This approach makes it simpler to introduce robots in schools right now, without requiring a profound reorganization of curriculums, like the one proposed by Scaradozzi et al. (2015). The objective of using robots in classes should be acquiring basic competences of such subjects through the programming of the robot and not just the programming itself.

As a first approach, the proposed methodology has been designed to be applied in the subject of mathematics, so the aim is that students learn specific mathematics contents with each robotics teaching unit. To reach such objective, they have to apply many different abilities from different disciplines, as will be explained later in detail. But, first of all, in the next section we will discuss why this practical approach to mathematics has been chosen.

Realistic Mathematics

There are numerous methodological perspectives on the teaching and learning of mathematics like those presented in Karampinis (2018), Karkazis et al. (2018), Daniela and Strods (2016), and Moro et al. (2018), but we are interested in highlighting realistic mathematics education (RME) (Freudenthal 1977). The aim is to move away from classical, memory-based, and abstract learning in which the teacher is limited to giving lessons and correcting written tests. Realistic mathematics proposes seeking, in an initial phase, real contexts for the meaningful construction of mathematical learning, trying to make more concrete the abstract contents of this subject. Realistic mathematics and cooperative learning are the two pillars on which the theoretical foundations of this chapter are based, and where robotics and mathematics walk hand in hand in primary classrooms.

According to Freudenthal (1971), in RME the teaching of mathematics is determined by an activity described as:

> An activity of solving problems, of seeing the problems, but it is also an activity of organizing a discipline. This can be an issue of reality that has to be organized according to the mathematical patterns if problems of reality have to be solved. It can also be a mathematical matter, of new or old results, ours or of others, that have to be organized according to new ideas, be better understood, in broader contexts or by an axiomatic approach. (Freudenthal 1971, 411)

Following Freudenthal ideas, later several authors (Alsina 2009; Alsina 2011; Bressan et al. 2004; Cobb et al. 2008; Martinez et al. 2002; Van den Heuvel-Panhuizen 2000) have described the RME from of the following principles:

1. *Activity*: Mathematics conceived as a human activity. The purpose of mathematics is to mathematize (organize) the world around us.
2. *Reality*: Mathematics is learned by doing mathematics in real or realistic contexts.
3. *Levels*: Students go through different levels of understanding: Situational (in the context of the situation); referential (schematization through models, descriptions, etc.), general (exploration, reflection, and generalization), formal (standard procedures and conventional notation).
4. *Guided reinvention*: A learning process that allows the reconstruction of formal mathematical knowledge through mediation.
5. *Interaction*: The teaching of mathematics is considered a social activity. The interaction between the students and between the students and the teachers can cause each one to reflect on what others contribute and thus reach higher levels of understanding.
6. *Interconnection*: Mathematical content blocks (numbering and calculation, algebra, geometry, and so on) cannot be treated as separate entities.

Based on these principles, Alsina (2011) includes the characterization of the most significant features of RME, and these are:

- Situations of everyday life or contextualized problems are used as a starting point to learn mathematics.
- These situations are mathematized to form more formal relationships and abstract structures.
- It is based on the interaction in the classroom among the students and between the teacher and the students.
- Students are encouraged to interpret mathematics under the guidance of an adult, rather than trying to transmit a pre-constructed mathematics to them.

Children must, therefore, learn mathematics in real and close contexts that have meaning for them, from which to develop concepts and apply rules. This way the need for mathematization arises: moving a problem from everyday life to the world of mathematics, solving it, and returning it to the real world, which familiarizes the student with the mathematical world.

Finally, it should be noted that, according to Freudenthal (Gravemeijer and Tewuel 2000), the strongest argument that supports and justifies the existence and importance of RME is that not all students will be mathematically mature, but almost all of them will use those mathematics that help them solve problems of daily life (Peters 2016). Robotics, as a support for teaching and learning mathematics, has obtained considerable contributions (Pinto Salamanca et al. 2010), making it ideal for learning by playing in an interdisciplinary fashion.

Proposed Methodology

To clarify how this methodology can be realized in practical terms, this section describes the specific experience carried out during the year 2018 in Spain. The sample of participants is composed of all elementary students of the Sigüeiro primary school, a total of 233 students, with an age range from 6 to 12 years, as shown in Table 6.1. All the gathered data from the participating students respect the ethical implications of the projects in the educational field, which refer to, generally, the establishment of an atmosphere of trust between the teaching staff and researchers, and to the adequate treatment of the data of sensitive nature. Both of these aspects are taken into account to be conveniently treated from the perspective of the socially responsible research (SRR). For the treatment of the information obtained through direct involvement with the students, authorizations were requested to the parents or legal guardians of the minor in order to collect the data through audio and video as well as in written form. In any case, it is maintained that the privacy of students is respected in the publications derived from this study, always identifying them under pseudonyms.

The proposed methodology for introducing educational robots in the existing mathematics curricula of primary schools starts from the two following general premises. First, on each primary school grade, some specific mathematics lessons from the official curriculum are selected to be reinforced, or taught, using the robot as a real-world application platform following the realistic mathematics methodology. Such lessons were organized in the form of a practical workshop. Second, these robotic classes were programmed in all the primary education grades, that is, the robot was used as a long-term didactical tool as students grow, so they were acquired technical knowledge about its operation in a progressive way.

Table 6.1 Details of the participants in the workshops that make up the proposed methodology

Course/Grade	1st	2nd	3rd	4th	5th	6th	TOTAL
N° students	48	18	24	48	48	47	233

Table 6.2 Specific topics selected for the workshop depending on the grade

Grade	Workshop 1	Workshop 2	Workshop 3	Workshop 4
1ST	Natural numbers	Sequential operations	Distances	Open and closed lines
2ND	Natural numbers: comparison	Time units	Straight and curved movement	Planar figures
3RD	Natural numbers: basic operations	Distance units	Angles	Following a path
4TH	Decimal numbers	Measuring distance and time	Angles	Basic algorithms
5TH	Decimal numbers	Measuring distance and time	Angles	Symmetry
6TH	Integer numbers	Measuring distance and time	Angles	Cartesian coordinates

According to the methodology, the main aim is to organize which topics of mathematics will be covered in each school year. In this sense, Table 6.2 summarizes the specific topics selected for each workshop in each of the six elementary grades (from 6 to 11 years of age). These topics are organized according to the existing curriculum. The table also includes a possible way to teach programming concepts. Each workshop lasts 2 hours, which is the minimum time required to administer a class like this. The number of robots per workshop depends on the number of students and the number of teachers who control the workshop.

Two different educational robots were used in the workshops: the MBOT (www.makeblock.com/STEM-kits/mbot) and the ROBOBO (http://theroboboproject.com/en/), to show that the methodology can be applied independently of the specific robotic platform the school has. This affects the specific challenge that can be carried out, Of course, it affects the specific challenges that can be carried out, and the teacher is the responsible to design the most appropriate ones according to the selected robot.

The MBOT (Fig. 6.1) is a small mobile robot based on Arduino, which is cheap and has many possible expansion options. It can be programmed using mblock, a programming environment created by Makeblock, which is based on the Scratch block-based language (http://scratch.mit.edu). It is equipped with two motors for the wheels, one ultrasonic sensor and a line sensor. Students have to construct the robot for the first time, which can be used as a part of the initial workshops. With this robot, three workshops have been carried out for the fourth, fifth, and sixth grades. Here, we will explain in detail of the one given in the fifth grade, focused on symmetries.

The ROBOBO (Fig. 6.1) is an educational robot based on the combination of a smartphone and a simple mobile base (Bellas et al. 2017). The smartphone is attached to the base as shown in Fig. 6.1 and linked by Bluetooth, so students can program both elements from the computer as if they were a single robot. ROBOBO has a much higher technological capability than the MBOT due to the smartphone's features, and the students can use advanced sensors like cameras, microphones,

Fig. 6.1 The two educational robots used in the workshops. The mBot (left) and the Robobo (right)

gyroscope, accelerometer, and so on. In addition, the speaker, screen, and base motors provide a large amount of interaction possibilities, so it is a very powerful robotic platform for teaching human-robot interaction topics. This is the reason why it was applied in the workshops for younger students, those in the first, second, and third grades (between 6 and 9 years old). ROBOBO can be programmed, as well, using the Scratch block-based language through the ScratchX environment (https://scratchx.org). Here, we will explain in detail a case study within the second grade, focused on geometry, specifically basic planar figures.

The workshop organization, which is an example of the methodology application, will be described in detail in the following sub-sections.

Didactical Basis

From a didactical perspective, the workshops have been designed considering a STEM project-based methodology.

Project Based There is a challenge to be solved with the robot that students must solve at the end of the workshop, and which is focused on the specific selected mathematics topic. This global challenge must be divided by the teacher into small robotic activities that lead to its completion in a progressive way. This division is important in our methodology because it is crucial that students understand how to face a complex problem in a hierarchical fashion. The mathematic concepts required to solve the challenge should be introduced in previous classes, and students can use in the workshop the sources of information that were provided.

To solve the challenge, it is required not only to solve the mathematical aspects, but many others related to robotics: students must build an experimental environment or arena where the robot operates, they must manipulate different accessories or tools like screws, insulating tape and so on, and, of course, they must program the robot using the computer. All of these activities are inherent to the project-based methodology, and it is very important that the teacher provides students with a general view of the tasks they must face, in order to carry them out in an ordered manner.

The solution to the activities, and global challenge, must be autonomously obtained by the students, with the guidance of the teacher mainly in the correct selection of steps to solve the problem, and not in the particular way it is solved.

STEM Although the final objective of the workshop is stated within the subject of mathematics, to solve it with the robot implies integrating knowledge from other disciplines, like programming, physics (kinematics), mechanics (design, manipulation), and, of course, robotics. A very relevant topic at this educational stage is that of learning the basics of programming, which can fit in the mathematics curriculum as well, as it trains logical reasoning. In this sense, the following considerations must be made:

- Programming knowledge will be introduced in a progressive way during the different workshops. This is a very important aspect of this methodology, as it does not require previous programming skills. They are acquired as the global challenge is addressed. Remember that this robotics methodology is opposite to the traditional use of robots just to learn programming, so these skills are acquired as they are required to solve the mathematics challenge, but they are not the main didactic goal.
- Each activity requires programming the robot, which must be introduced by the teacher following an adequate order, with the objective of teaching a complete set of programming skills in a long-term setup, that is, during the whole primary education. As a consequence, the proposed challenges must be adapted to the programming complexity.
- The programming language at this age should be based on blocks, as it is simpler for students and the learning stage is short.
- The programming concepts can be explained in different order, but here we propose adhering to the following one. We also indicate an optimal learning age:
  - Programming basics: sequential operation, logical thinking, and basic blocks usage (from 7 years old)
  - Sensors and actions (from 7 years old)
  - Conditionals (from 9 years old)
  - Loops (from 9 years old)
  - Variables (from 10 years old)
  - Expressions (from 10 years old)
  - Functions (from 10 years old)

Regarding robotics itself, there are many concepts that are specific to this discipline and that will be introduced during the different learning stages. Specifically:

- *Sensors*: Understanding basic concepts of sensing like the magnitude to be measured, the data processing, the calibration, or the noise.
- *Actuators*: Understanding how the robot can act in the real world, mainly in terms of motors and how they work.
- *Reality gap*: It is a key aspect when learning robotics, because students have to understand that the real world where robots operate is complex and the translation between the program logic and its real consequence is not direct.

In addition to these general aspects, students must perform many physical manipulation tasks, both with the robot but also with the experimental environment where the robot performs the task, so teaching them basic manipulation skills like screwing or gluing is very important. In fact, some challenges may imply a more elaborate environment for the robot that students should construct in previous classes, for instance, a small city created with streets the robot must travel.

Evaluation

The evaluation of the workshops is based on the analysis of the student's notebooks and on rubrics. An example of the used rubrics is that of Table 6.3, which allows the teacher to evaluate the student's competence and motivation in different aspects of the workshop, as well as their knowledge in specific questions about the mathematics concepts treated during the workshop. Each student has their own notebook (Fig. 6.2) where they must take notes about the steps followed to achieve the objective of the workshop, mainly those related to the challenge and activities proposed by the teacher, but all they consider important. This notebook can be used to assess what data each student collects and how they do it when they are doing the workshops. Likewise, it serves to complete, in the rubric, the aspect about the attention placed in the classroom.

In addition, with the aim of evaluating the student's motivation when working with robots, a questionnaire with 12 items in a five-level rating scale (nothing, little, something, enough, a lot) was created. The specific items are included in Appendix 1. This motivation questionnaire was applied at the end of each session so that each student could fill it individually in the classroom. With respect to its analysis, the first thing that we are going to indicate is the degree of satisfaction of the students with the robotics session in which they participated. This is observed in the second item (did you find the class fun?) and in the twelfth one (would you like to continue learning with robots?). In general, the results of the class were positive for all levels, which are discussed in the section "Motivation Questionnaire Analysis."

Table 6.3 Proposal of possible rubrics that can be used for each student

Level (score)/ Aspects to be evaluated	Expert (4)	Competent (3)	Partially competent (2)	Not yet competent (1)
Time management	Satisfactory use of time during the entire workshop	Uses time well but can be delayed in some aspects	Has issues with time management and can cause delays to the team	Has serious issues with time management
Design and construction of the solution: ability to understand the objective	Understands the objective of the workshop, and the path to reach it, and to obtaining the solution of the activities	Understands the objective of the workshop, but the path to reach the solution is unclear	Has doubts when understanding what is the objective of the workshop	Has great difficulties to understand the objective of the workshop
Mathematics knowledge	Recognizes and relates the mathematical concepts involved in the workshop	Recognizes the concepts appropriately, although has trouble establishing relations between them	Has difficulties to recognize some concepts involved in the workshop	Does not recognize the majority of the mathematical concepts involved in the workshop
Attention in the classroom	Always pays attention to the teacher's explanations and to everything discussed in the classroom	Pays attention to the teacher's explanations and to everything discussed in class most of the time	Pays attention but is frequently distracted	Does not pay attention to the material discussed in the classroom, focusing on things that have no relation to the teacher's explanation
Attitude: active participation	Always participates in an active and voluntary manner	Usually participates in an active manner in the classroom	Often participates, only when asked to	Does not participate in class, not even when asked to
Problem solving: practical ability	Contributes with information and abilities when solving problems, showing initiative, and fomenting other's work	Usually contributes with information and abilities when solving problems showing initiative	Contributes with information and abilities when solving problems, only if asked to	Hardly ever contributes with information or abilities when solving problems

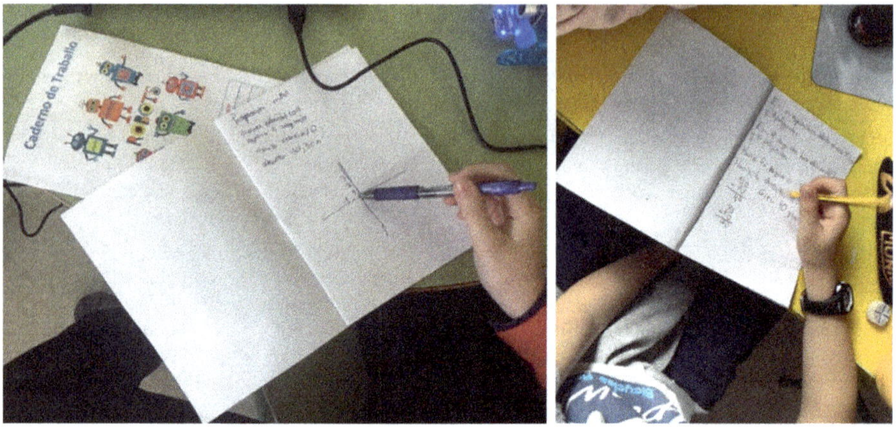

Fig. 6.2 Examples of students' personal notebook that they must use in the workshop

Classroom Organization and Equipment

The classroom organization and features where the workshops are carried out are very important in this methodology. A properly organized teaching space as well as different tool and practical elements are required for each group. In this sense, Table 6.4 contains a specific list of elements that should be present in the classroom:

The students of each class were divided into groups of four members per group. Each group contemplates the following roles:

- *Programmer*: Responsible for programming the robot using the computer
- *Robotician*: Responsible for manipulating the robot (turning it on and off, moving it from the table to the moving area and so on) and taking care of it (controlling that is has enough battery charge for the workshop, that it is not damaged during the class, etc.).
- *Technician*: Responsible for all the external elements and devices required to carry out the workshop, for instance, measuring tape, obstacles, etc. If any element must be constructed to carry out the lesson, it should be made before the workshop period in order to optimize the existing time.
- *Organizer*: Responsible for managing the group activity, controlling the time used on each activity, and interacting with the teacher in case of questions or comments.

Each student can help others in a different role in case of necessity, with the aim of all of them being active during the whole class. The teacher must assign to each student in a group one of the previously mentioned roles before the workshop and explain to them the main responsibilities associated with it. The roles must be interchanged during the four workshops that will be carried out during the school year, so that each student in the group assumes each role at least once.

Table 6.4 Elements that should be present in the room and basic equipment

Room	Equipment for each group
Round tables where students can work in teams of 4 or 5	1 educational robot, mBot or Robobo
Flat open space, in the floor or in an additional table, where robots can move freely	1 computer (ideally a laptop)
Workshop space	Connection cables and power supply (power strip)
Wi-Fi connection available	Screw, measuring tape, insulating tape, and scissors
1 projector to show slides of the workshop contents	
1 computer or laptop, for the teacher	

Workshop Description

In the following two subsections, two specific workshops for each robot will be explained in detail, in particular those marked in red in Table 6.2 corresponding to the fifth and second grades.

Fifth Grade Workshop

First of all, students were organized before the workshop into groups of four members with their specific roles previously assigned. Each group had its own table and chairs, with one laptop and one robot, as can be observed in Fig. 6.3. The class starts with the teacher presenting the robotics challenge they must face, in this case summarized in the diagram displayed in Fig. 6.4. They must implement a program in Scratch so that the robot can *avoid a rectangular obstacle ahead*. The specific obstacle was the mBot box, which could be located in any position in front of the robot, so it has to detect it using the sensors and then perform the movements displayed in Fig. 6.4.

Once the challenge is clearly understood, the student responsible for each role starts preparing their own part: turning on the computer and launching the programming environment (mBlock software in this case, which uses standard Scratch blocks and additional blocks specific for the mBot robot), turning on the robot, preparing the space on the floor, and preparing the additional elements, like the measuring tape. To solve the challenge, each group must have the following additional elements: measuring tape, protractor, mBot box, adhesive tape, and scissors.

To guide students toward the completion of the challenge, the teacher proposes the steps to be followed in the form of activities, and gives time to students in order to carry them out. In this workshop, seven small activities were proposed:

1. *Moving the robot to a certain distance*: To move the mBot a certain distance, students have to make a small program because this robot does not have any

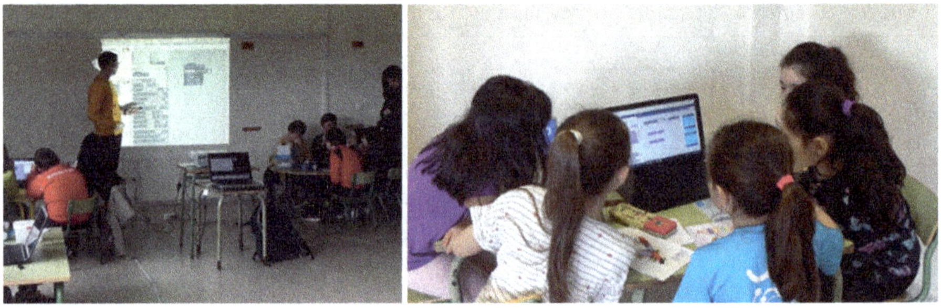

Fig. 6.3 Classroom and groups organization

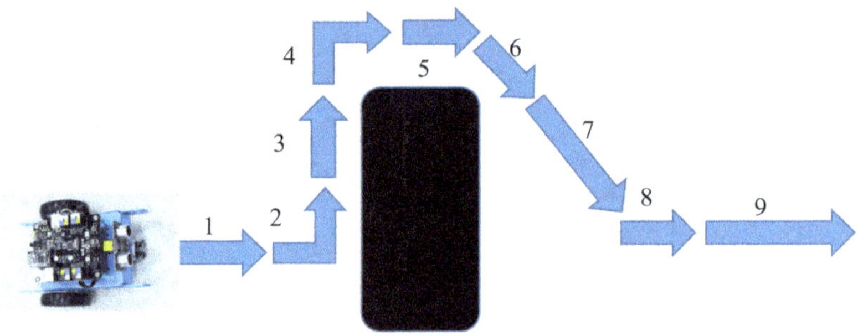

Fig. 6.4 Path that the robot must follow in the fifth grade workshop

predefined block to do that. Following the proposed methodology, first the teacher shows students a program with a preliminary solution to this problem, displayed in Fig. 6.5 left. Before they copy the program on to their computer and execute it on the robot, the teacher must explain the behavior of the blocks *if it is the first time students use them.* It is important to remember that, although the objective of the workshop is not on the robotics part, students must understand its basic operation. For instance, in this case the first block "- at speed -" makes the robot move in different directions and with different speeds using the wheel motors. There are two fields in this block the teacher should explain showing their effect in the robot movement, and the physical reason of such effect. The left field allows to select the robot direction between 4 options: "run forward", "run backward", "turn right", or "turn left". The teacher should explain that behind these pre-defined directions, the motor speeds are different for each wheel, obtaining this way a different direction. The right field allows to choose the robot speed, and it ranges from −100 to 100. The teacher should explain that this is an arbitrary unit, it does not correspond to any standard speed unit like cm/s or m/s. Moreover, the difference between using positive or negative values must be remarked too, highlighting how the wheel turning direction creates forward and backward movements. Summarizing, explaining the details of each

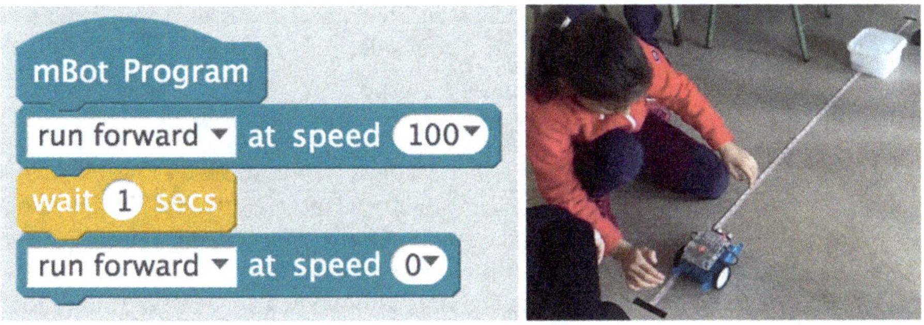

Fig. 6.5 Preliminary solution (left) and distance measurement procedure (right)

block and its relation with the robot response is very important in this methodology to allow students to understand the robotics background and get used to it.

The third block in the program displayed in Fig. 6.5 is a "wait - secs" block, which is a basic block in any programming language and has the effect of pausing the program a predefined time. This time can be an integer or decimal value, and it is important that students understand this difference. The fourth block is, again, a movement block, in this case "run forward at speed 0", which makes the robot stop. So, once all the blocks are clear, the teacher must carry out an overall explanation of the program logic before students try it: the robot starts moving forward at speed 100, and 1 second later it stops.

To execute this particular program, students must first know how to download the program to the robot using the USB cable or by Bluetooth. This is part of the *Robotician* role in the group who, in addition, must put the robot on the floor and leave free space in front of it. Moreover, they must fix a measuring tape to the floor and make the robot move next to it, as displayed in Fig. 6.5 right. After executing the program, the robot moves straight for 1 second at a speed of 100, and the students must write down the distance covered by the robot in their notebooks. In this specific case, the distance covered by the mBot was about 5 cm. It is important that the teacher emphasizes that the measurement must be reliable, so the robot must start in the zero value of the tape, and they must be precise with the measurement of the final position. Moreover, the program execution should be carried out more than once in order to avoid punctual fluctuations. All of these tips are very important to introduce students to the relevance of being technically formal.

Next, students had to measure the box sides with the measuring tape and annotate them again. Considering the distance covered by the robot when executing the program shown in Fig. 6.5, and without changing the speed, students had to adjust the time the robot moves in the "wait – secs" block in order to make it advance these two distances (in the case the box is not a rectangle but a square, they will have only one distance). Students at this level know the mathematical concept of rule of three, so instead of trying different time values, they have to calculate the right one, put it on the wait block, and test if the calculation was right. Specifically, for a box of 22 × 22 cm side:

$$5 \text{ cm} \rightarrow 1 \text{ second}$$

$$22 \text{ cm} \rightarrow x$$

So, the time they should try is 4.4 seconds. If the distance covered is not exactly the expected one, students can slightly adapt it. Notice that in this initial activity they worked with time and distance measurements, integer numbers, and decimal numbers in an integrated fashion, as proposed.

2. *Turning the robot 90°*: Once the students know how to make the robot move frontally the predefined distances, the second step toward the completion of the challenge is to make it turn 90°. If you see the path displayed in Fig. 6.4, the mBot must perform two 90° turns, one to the left and another to the right, in order to avoid the obstacle. Again, the mBlock software does not have any block that allows the robot to turn a specified value, so students must create a program to do it. In this case, the teacher presents the program displayed in Fig. 6.6, explaining that the only difference with respect to the previous one is in the first block, where now the specified movement is "turn right", so the logic would be: the robot starts turning right at speed 100, and 1 second later it stops.

   Students copy this new program in the mBlock software, or modify the previous one, and then they download it and execute it on the robot. To do it, the one with the *technician* role must fix the protractor on the floor using the adhesive tape, as shown in Fig. 6.6 right, and the *robotician* puts the robot on top of the protractor. Students must measure the degrees rotated by the robot in this specific case, and annotate this value in the notebook. From this value, and using again a rule of three, students must now calculate the time required in the "wait – secs" block to make the robot turn 90° right. This value was around 1.2 seconds with the selected speed. Finally, they must change the program to make the robot turn 90° left, which implies changing the first block and selecting "turn left at speed 100", using the same value of 1.2 seconds for the wait block.

3. *Turning acute and obtuse angles*: Now students have three small programs that allow them to move straight a predefined distance and turn 90° right and left. The next step to solve the challenge displayed in Fig. 6.4 is to perform a small turn to the right and then to the left to return the robot to the original path. To do it, students must understand the concept of acute angle. In addition, we introduce here the concept of obtuse angle although it is not necessary in order to solve this particular challenge (the diagram displayed in Fig. 6.7 is shown through the projector). So, in this activity, students must modify the previous program to make the mBot turn an acute angle and then an obtuse angle (the specific values must be selected by them) by changing the time in the "wait – secs" block.

   They test their solution using the protractor and annotate the time in their notebook. There are many possibilities on each case but, for instance, in the case of the acute angle the time used in the block must be lower than 1.2 seconds.

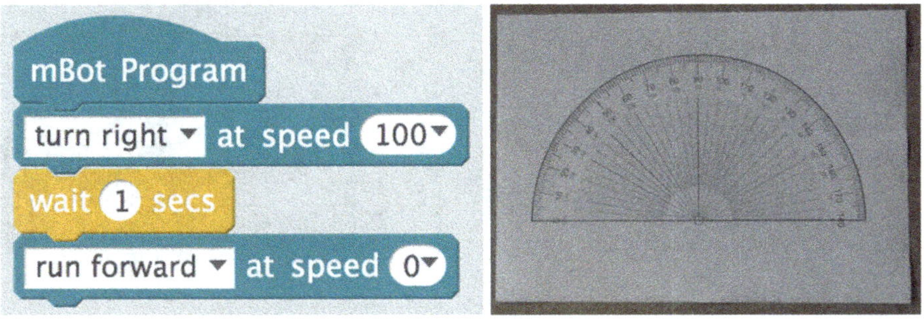

Fig. 6.6 Preliminary solution (left) and protractor fixed to the floor (right)

Fig. 6.7 Example of acute and obtuse angle turned by the robot

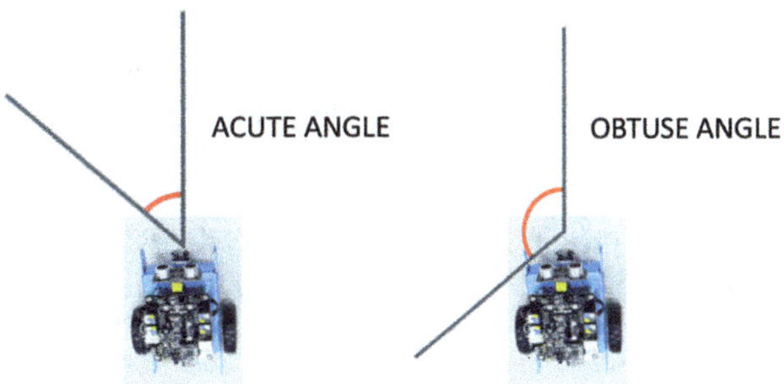

4. *Stopping the robot in front of the obstacle*: The previous activities create small programs that move or turn the robot a predefined value. This type of program is not very useful in robotics, because the actions should rely on the sensing, that is depending on what the robot perceives, it moves or turns in a different fashion. To show that to students, they put the robot on the floor with the box in front of it at a distance of 15 cm and they execute, again, the program displayed in Fig. 6.5 using a time of 4.4 seconds, which corresponds to covering 22 cm. The result is that the robot crashes with the box. Next, they put the box at a distance of 40 cm from the robot and try the same program. The result now is that the robot stops far away from the box. What the teacher must point out is that this program depends on a predefined distance to the box, which is not useful in many real cases, where the robot does not know, beforehand, where the obstacle will be placed.

   The solution is using a sensor that provides the distance to the box, in this case, the ultrasonic one that is placed on the frontal part of the mBot. The teacher shows the program displayed in Fig. 6.8 left, which makes the robot start moving straight at speed 100, then it waits until the distance returned by the ultrasonic sensor is lower than a threshold (the robot keeps moving), and then it stops. Students copy and try this program placing the robot in front of the box at an arbitrary distance. In fact, they should try the program with different distances to realize that now the robot is really autonomous, that is, it stops without knowing

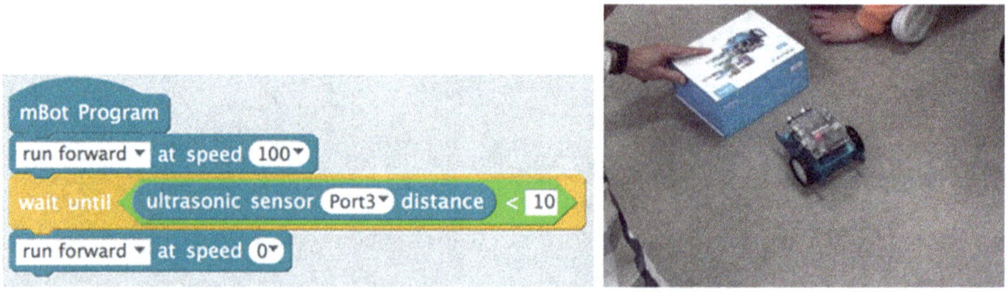

Fig. 6.8 Preliminary solution (left) and obstacle used in the workshop (right)

the distance to the box beforehand. The threshold value (10 cm in the example shown by the teacher), must be adjusted by each group considering that the robot must have enough free space to turn without crashing with the box (see Fig. 6.8 right). Once chosen, students must annotate it in their notebook.

5. *Stopping the robot and avoiding the obstacle*: At this point, students have all the components of the global program, so they have to join them to create the solution to the challenge. Thus, starting from the program shown in Fig. 6.8 with the threshold distance adjusted by each group, the first step is to concatenate it to the program that performs a 90° left turn developed in activity 2 (step 2 in Fig. 6.5). Second, students must add the program developed in activity 1, which moves the robot straight for a distance equal to the box width, in order to overpass it, as represented by step 3 in the diagram of Fig. 6.4. The third step is to use again the program developed in activity 2 but now to perform a 90° right turn (step 4 in Fig. 6.4). The solution to this activity is displayed in Fig. 6.9, and it must be found by the students, which may require a period of thinking and reflection before trying it on the robot. When executed, the robot finishes on the left side of the box, which must be tested by all groups before moving to the next activity. Although it is not mandatory to stop the robot after each single movement with the block "run forward at speed 0", it is interesting to do that at this level, in order to show that the global movement is composed by discrete steps that are easier to compose and control.
6. *Returning the robot to the original path*: The steps required to complete the program are those shown in Fig. 6.4 as 5, 6, 7, 8, and 9: moving the robot straight until it overpasses the left side of the box, turning right at an acute angle, moving straight until it reaches the original path, turning left at the same acute angle, and finally moving straight a predefined distance. This final step has been included just to show that the robot has avoided the obstacle and it can keep on moving. These four steps can be carried out using the programs developed in previous activities, but this is part of the student's job, that is, it is important that they understand the objective and how it is related to the previous steps, so they can divide the whole problem into small ones by themselves in the future.

A possible solution to this activity is shown in Fig. 6.10, but each group can perform their own variation. The execution of this program solves the global

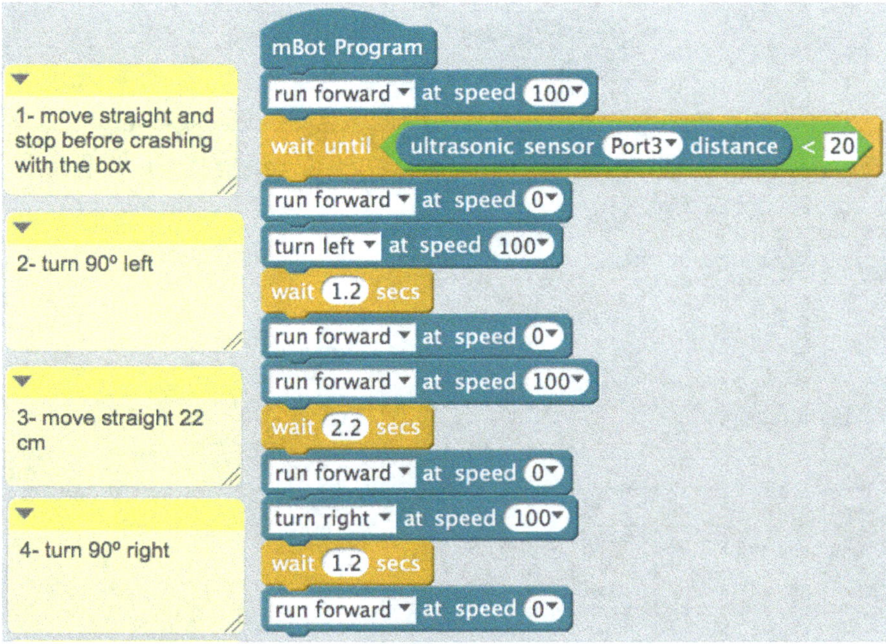

Fig. 6.9 Partial solution to the fifth grade workshop

challenge, so it is important that the teacher emphasizes that it is important to reach a valid solution in practice. They must recognize whether their solution is successful, that is, if the robot returns to the original path or not, although high precision is not required. The final movement of the robot could be recorded on video by the students.

7. *Symmetric movement*: The main objective in this workshop was to solve the challenge displayed in Fig. 6.4, but with the aim of understanding the concept of symmetry. So, at this point, the teacher can pose the following question to students: why do you avoid the obstacle on the left part of the box? Why not on the right? The typical answer is that, of course, it is possible to do that and it would be a symmetric movement, as shown in Fig. 6.12 in red color. So now, the students have to create a copy of the final program, and change it so that the robot avoids the obstacle on the right (we do not show this solution because it is equal to that of Fig. 6.11 but changing steps 2, 4, 6, and 8).

Summarizing, the proposed methodology has been clearly shown with this workshop example. The main didactical objective was understanding the concept of symmetry from a practical point of view, and it has been clearly achieved. To reach it, many other mathematical concepts have been used: integer and decimal numbers, time and distance measurements, rule of three, and angles. From an algorithmic perspective, students have created a simple solution based on the sequential combination of small programs, which is very important in programming. Regarding specific

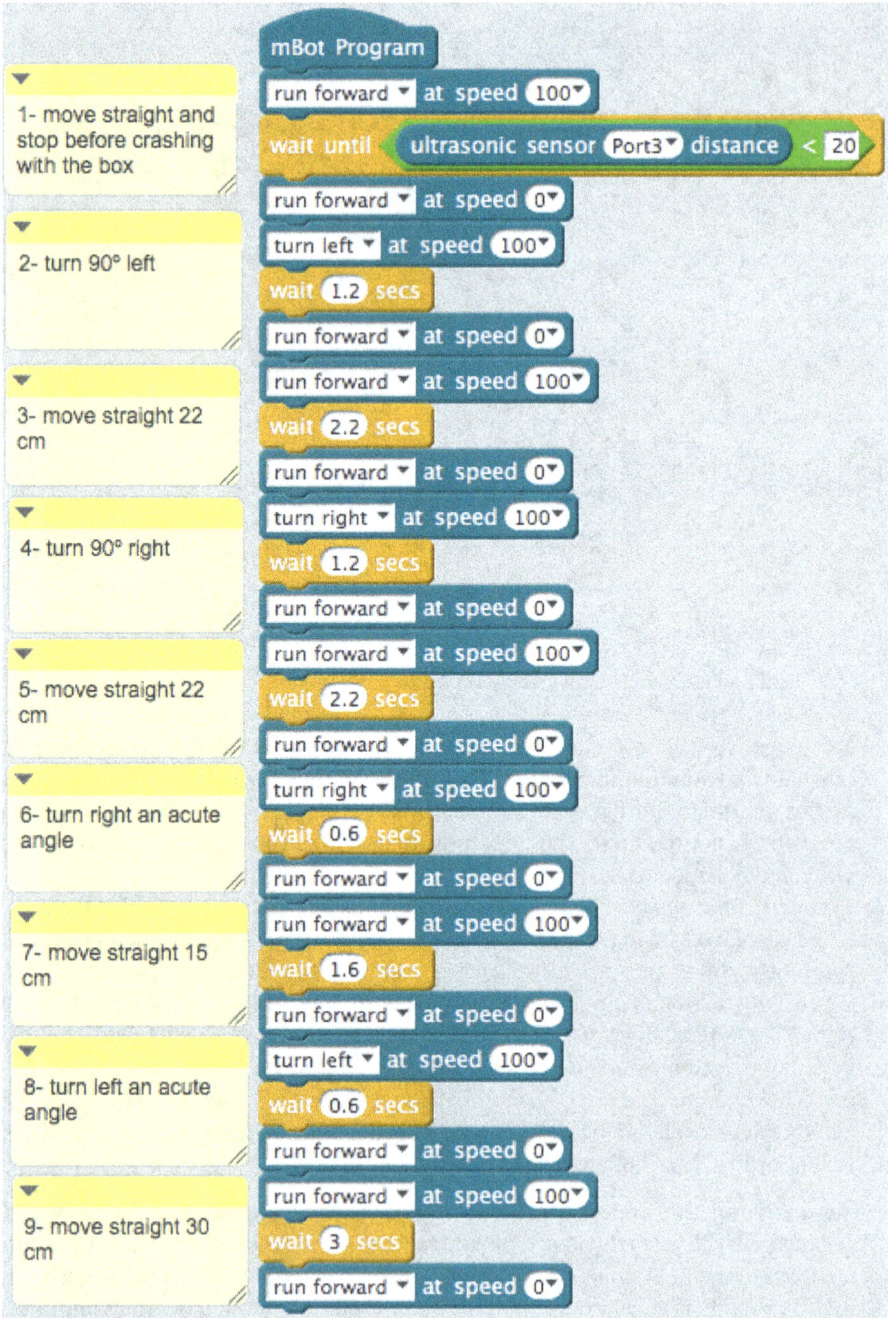

Fig. 6.10 Final solution to the fifth grade workshop

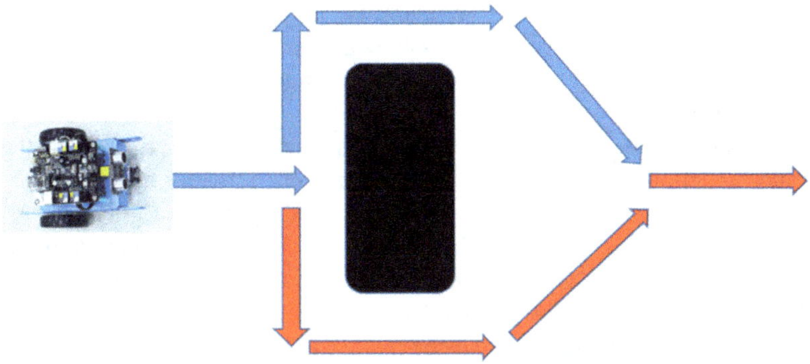

Fig. 6.11 Schematic view of the symmetric movement

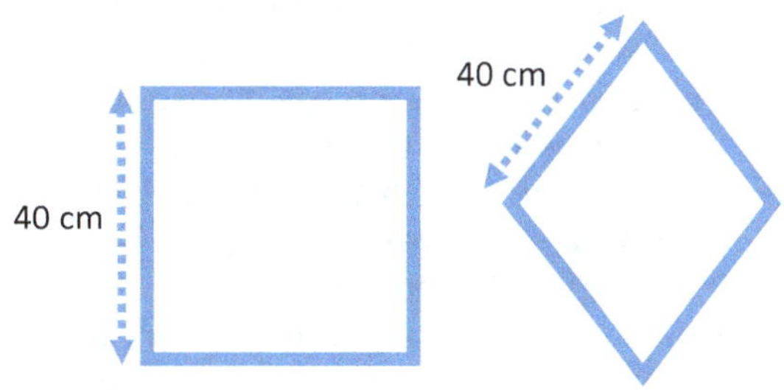

Fig. 6.12 Representation of the type of figure the robot must describe in its movement

programming topics, students have learned to reinforce basic blocks as "wait" or "wait until". Finally, from a robotics point of view, basic concepts of motor movement and ultrasonic sensing have been used. As it can be seen, the STEM methodology is clearly exploited in this type of workshop.

Second Grade Workshop

According to the proposed methodology, students were previously organized into groups of four members with the specific roles previously assigned. Again, each group used a round table with chairs, one laptop, and one robot. The class starts with the teacher presenting the robotics challenge, in this case summarized in the diagram displayed in Fig. 6.12: they must implement a program in Scratch so the robot can *move describing two simple planar figures, a square, and a diamond.* Both figures will be drawn on the floor using masking tape, so the robot must follow this path. Take into account that these students are younger than those of the previous workshop, so the challenge is simpler. Following the STEM approach, to reach this final didactic objective, many other topics will be necessary: natural numbers, distance and time measurements, simple sequential algorithms, or angles.

As in the previous workshop, once the challenge is clearly understood, the student responsible for each role starts preparing his/her own part: turning on the computer and launching the programming environment (ScratchX in this case, which uses standard Scratch blocks plus specific Robobo blocks), turning on the robot, preparing the space on the floor, and preparing the additional elements, in this case, measuring tape, masking tape, a protractor, and scissors. Considering the student's age, the teacher must organize the workshop into very simple and clear activities so the way toward the completion of the challenge can be easily followed. In this workshop, five small activities were proposed:

1. *Moving forward and backwards*: In this case, due to the students' age, their programming skills were very limited. As a consequence, the workshop does not use the original Scratch blocks but custom blocks that the teacher must first create. In this first activity, the goal was to move the robot forward and backward by using the blocks shown in Fig. 6.13. These blocks are custom blocks defined by the teacher in ScratchX (following the same procedure as in Scratch) as can be seen in the bottom part of this figure. For instance, the "move forward – seconds" block makes the robot advance in a straight line per the time specified in the field. Internally, this custom block contains many interactive elements that Robobo allows to use. Thus, the robot first says "forward" using the smartphone's speaker, then it changes the robot emotion (facial expression) to "laughing", and finally it turns on the frontal LEDs in magenta color. All of these actions are performed before the robot starts moving with the command "move

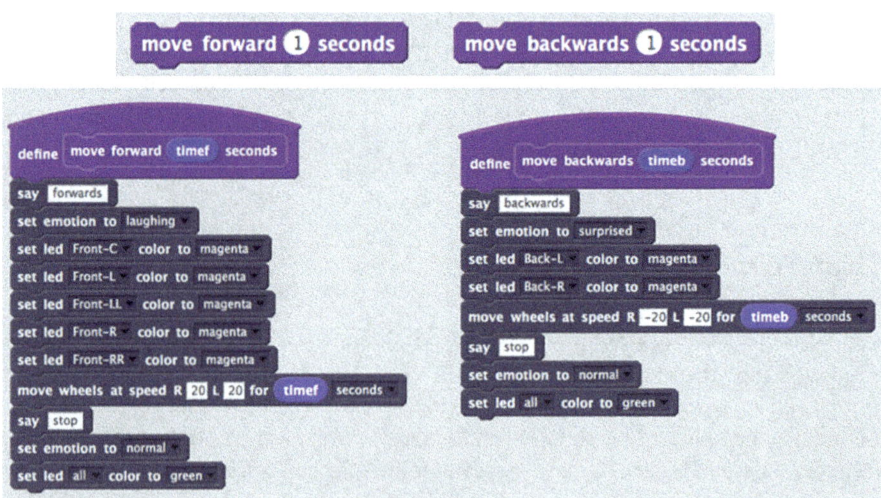

Fig. 6.13 Custom blocks for activity 1 created by the teacher (top) and their internal code (bottom)

wheels at speed R – L for – seconds", which is responsible for moving the robot wheels. When the movement is finished, the robot says "stop", changes its emotion to the normal state, and all the LEDs back to green. The "move backwards – seconds" block is similar to this one and it can be observed on the right part of Fig. 6.13. The specific interactive actions that have been included in this workshop are not relevant, and many others could be used. The most important aspect here is that students perceive the change in the robot state when it moves or when it stops.

Regarding the activity itself, students must execute the "move forward 1 seconds" block and measure the distance covered by the robot. As in the previous workshop, to do this they fix a measuring tape on the floor and place the robot at the beginning, as displayed in Fig. 6.14, so they can measure the real displacement of the robot. This value was annotated by students in their notebook (see Fig. 6.14), and then the objective was to adapt the time field in this block to reach a 40 cm displacement. At this grade level, students may not know the concept of centimeters, but it is not relevant because the key aspect here is that of the measurement unit. That is, the teacher must emphasize that, to compare different distances, it is required to have a reference one, and the measuring tape has some of them (m, cm, and mm). So, although they do not understand the difference between these units, they can use the centimeter marks in order to compare the robot movements. In this case, as students did not know the rule of three as yet, this adjustment was carried out by using a simple proportionality rule. In 1 second, Robobo covered 10 cm approximately, so students easily find out that they must use 4 seconds in order to advance 40 cm. In this grade level, only natural numbers can be used, so a more precise adjustment through decimal numbers is not possible.

2. *Turning left and right*: Once the students know how to move the robot 40 cm in a straight line, they learn how to turn the robot left and right an arbitrary angle. To do it, again, the teacher must prepare simplified blocks that allow the robot to rotate in place an angle that is specified as a parameter. The blocks used in this activity were those displayed in Fig. 6.15. On the top, the custom blocks are shown, with their corresponding internal blocks on the bottom. It can be seen that now the robot says that it is turning left or right, and LEDs corresponding to

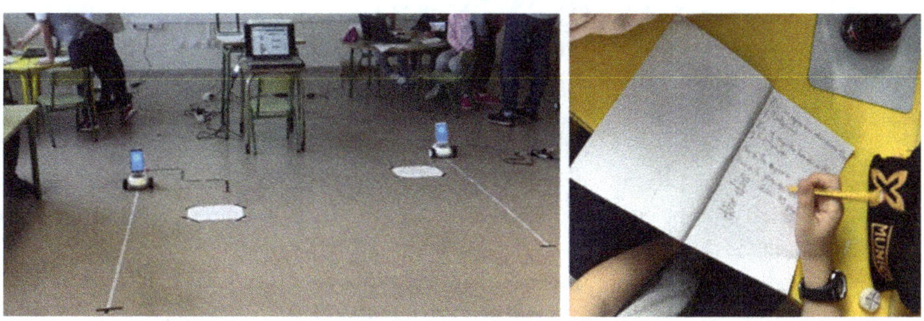

Fig. 6.14 Setup created by students to solve the challenge (left). Students writing results in their notebooks (right)

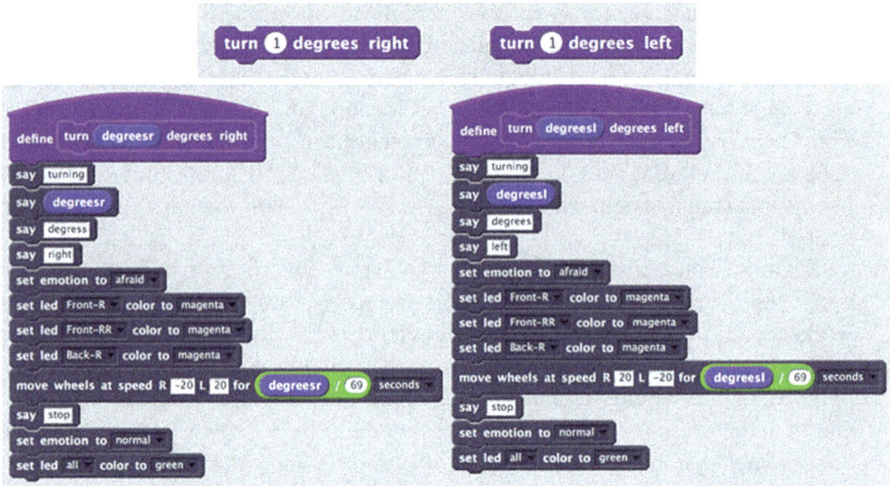

Fig. 6.15 Custom blocks for activity 2 created by the teacher (top) and their internal code (bottom)

this side are turned on. In the "move wheels" block, the time has been adjusted using a simple rule of three, so the robot moves a time proportional to the specified turning degrees at a speed of 20 on each wheel.

In this activity, students must try 90° right and left, and annotate what happens, that is, how the robot finishes with respect to its initial orientation. To do it, as in the previous workshop, each group must fix a protractor on the floor and put the robot on top of it, as shown in Fig. 6.16. In the specific workshop carried out at Sigüeiro school, it was the first time the students saw a protractor, and the concept of rotation degrees was also new to them, but this was not a problem, and all of them could follow the activity without trouble. As in the previous case, the specific concept of degree is not as important as the concept of measurement unit, and how the turns can be compared using it. Once the 90° rotation was understood, the teacher explained the concept of acute and obtuse angle, and students had to select a value to obtain such rotations in the robot, one larger than 90° and other smaller than 90°. These specific values were annotated by the students at the end of this activity.

3. *Following a square*: At this point, students know how to move the robot 40 cm in a straight line and how to perform different types of rotations. In this activity, they have to compose these two custom blocks in order to make Robobo follow a square drawn on the floor with masking tape. Each group must create its own square of 40 cm per side, implement the program in Scratch, and modify it until they reach the solution, shown in Fig. 6.17. It is a simple solution that implies repeating the same pattern of moving and turning four times. Once it is achieved, students must annotate this solution and the teacher can record the real execution on video. Figure 6.17 shows the same solution but using a very simple loop with four repetitions. This program can be explained to the students so they have a simple and clear introduction to the concept of loop in programming.

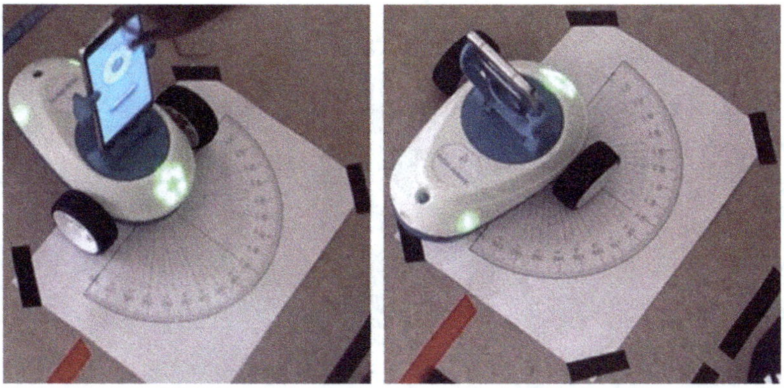

Fig. 6.16 Measurement of turned angles using the protractor

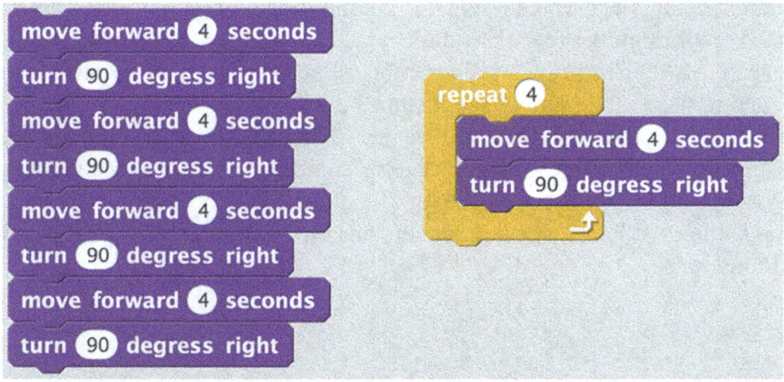

Fig. 6.17 Solution that makes the robot describe a square using (right) and not using (left) a loop

4. *Following a diamond*: With the square activity already finished, students must solve the last activity, which is making Robobo follow a diamond drawn on the floor with masking tape again. In this case, the angle that must be turned on each vertex must be adjusted by measuring it with the protractor or by simple trial and error. What is relevant is that students understand that the diamond requires two turns larger than 90° (obtuse angles) and two smaller than 90° (acute angles). Figure 6.18 shows the solution obtained by one of the groups, where the different turns created by the students can be observed.
5. *Optional (new figures):* As an optional activity, in case the workshop still has time, or some groups finish the diamond before the class ends, they can draw a more complex planar figure on the floor and implement the Scratch program to follow it. For instance, students can try to follow a pentagon, hexagon, etc.

This second workshop example is interesting to show how this robotics methodology can be introduced in early stages easily by adapting the topics to the level. In

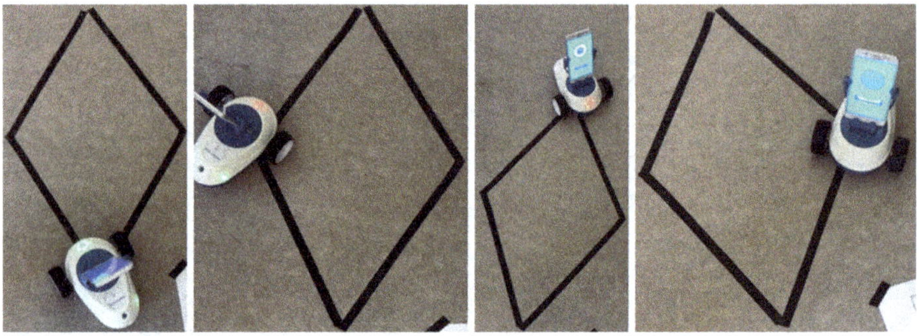

Fig. 6.18 Example of a final diamond movement obtained by students

this sense, the mathematical concepts are imposed by the official curriculum and the natural development at this age, so the main work of the teacher lies in adapting the programming language and in seeking a simple challenge that does not require advanced programming skills. The main didactical objective in this case was around the concept of planar figure, which can be reduced to work with linear displacements and turns, so the main concepts that students reinforce from a practical perspective are those of distances and angles. The programming topics were very simple, focused on the use of simple motor commands and sequential operations. Finally, regarding robotics, at this level, the most important aspect is that students become familiar with this new tool, understand how to interact with it, and see some of its limitations.

Motivation Questionnaire Analysis

The objective of this point is to verify whether robots can be considered or not a motivating tool for the classrooms and the development of mathematical contents in primary education. Motivation largely determines the performance of students. It can be said that improving motivation is one of the two main purposes of schooling as it can influence how and when they learn (Schunk 2001). There is a reciprocal relationship between motivation, learning, and execution, so motivation influences learning and execution and what students do and learn affects their motivation (Pintrich and De Groot 1990). The sample of participants in this case was composed of all elementary students of the CEIP school in Sigüeiro, a total of 233 students, with an age range from 6 to 12 years. The instrument used was the questionnaire presented in Appendix 1.

The analysis is presented through the components of motivation proposed by Pintrich and de Groot (1990): (1) the value component, (2) the expectation component, and (3) the affective-emotional component. The value component would be related to the question "Why do I do this task?" It would include those motives, purposes, and reasons why the student would carry out this activity. This is very much linked to motivation since, depending on the weight of that reason for oneself, the motivation will be greater or lesser. The expectation component is related to the question "Am I capable of performing this task?" It would fit in with individual perceptions and beliefs about one's ability to perform the task. If a student believes that he can do the task and that he hopes to do it well, he is likely to obtain good performance, involving himself cognitively and persisting for a long time in the task (Pintrich and Schunk 2006). The affective-emotional component is related to the question "How do I feel when performing this task?" refers to the feelings and emotions that arise when the activity is performed.

The Value Component

The value component is included in items 1, 8, 10, and 11. The following dimensions are differentiated within the value component: the intrinsic value, the utility value, and the cost value. The intrinsic value is related to the satisfaction that is obtained during the activity. Many of the experiences on robotics in the classroom coincide in that this methodology achieves a high degree of involvement in children, pointing out the satisfaction that children obtain when carrying out the challenges as one of the main reasons. From our observations, in general the boys and girls were very committed to the task and many were implicated in the importance of correctly carrying out the challenges. This could be observed every time they checked their experiments and robots, as they placed themselves around or inside the circuit, attentive to the robot, often nervous. Within this item, Krapp, Hidi, and Renninger (1992) distinguish situational interest, influenced by factors such as novelty or intensity, and topical interest. Without any doubt, the context created can be considered as an important motivational factor. The novelty, the playful nature, and the freedom and responsibility that was perceived generated great interest in the children and thus great motivation. This was clearly expressed in the questionnaires, for instance, analyzing the responses of item 10 shown in Fig. 6.19. It displays a bar graph where the colored scale indicates the grades (from first to sixth) and the y-axis corresponds to the average value for each grade considering the previously explained scores (1-nothing, 2-little, 3-something, 4-enough, 5-a lot). So, in this case, Fig. 6.20 clearly shows that students feel they put interest when working with robots, a little more as the age increases.

Regarding topical interest, it is related to the preferences of people for topics such as educational robotics, tasks, or contexts. The first question of the questionnaire did not directly ask if robotics was among their interests or tastes, although we can get an idea about that relationship assuming that those who had robotics among their interests would consider themselves more knowledgeable about the subject. As for the results in this case, they were those displayed in Fig. 6.20. The average response to this item is 2.6, which translated into the established variables would be between "something" and "a little," meaning that most students do not have a clear previous experience.

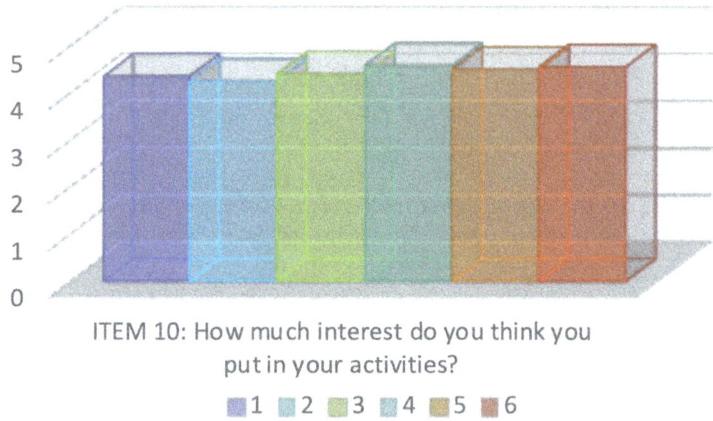

Fig. 6.19 Results item 10

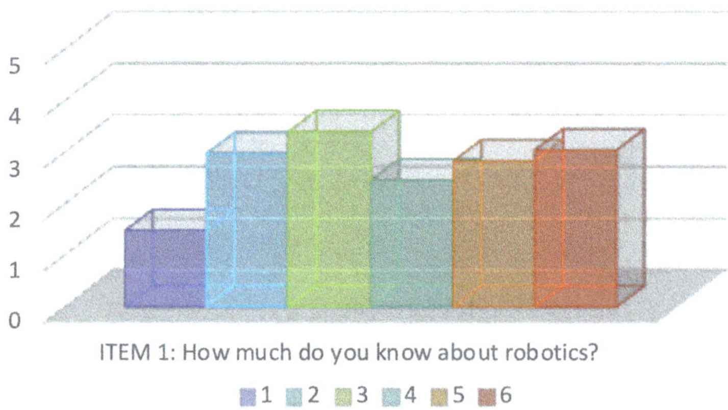

Fig. 6.20 Results item 1

Regarding the utility value, in item number 11, the results displayed in Fig. 6.21 were obtained. The average value of the answers is close to 5; specifically 89.27% of the all the children marked "a lot" in their questionnaire.

Finally, we consider the cost value, linked to the negative aspects that imply commitment to the task. These trade-offs include anticipated negative emotional states (e.g., anxiety and fear of both failure and success) as well as the amount of effort needed to succeed in different tasks or activities (Wigfield and Eccles 2000). In robotics, the realization of challenges is often hindered by the lack of precision of the robots or difficulty. During the sessions, it is surprising to see that in spite of the number of mistakes made, the children are still motivated.

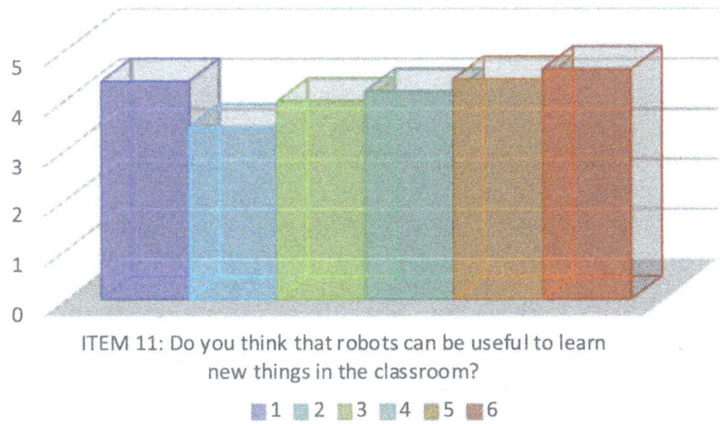

Fig. 6.21 Results item 11

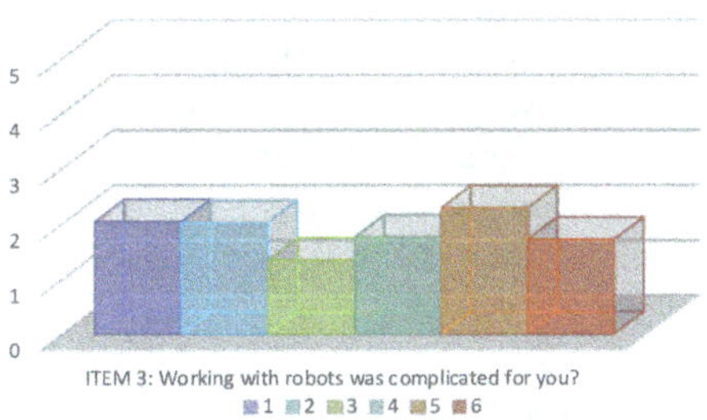

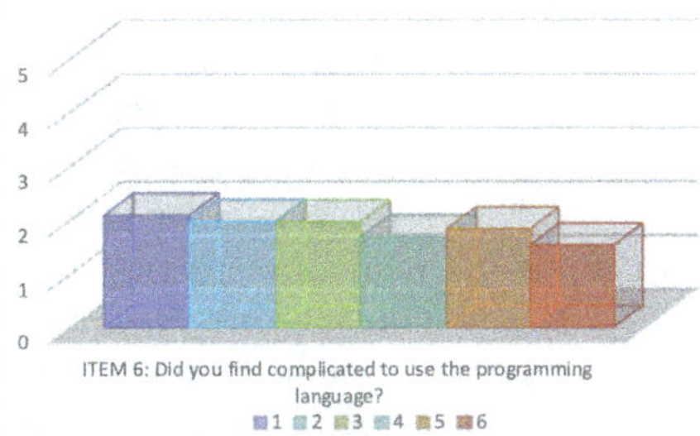

Fig. 6.22 Result items 3 (left) and 6 (right)

The Expectation Component

This component analyzes the perception of their own competence. Associated items are 3, 4, and 6 in the questionnaire. Many studies, like Harter (1981), state that "students with a positive perception show greater interest in learning, like challenges and, in general, obtain better results in their academic performance." In order to analyze this component, we first consider the perception of the students regarding the difficulty of the challenges. Such difficulty has a great impact in the academic motivation and it can lead to a higher or lower motivation in the student. In this case, items 3 and 4 refer to the difficulty of handling the robots, and item 6 refers to the specific programming language. In both items the results were similar, the great majority of the students answered to these two questions between "little" and "nothing" (see Fig. 6.22).

The Affective-Emotional Component

The items related to this component are 5 (Fig. 6.23) and 9 (Fig. 6.24). By analyzing the results, it can be seen how the vast majority of children felt comfortable, that is, had a positive emotional response, although they felt "something" or "a little" confused at some point in the class.

Going in depth into these responses of the students, a link can be established between this item 9 and number 7 ("Do you think that programming robots is boring?"), shown in Fig. 6.25. Eighty-five percent of the children think that programming robots is not boring. The relation of this with the previous item 9 is that 63.33% of those that concluded in their answer that to program the robots was between a little and much of a pain also felt confused within that interval, that is, they were between "a little" and "very confused." So, it can be considered that the programming process is one of the factors that lead children to feel confused and, therefore, influence their motivation.

In general, the results show a high motivation of students, although there are individuals who are not attentive to the task, who let themselves be carried away by the ludic atmosphere of the classroom. The degree of satisfaction of the students with the robotics session in which they participated was high. This is observed in the results of items 2 and 12, displayed at Fig. 6.26. In general, the result of the class was positive for all levels. The same goes for the twelfth item. In the last question of the questionnaire, the answers also had a high degree of uniformity, so the data was concentrated between "enough" and "a lot."

Within the questionnaires, the students in the first grade were suggested to add a small phrase to summarize their feeling about it. The great majority of the answers were related to the questions we have just analyzed. Some of them are shown in Fig. 6.27, and they clearly reflect the motivation of these young students in favor of the robot.

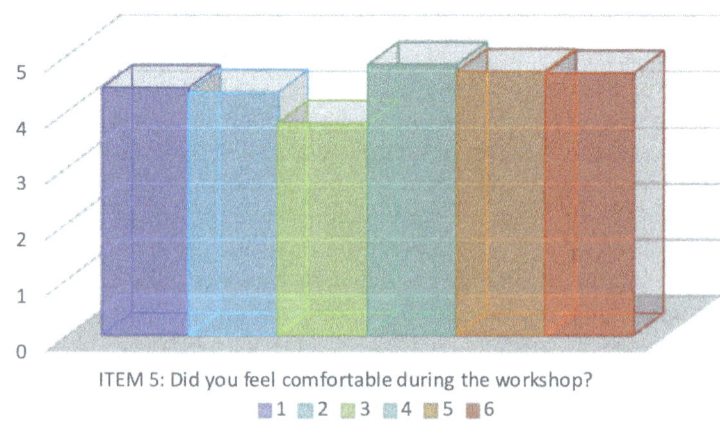

Fig. 6.23 Results item 5

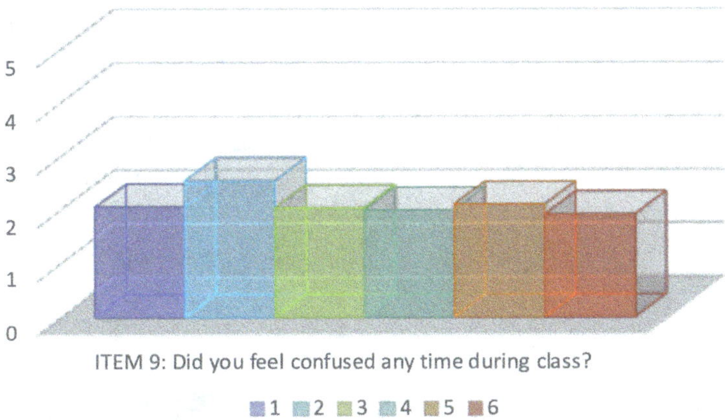

Fig. 6.24 Results item 9

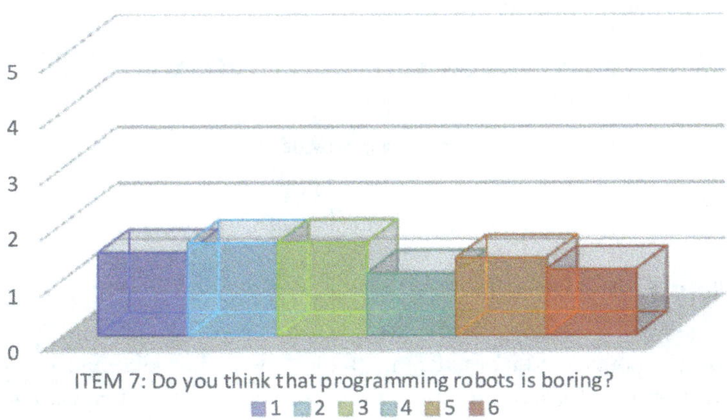

Fig. 6.25 Results item 7

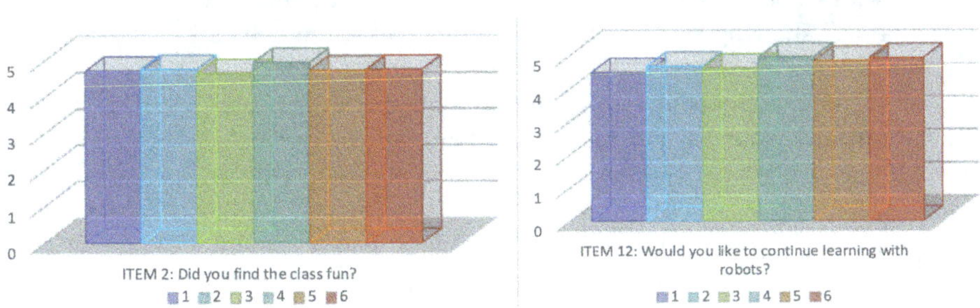

Fig. 6.26 Results of items 2 and 12

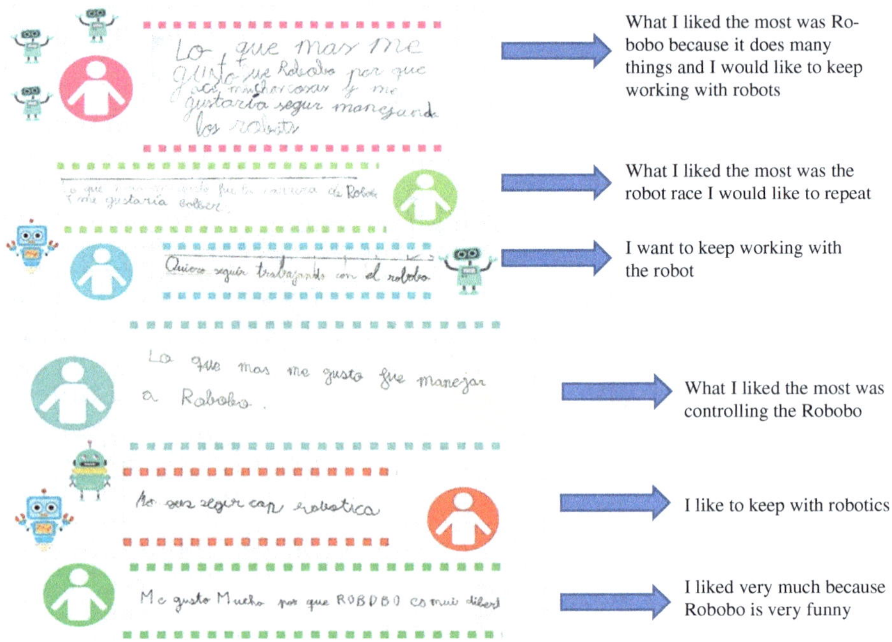

Fig. 6.27 Some impressions about the workshop provided by first grade students

Conclusions

This chapter has presented a practical methodology for introducing robotics in primary education in a formal way through the subject of mathematics, and using a realistic mathematics approach, as explained in section "Realistic Mathematics". In the section "Proposed Methodology," the methodology has been detailed in terms of didactical premises, evaluation, and class organization. Two specific workshops carried out with second and fifth grade students were presented in the section "Workshop Description," showing specific challenges that have been solved by students with high success.

One of the main conclusions of this study is that educational robotics has two main motivation sources. The first one is the robot itself, which makes students to be highly interested and curious, as shown in the results of the section "Motivation Questionnaire Analysis." But we must be careful with this result, because that motivation can be derived for using a new element in classes, and not by the element itself. The second source comes from the learning environment used to carry out the workshops. It must be a comfortable and open space, where students can interact between them and build their knowledge in an autonomous way.

After the implementation of this pilot experience in the Sigüeiro center during the last academic year 2017–2018, the future perspective is very positive with regard to robotics. The center managers, supported by the teaching staff, will create a STEM classroom in the main building of the school and will provide it with non-expendable material (tables, stools, computers, screen, projector, and others) as well as an Internet and Wi-Fi connection. On the other hand, for the next academic year, they aim to teach robotics workshops throughout the course (every two weeks more or less) in three educational levels: sixth grade infant education, fourth grade primary education, and sixth grade primary education. The reason for establishing the workshop in three levels is to guarantee in the long term the opportunity for all the students of the center to learn about, with, and through robots.

Moreover, the teaching staff of the center, considering the students' enthusiasm, supports the continuity and immersion of robotics in the school, as they believe in the potential of the robot as an educational tool (Badía et al. 2015). To do this, they propose to continue with robotics in the training plan of the school, thus training teachers to be able to respond to student demand. In addition, the school manager decided to request the regional government, XUNTA de Galicia, the increase of the endowment of educational robots in the school, which at the moment has six mBot and two Robobo.

Finally, it should be pointed out that, although the workshops were programmed in coordination with the mathematics tutors, it is not stated whether the experience had repercussions on the abstraction and comprehension of the mathematical contents. For this reason, with a future perspective, evaluation is highlighted as a priority element in order to justify the final introduction of this tool in the center to improve the mathematical knowledge of students.

Appendix 1

The specific questionnaire presented to the students at the end of the workshops is included here.

Questionnaire on the motivation of students working with robots

		Nothing	Little	Something	Enough	A lot
1	How much do you know about robotics?					
2	Did you find the class fun?					
3	Working with robots was complicated for you?					
4	Do you consider yourself skilled with robots?					
5	Did you feel comfortable during the session?					
6	Did you find it difficult to use the programming language?					
7	Do you agree that programming robots is boring?					
8	Do you agree that programming robots is boring?					
9	Did you feel confused at some times during class?					
10	How much interest do you think you put into the activities?					
11	Do you think that robots can be useful for us to learn things in the classroom?					
12	Would you like to continue learning with robots?					

References

Alsina, A. (2009). El aprendizaje realista: una contribución de la investigación en educación matemática a la formación del profesorado (Realistic learning: a contribution of research in mathematics education to teacher training). In M. J. González, M. T. González, & J. Murillo (Eds.), *Investigación en Educación Matemática XIII* (pp. 119–127). Santander: SEIEM.

Alsina, A. (2011). *Educación matemática en contexto: de 3 a 6 años (Mathematical education in context: from 3 to 6 years) Cuadernos de educación n 62*. Barcelona: Horsori.

Badia, A., et al. (2015). Factores que influyen en la percepción de los profesores de los beneficios instruccionales de los medios educativos digitales (factors that influence teachers' perception of the instructional benefits of digital educational media). *RELIEVE, 21*(2), art. 1. https://doi.org/10.7203/relieve.21.1.7204.

Bellas, F., et al. (2017). The Robobo project: Bringing educational robotics closer to real-world applications. In W. Lepuschitz, M. Merdan, G. Koppensteiner, R. Balogh, & D. Obdržálek (Eds.), *Robotics in education. RiE 2017. Advances in intelligent systems and computing* (Vol. 630). Springer. Switzerland.

Bressan, A., Zolkower, B., & Gallego, F. (2004). *Los principios de la educación matemática realista. (The principles of realistic mathematical education)* En Reflexiones Teóricas para la Educación Matemática. Compilador: Alagia, H. y otros. Edito-rial Libros del Zorzal, Buenos Aires, Argentina.

Cobb, P., Zhao, Q., & Visnovska, J. (2008). Learning from and adapting the theory of realistic mathematics education. *Éducation et didactique, 2*(1), 105–124.

Daniela, L., & Strods, R. (2016). *The role of robotics in promoting the learning motivation to decrease the early school leaving risks*, ROBOESL Conference, November 26, 2016. Athens, Greece: Proceedings Athens, 2016. p.7–13.

Freudenthal, H. (1971). Geometry between the devil and the deep sea. *Educational Studies in Mathematics, 3*, 413–435.

Freudenthal, H. (1977). Antwoord door Prof. Dr H. Freudenthal na het verlenen van het eredoctoraat. (answer by Prof. Dr H. Freudenthal after granting the honorary doctorate). *Euclides, 52*, 336–338.

Gravemeijer, K., & Tewuel, J. (2000). Hans Freudenthal: A mathematician on didactics and curriculum theory. *Curriuculum Studies, 32*(6), 777–796.

Harter, S. (1981). A new self-report scale of intrinsic versus extrinsic orientation in the classroom: Motivational and informational components. *Developmental Psychology, 17*(3), 300–312.

Karampinis, T. (2018). Activities and experiences through RoboESL project opportunities. *International Journal of Smart Education and Urban Society, 9*(1), 13–24. https://doi.org/10.4018/IJSEUS.

Karkazis, P., Balourdos, P., Pitsiakos, G., Asimakopoulos, K., Saranteas, I., Spiliou, T., & Roussou, D. (2018). To water or not to water. The Arduino approach for the irrigation of a field. *International Journal of Smart Education and Urban Society, 9*(1), 25–36. https://doi.org/10.4018/IJSEUS.

Krapp, A., Hidi, S., & Renninger, K. A. (1992). Interest, learning, and development. In K. A. Renninger, S. Hidi, & A. Krapp (Eds.), *The role of interest in learning and development* (pp. 3–25). Hillsdale: Lawrence Erlbaum Associates, Inc.

Martínez, M. A., Da Valle, N., Zolkower, B., & Bressan, A. (2002). La relevancia de los contextos en los contextos en la Resolución de Problemas de Matemática: una experiencia para docentes y sus capacitadores (The relevance of contexts in Mathematical Problem Solving contexts: an experience for teachers and their trainers). *Paradigma, 23*(1), 59–94.

Moro, M., Agatolio, F., & Menegatti, E. (2018). The development of robotic enhanced curricula for the RoboESL project: Overall evaluation and expected outcomes. *International Journal of Smart Education and Urban Society (IJSEUS), 9*(1), 48–60. https://doi.org/10.4018/IJSEUS.

Peters, B. I. (2016). *Realistic mathematics education and professional development: A case study of the experiences of primary school mathematics teachers in Namibia* (Doctoral dissertation). Stellenbosch University, Stellenbosch.

Pinto Salamanca, M. L., Barrera Lombana, N., & Pérez Holguín, J. (2010). Uso de la robótica educativa como herramienta en los procesos de enseñanza. *Ingeniería Investigación y Desarrollo (Use of educational robotics as a tool in teaching processes. Engineering Research and Development), I2+D, 10*(1), 15–23.

Pintrich, P. R., & De Groot, E. (1990). Motivational and self-regulated learning components of classroom academia performance. *Journal of Educational Psychology, 82*, 33–40.

Pintrich, P. R., & Schunk, D. H. (2006). *Motivación en contextos educativos: teoría, investigación y aplicaciones (motivation in educational contexts: Theory, research and applications).* Madrid: Pearson Educación S.A.

Scaradozzi, D., Sorbi, L., Pedale, A., Valzano, M., & Vergine, C. (2015). Teaching robotics at the primary school: An innovative approach. *Procedia-Social and Behavioral Sciences, 174*, 3838–3846. https://doi.org/10.1016/j.sbspro.2015.01.1122.

Schunk, D. H. (2001). *Motivation in education: Theory, research, and applications* (2nd ed.). Pearson Education. Essex, UK.

Van den Heuvel-Panhuizen, M. (2000). *Mathematics education in the Netherlands: A guided tour. Freudenthal Institute Cd-rom for ICME9*. Utrecht: Utrecht University.

Wigfield, A., & Eccles, J. S. (2000). Expectancy–value theory of achievement motivation. *Contemporary Educational Psychology, 25*(1), 68–81.

Chapter 7
Crab Robot: A Comparative Study Regarding the Use of Robotics in STEM Education

Icleia Santos, Elaine Cristina Grebogy, and Luciano Frontino de Medeiros

Abstract Educational robotics has proved to be a propulsive pedagogical tool for the recovery of students' interest, creativity, imagination, and logical reasoning, as well as assisting in the development of critical thinking, motor coordination, teamwork, and problem solving through hits and errors. The objective of this study was to verify the effectiveness of the use of Robotics as a pedagogical tool through a qualitative–quantitative research study with students in the fourth year of primary education, in the science discipline in an elementary public school in São José dos Pinhais (South of Brazil), with content on invertebrate animals. The study was carried out in two groups of the fourth year. Both underwent initial diagnostic evaluations, and one received the content in a traditional way, while the other followed the same planning but with the intervention of Robotics in parallel. The qualitative analysis considered the students' reports after the development of the activity. Regarding the quantitative analysis, we used the statistical tests of hypotheses, Wilcoxon–Mann–Whitney medians and Student's t-test, for comparison of means, from which it was possible to show that the group that received the test content with robotics had a better performance in relation to the appropriation of the content.

Keywords Educational robotics · Sustainable robotics · Meaningful learning · STEM education · Elementary school

Introduction

Facing the need to fulfill a list of contents prescribed in the curricula, in many cases, traditional teaching activities can limit the thinking, creating, and imagining that is so necessary in elementary schools. In this scenario, the school must use some means to encourage students' interest with respect to the contents that are part of the curriculum, always trying to make the most motivating classroom.

I. Santos · E. C. Grebogy · L. F. de Medeiros (✉)
International University Center UNINTER, Curitiba, Paraná, Brazil
e-mail: luciano.me@uninter.com

L. Daniela (ed.), *Smart Learning with Educational Robotics*,
https://doi.org/10.1007/978-3-030-19913-5_7

Motivation is one of the main factors that determine the behavior of human beings. For Perrenoud (2000), the first step to learning is motivation. Classes that include play activities are more enjoyable, captivating, and enable meaningful learning by allowing the expression of the wishes and desires of students.

In this perspective, we sought to achieve this motivational aspect with the use of robotics, in addition to the other benefits mentioned previously, so that there is a better performance of the students and a greater involvement of these in the proposals of classes brought by the teachers. Robotics consists of an educational tool that has a great and important potential to learn through the construction and testing of educational robots (Romero 2016). STEM-based robotic projects can be found in the literature (STEM–Science, Technology, Engineering and Math), exploring low-cost solutions, closer to real-world applications, from primary to high school levels (Bellas et al. 2018; Daros et al. 2016; Karkazis et al. 2018; Moro et al. 2018; Saleiro et al. 2013; Santos and De Medeiros 2017).

However, despite the wide availability of papers reporting the application of robotics in the classroom, few investigations seek to show the effectiveness of interventions in a quantitative way, analyzing the evaluation of learning by the students involved and applying tests statistical tools or analysis tools. A preliminary research study in the CAPES[1] Journal Portal with the keywords "robotics", "sciences", and "education" revealed only 16 double-reviewed articles, none of which is related to the application of robotics in science education.

Another fact pointed out by some authors is the lack of research that addresses the use of some kind of quantitative methodology in the research community on robotics. Although a number of benefits are achieved in terms of educational and motivational aspects, more rigorous research on the quantitative approach is lacking (Campos 2017, p. 2117). Methods of data analysis involving numbers can be quite useful in understanding various educational problems. It can enrich the understanding of events, facts, and processes by combining qualitative data with the quantitative (Gatti 2004, p. 13).

Thus, this chapter details a causal–comparative research study in a 4th grade elementary school class in the science discipline, developed in a municipal public school in São José dos Pinhais (South of Brazil), analyzing the activities developed with the "Crab" robot and verifying the effectiveness of such practices. To achieve this goal, we present a preamble on robotics applied to education, followed by a description of the methodology adopted for the development of this work, the presentation and discussion of the quantitative and qualitative results, and the final considerations.

Educational Robotics and Meaningful Learning

According to Mataric (2014, p. 19), a robot is an autonomous system that exists in physical reality, being able to sense its external environment as well as act on it to reach objectives. Therefore, robotics is considered the science that studies the

[1] http://www.periodicos.capes.gov.br/

assembly and programming of robots, which requires and combines knowledge from several areas, thus making it a multidisciplinary science. Robotics began to be used in the first quarter of the twentieth century, from the need to increase productivity and improve product quality. It is important to note that robotics is a very useful tool that can be considered for other types of tasks. Its use has surpassed industrial applications and expanded to other areas, including education (Lima et al. 2012).

As an effort to understand how children learn and think, Seymour Papert applied the principles of robotics in education, using the computer to provide an attractive facilitator in the process of learning mathematics for children. From the perspective of Papert's Constructionism, the student is seen as the protagonist of his learning, building his own knowledge through some tool (Papert 2008). In this way, the intention of eventually bringing automated devices into classroom learning is that they are cognitive artifacts, which students use to explore and express their own ideas. They would be "an-object-to-think-with" (Papert 1985).

In education, the use of robotics tends to generate a number of benefits, such as improving the quality of learning and new participatory methodologies. For students, it is an excellent tool for exercising creativity, studying, and practicing concepts related to different disciplines (Torcato 2012).

As a complementary tool to teaching, it can be said that there is a better significance of the contents when they are combined with the use of robotics. When the student experiences the theoretical concepts in practice, they become more meaningful, giving him greater intellectual gain, as well as being a fun way to get students to discover the workings of technology (Santos 2017).

In order to provide such content significance, this work is also based on David Ausubel's theory of meaningful learning. From a general concept, pre-existing in the cognitive structure of the student, the construction of new knowledge relevant to the relationship of new information is tied to the previous concepts. These prior concepts are called anchors, anchorage points or anchorages. The result of such a relationship of the material to the cognitive structure is reflected in the psychological meaning that also embodies elements of the student's way of being. Therefore, for meaningful learning, the material of support must have the potential of signification for the student (Ausubel 2000, p. 1). Robotics is inserted in a peculiar way in this context because it manifests high potential of content significance for the apprentices.

For the development of the robotics application in the context of sustainability advocated in this work, new and reused materials, such as bottle caps, EVA (ethyl-vinyl-acetate rubber sheets), wires, DC motors, switches, and 3 V batteries, were used. The engine to provide the Crab robot's motion effect was removed from old or non-operational cell phones and tablets.

The pedagogical goal for the construction of these robots was to awaken students' attention and creativity, promoting the motivation to learn certain content in a pleasant and enchanting way, as well as encouraging the reuse of materials that would simply be discarded. Thus, in addition to developing the aforementioned capacities, we also take advantage of the opportunity to raise awareness of the preservation of the environment. In addition to the appropriation of the proposed content, one can teach the principles of robotics and introduce students to some technological knowledge through which they can learn the operation of motors, batteries, and on-off switches. The democratization of technological knowledge is promoted by using low-cost materials.

Methodology

This chapter considers a descriptive and causal–comparative research study (Moreira and Caleffe 2008), in order to analyze the association between learning and the use of robotics as a pedagogical tool for teaching science, in the content “Invertebrate Animals”, in two classes of the 4th year of Elementary School of a Municipal Public School in São José dos Pinhais. The classes A (experimental group) and B (control group) have respectively 24 and 27 students, with ages between 9 and 11 years old. The research followed ethical regulations, with all students involved having acceptance terms signed by their parents.

Based on a qualitative and quantitative approach, besides the use of hypothesis tests to compare the means of the evaluations of the studied groups, it was sought to capture the phenomenon under study from the perspective of the people involved. The relevant points of view were considered, such as the importance given to the process and not only to the results, as well as the interpretive search for meanings from the subjects’ perception in the observed context (Godoy 1995).

Table 7.1 shows the stages of development for the Crab robot project, including the application related to the experimental and control groups.

Phase 1 contemplated the collection of the necessary materials for the construction of the robots. There was a successful collective action at the school for unused cell phones, with strong community participation, from which the vibration motors were removed and later used to give movement to the robots. In phase 2, a diagnostic evaluation in both classes A (experimental group) and B (control group) was applied, in order to investigate students’ previous knowledge about the proposed content by means of a diagnostic evaluation. In phase 3, both classes received the content in the traditional way from the teacher, using texts on blackboards, in

Table 7.1 Phases of pedagogical proposal for the Crab robot

Phase	Experimental group (class A)	Control group (class B)
1	Collective action in the school to collect unused waste or recyclable material needed for the project.	
2	Diagnostic evaluation to measure the previous knowledge on invertebrate animals of the students through a written exam.	
3	Teaching of invertebrate animals and proposal of Crab robot construction.	Teaching of invertebrate animals.
4	Evaluation of learning through a written exam.	
5		Proposal of Crab robot construction
6	Reports and testimonials about students’ experiences.	

textbooks, and illustrative images, besides talking and discussion sessions about the subject. However, for group A, the proposal to construct a robot of an invertebrate animal (a Crab) was presented and developed. Phase 4 included a subsequent assessment in class A and class B to identify student learning differences. In phase 5, with the intention of leveling both classes, group B was also exposed to the construction of the same robot, but at this moment, without the intention to measure quantitative results. In the last phase, the reports and testimonials were collected from the students to do the qualitative analysis.

The work with the robots lasted approximately 12 h, distributed in four classes. Each robot used by the students needed the following materials for its assembly:

- EVA circles
- Flexible wire
- Two black beads
- One 3 V vibration motor (taken from unused cell phones)
- One 3 V button battery
- 20 cm of 0.8 mm^2 electric wire
- One on/off switch

It was also necessary to use the following tools: soldering iron, soldering tin, hot glue gun, hot glue sticks, pliers, and scissors.

The assembly of the prototype was partially done by the teacher, especially at the times using tools, such as hot glue, soldering iron and pliers, always thinking about the student's safety. The teacher set up the electric circuit outside of class hours, following the electrical scheme shown in Fig. 7.1. After assembly and testing, the electric circuit was brought to the classroom and was available for the students to manipulate.

The students manipulated the materials and had the opportunity to see the operation of the vibration engine outside a cell phone and understand its function. They assembled the prototype body, using EVA circumferences and pieces of flexible wire for the legs, tweezers to put eye beads, as well as inserted the electric assembly into the body of the prototype ensuring that he stood balanced on his paws. At the end, the students were also encouraged to baptize their creation with a unique name,

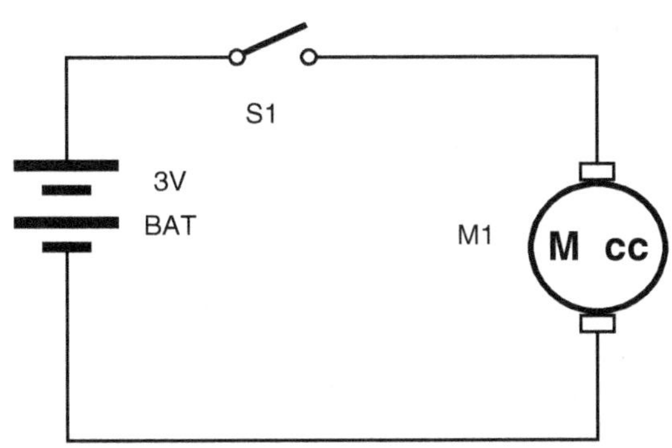

Fig. 7.1 Electric circuit used in the Crab robot

Fig. 7.2 Photo of the Crab robot totally assembled

which provided a feeling that the robots belonged to their creators. The assembled and ready-to-use prototype is shown in Fig. 7.2.

Qualitative and Quantitative Analysis

Regarding qualitative results, according to Table 7.1, in phase 6 the students were asked to write a text reporting their own experience with the Crab robot. There was a unanimous request that more classes using robotics should be presented in the school and that these would be extended to other classes and contents. In addition, it is worth mentioning the students' interest in assembling new creations based on the knowledge they had acquired. The students brought several suggestions for future work.

The contents of student's reports were processed to create a "word cloud" in order to show the most significant terms used in their writings. The result with the words present in 80% or more of the texts is shown in Fig. 7.3.

For the students, the experience of "giving life" to their creation was remarkable and a lot of fun "it's cool to play with my crab robot, the vibration makes him move in a very fun and very creative way" (real testimony of a student). Another issue noticed in the students' report was the reinforcement in their self-esteem: "I always wanted to do a robot, and I still do not believe I got it" (a student's actual testimony). Another student reports that "before he was just a model, but I made him walk and turn it into a robot."

Figure 7.3 is a word cloud extracted from the student's reports. One can notice a series of words related to positive reception from the students. Another interesting finding is the relationship between the words "robot" and "robotics" with those related to the concepts of the theme worked ("crustacean", "crab", "crab-uçá" – a local specimen of crab), reinforcing the context of meaningful learning.

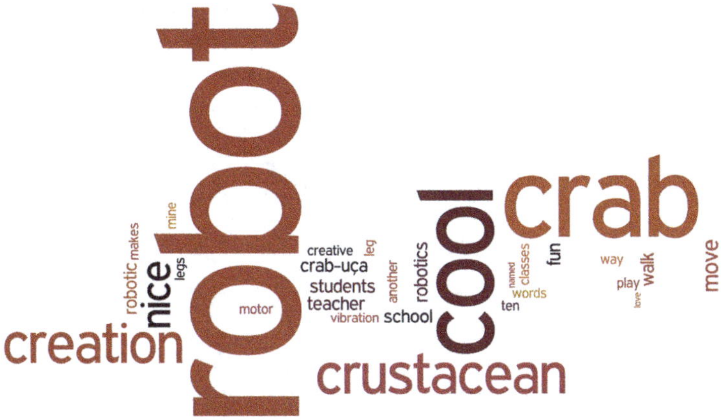

Fig. 7.3 Word cloud of student's written reports

For the quantitative analysis, the individual results of each student (the grades obtained from the written exams) were collected and tabulated during evaluation. The information was recorded and processed using various Microsoft® Excel software tools and R statistical software. From the construction of the histograms of the classes and the boxplot graph of the statistical quantities, the Shapiro–Wilk Normality Test was applied to verify that the data have normal distribution. Subsequently, Hypothesis Tests were applied to identify if the means calculated from the grades were equal between the classes and internally in the classes. Figure 7.4 shows the flow of application of the Hypothesis Tests used here. In the case of the distribution being normal, Snedecor's F-test was applied to determine the homogeneity of the variances and then the appropriate Student's t-test was chosen.

Aiming to reduce uncertainty on the statistical techniques used and to validate the process, an incremental search was performed involving the terms related to the techniques, strictly in the educational area. From this search, nine peer-reviewed articles were considered near the statistical methodology planned here. These works are described in Fig. 7.2, detailing the objective, method or tools, statistical techniques, and the sample size.

Table 7.2 shows that the methodology in most of the studies used diagnostic and subsequent evaluations, adequately characterizing the comparative causal study, combined with the use of experimental and control groups and the use of a questionnaire. As for the statistical techniques employed, the Student's t-test is the most used, from which the adoption of the normal distribution in a broader manner is assumed. Wilcoxon–Mann–Whitney and Shapiro–Wilk tests are also mentioned in some cases. Sample size also varies, in a range from 31 to 359 participants. Therefore, the proposal presented here is close to these works, regarding the quantitative methodology employed and properly grounded by the use of statistical techniques adopted in this work.

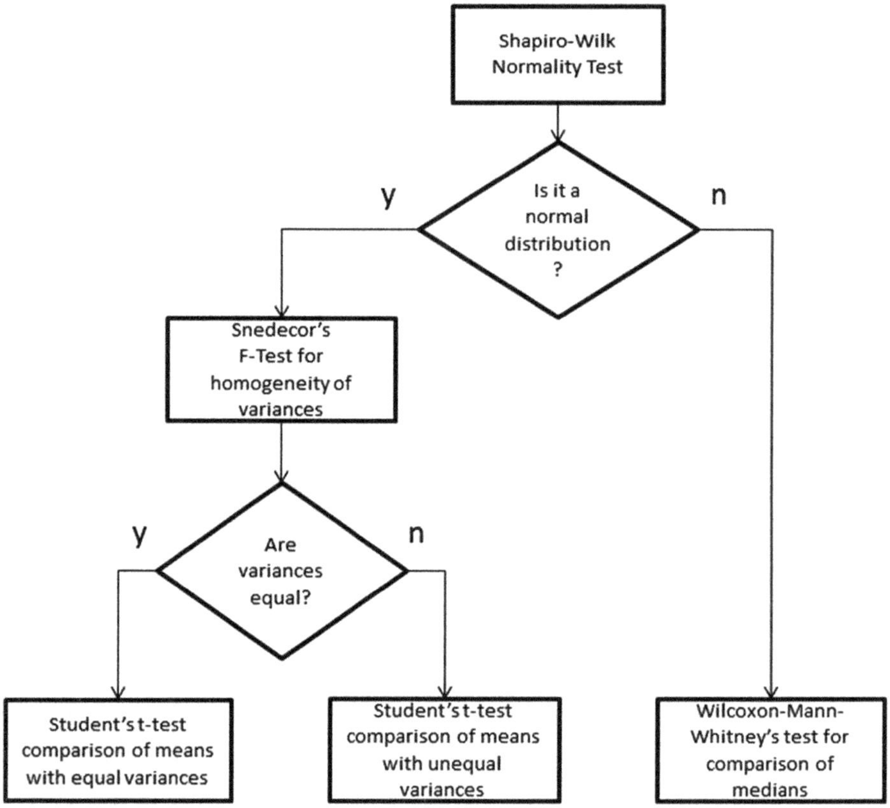

Fig. 7.4 Flow adopted for Hypotheses Tests

The comparison of the evaluations results before and after the content's presentation is visualized in the *boxplo*t graph of Fig. 7.5. This type of graph allows a visual comprehension of position, dispersion, asymmetry, tails, and discrepant data (Bussab and Morettin 2017, p. 57). Students' means grades are normalized in the range [0,1]. It can be noticed that there is a greater dispersion of the notes in the previous evaluations and smaller in the posterior ones in both classes. It is also noteworthy that in the second evaluation, class A has an average value higher than class B. Table 7.3 shows the means and standard deviations obtained for each class at each moment.

Knowing that the mean grades of the written exam for class A were higher than those for class B, the next step is to determine if this increase obtained in the means of the evaluations is statistically significant. The hypothesis tests is applied for this purpose, first referring to the determination of the data distribution's characteristic and then making the comparison tests themselves.

Before applying a hypothesis test for the comparison between classes A and B, it is necessary to verify whether the scores constitute a normal distribution or not.

Table 7.2 Kinds of statistical tests used in some articles

Article	Objective	Methods	Techniques	Sample size
Bhagat et al. (2016)	Effectiveness of the inverted classroom in the learning of mathematical concepts	Diagnostic and posterior evaluation; control and experimental groups	Student's t-test	82
Johnson and Mighten (2005)	Comparison of teaching strategies: Class notes combined with group discussion × class only	Control and experimental groups	Student's t-test	169
Dilshad et al. (2016)	Study of the impact of the use of computer simulation in the learning of secondary school biology students	Diagnostic and posterior evaluation; control and experimental groups	Student's t-test	60
Pérez-Marín et al. (2016)	Comparison of academic performance and level of engagement in active learning and collaborative knowledge construction	Diagnostic and posterior evaluation; control and experimental groups	Student's t-test	162
Wolbers et al. (2018)	Comparison of writing performance of elementary students receiving interactive and strategic writing instructions	Diagnostic and posterior evaluation; control and experimental groups	Wilcoxon–Mann–Whitney's test	31
Ramezani-monfared et al. (2015)	Effectiveness of math teachers in student learning in terms of knowledge and understanding	Questionnaire	Wilcoxon–Mann–Whitney's test	359
Hohman et al. (2015)	Determination of students' basic skills through the use of motivational interviewing	Diagnostic and posterior evaluation; questionnaire	Student's t-test	137
Mayer et al. (2018)	Research of the active involvement of biostatistics students in the data collection process for improvement in tests and motivation	Control and experimental groups	Student's t-test, Wilcoxon–Mann–Whitney's test	70
Sungur and Tekkaya (2006)	Effectiveness of problem-based learning in relation to traditional instructional approaches	Questionnaire	Shapiro–Wilk's test	61

For this, the normality test of Shapiro–Wilk (Bielefeldt et al. 2012) is used. If the distribution is normal, then parametric Student's T-test can be used to compare medians. Otherwise, the nonparametric Wilcoxon–Mann–Whitney test should be used for comparison (Larson and Farber 2015). By means of the Shapiro–Wilk normality test, the hypotheses are defined as follows:

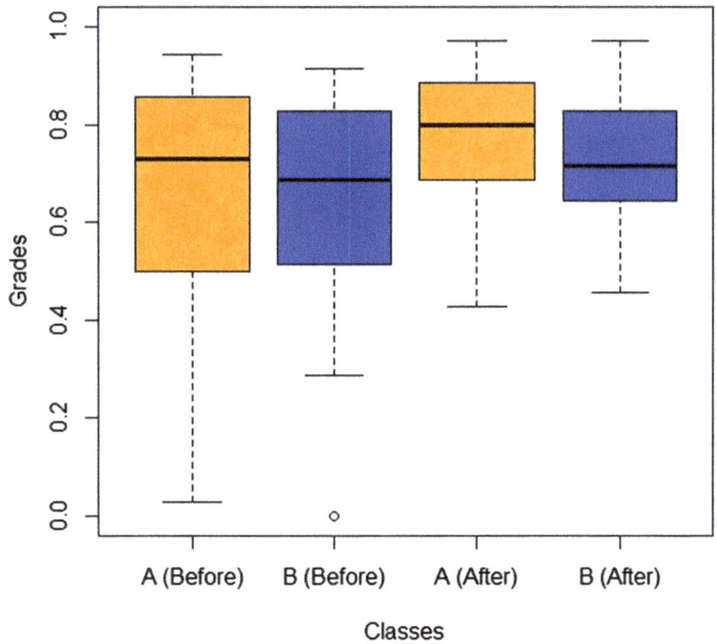

Fig. 7.5 *Boxplot* comparative graph with the grades of classes A and B, according to the results obtained in the diagnostic evaluations before and after the content's presentation

Table 7.3 Summary of statistical measurement of the classes

Classes	A (Before)	B (Before)	A (After)	B (After)
Average	0.6583	0.6212	0.7821	0.7259
Standard deviation	0.2416	0.2742	0.1389	0.1409

- *Null Hypothesis*: The sample comes from a normal distribution, with significance level $p \geq 0.1$.
- *Alternative Hypothesis*: The sample does not come from a normal distribution, with significance level $p < 0.1$.

The results are shown in Table 7.4. As one can check, it is not possible to state that the diagnostic evaluation follows a normal distribution, as well as in the evaluation after the written exam, there is no evidence that justifies the rejection of the null hypothesis.

The Shapiro–Wilk test can be supplemented with a visual analysis of the histograms as shown in Fig. 7.6. The histograms obtained from the diagnostic evaluation (class A and B, before) do not represent an approximate distribution of the normal, whereas the data related to the posterior evaluation seem to indicate such approximation. Figure 7.7 shows another graph that can aid in the interpretation of the data, regarding the approximation of a normal distribution, the Q–Q (quantile–quantile)

Table 7.4 Shapiro–Wilk normality tests

Shapiro–Wilk test	Class A (before)	Class B (before)	Class A (after)	Class B (after)
Statistics W	0.91224	0.84143	0.94795	0.97542
p-value	0.03946	0.00078	0.24460	0.74770
Result	Alternative hypothesis	Alternative hypothesis	Null hypothesis	Null hypothesis

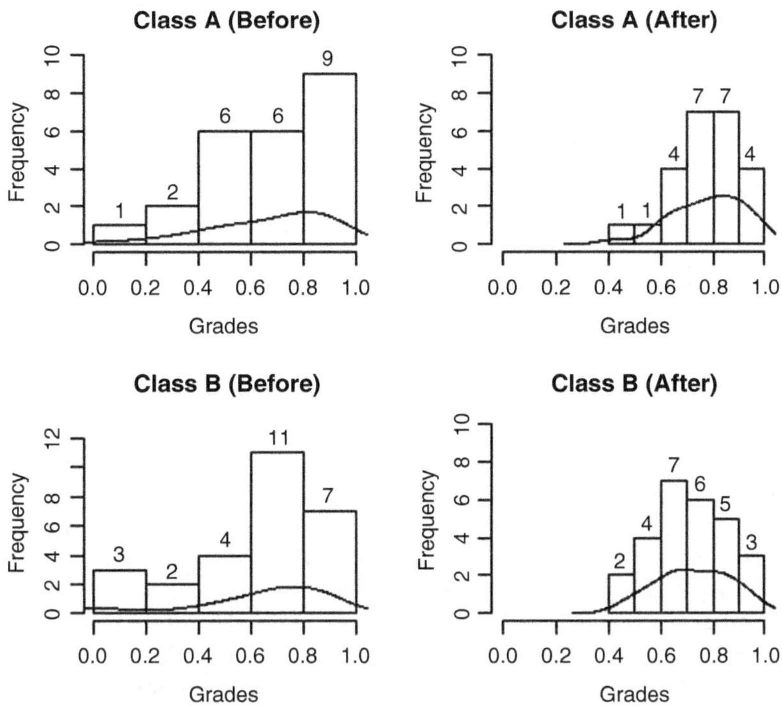

Fig. 7.6 Classes' histograms

plot. These graphs show the characteristic of normality if the samples are close to the diagonal line representing the normal curve (Bussab and Morettin 2017, p. 55). One can notice the greater approximation of classes' grades in the posterior evaluation than in the diagnosis phase. Thus, for the hypothesis tests, the diagnostic evaluations were evaluated from non-parametric tests, whereas in the later evaluations, parametric tests were used.

In this way, the next step is to perform the hypothesis tests for the comparisons. Two types of comparison were made: (i) internal comparison: the class (A or B) is compared to itself, whether at the diagnostic evaluation (before teaching the contents) or at the written exam (after teaching the contents); (ii) external comparison:

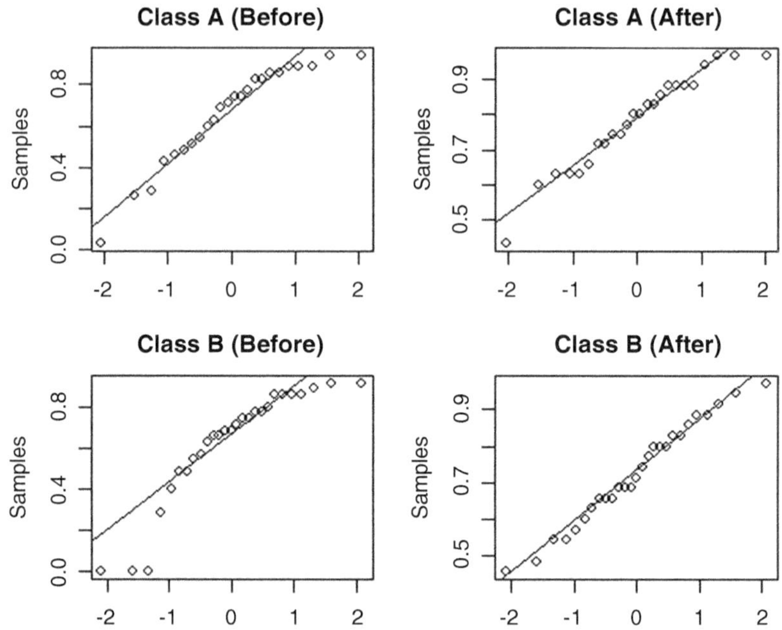

Fig. 7.7 Q–Q plots

class A is compared with class B, considering both evaluation events (before and after teaching the contents).

For the internal comparison, because one of the distributions for each class (A and B) was characterized as not normal, for medians comparison the Wilcoxon–Mann–Whitney's (WMW) non-parametric test was used (Larson and Farber 2015).

For the external comparison, for the diagnostic evaluation of classes A and B (before the application of the contents), the Wilcoxon–Mann–Whitney test was also used. For the comparison related to the posterior evaluation, since the distribution is assumed to be normal, the Student's t-test was used for comparison of means (Bussab and Morettin 2017), assuming that the average of class A is higher than the average of class B (as shown in Table 7.3, 0.6583 against 0.6212).

The hypotheses for the WMW Test, in this case are:

- *Null Hypothesis*: medians equal, with significance level $p \geq 0.1$.
- *Alternative Hypothesis*: medians not equal, significance level $p < 0.1$.

For the Student's T Test, the hypotheses are defined as follows:

- *Null Hypothesis*: the average of class A is not bigger than the average of class B, with significance level $p \geq 0.1$.
- *Alternative Hypothesis*: the average of class A is bigger than the average of class B, with significance level $p < 0.1$.

Two types of Student t-tests can be performed; one assumes the sample variances are the same and the other that they are not. The Snedecor's F-test is used to verify the homogeneity of the variances (Bussab and Morettin 2017, pp. 379–380). After calculating the W Statistic, the hypotheses tests for the comparison between classes A and B become:

- *Null Hypothesis*: the variances for classes A and B are homogenous, with significance level $p \geq 0.1$.
- *Alternative Hypothesis*: the variances for classes A and B are not homogenous, with significance level $p < 0.1$.

Table 7.5 shows the result for Snedecor's F-test, with value $p = 0.9504$, that attests the homogeneity of variances between both classes. Thus, the Student's t-test is applied comparing the averages supposing the same variances.

Table 7.6 shows that the results of the internal comparisons of grades' medians demonstrated the alternative hypothesis, that is, even in the diagnostic evaluation there is no way to affirm that the medians of the grades were equal according to statistical criteria. Regarding the external comparison after the diagnostic evaluation, there is no evidence to reject the null hypothesis, with the medians being statistically equal. Regarding the external comparison after the posterior evaluation, the last column of the table shows that the average of class A is higher than the average of class B, statistically evidencing the increment in the evaluation due to the application of robotics with the pedagogical proposal, with a level of significance of 0.1.

A final point to be made regarding the quantitative analysis refers to the value adopted for the level of significance (the *p*-value item) of 0.1. According to Cramer and Howitt (2004, p. 151), the probability level of 0.05 was historically adopted as an arbitrary choice and has been accepted as a reasonable choice in most circumstances of use. However, the same authors state that if there is a reason for varying the level of significance, it is acceptable to do so. Thus, a first justification for the use of a value of 0.1 is given by the fact that any gain related to learning in the Brazilian public educational context, in a causal–comparative research study, is always desirable under any circumstance.

Another justification may be because quantitative, causal–comparative research involving educational robotics is still incipient and that it would be necessary to establish a larger body of related research for an in-depth analysis, including discussion of a value for the level of significance for broad use. Therefore, establishment of any criterion for assessing the level of significance that should be properly adopted is still in the early stage.

Table 7.5 Results of Snedecor's F-test for homogeneity of variances

Snedecor's F-test	Classes A × B (after)
Statistic F	0.97156
p-value	0.9504
Result	Null hypothesis

Table 7.6 Results of Wilcoxon–Mann–Whitney and Student's t hypotheses tests

Comparison	Internal Class A (before vs. after)	 Class B (before vs. after)	External Class A × B (before)	 Class A × B (after)
Test	WMW	WMW	WMW	Student's t
Statistic W	205	306	347.5	–
Statistic t	–	–	–	1.4315
p-value	2.867e-05	1.886e-05	0.6637	0.0793
Results	Alternative hypothesis	Alternative hypothesis	Null hypothesis	Alternative hypothesis

Conclusion

The use of robotics, as a pedagogical tool, has an inherent potential to motivate students in the classroom when combined with the contents to be taught. The work described here enabled us to highlight the improvement obtained, both in a qualitative way, evidenced from the positive reports of the students, as well as in a quantitative way, from the result of statistical tests. One of the results obtained with this work was to define an evaluation methodology that also contemplates the quantitative aspects and that could be used later in new interventions.

It is noteworthy that the Crab robot design used has very low complexity, is really inexpensive, has quick application, and enables students to engage in a project that also reuses materials, making an appropriate connection between the science discipline and environmental education. One can also identify that the construction of new knowledge in this context occurs in a playful and affective way, as can be noticed from some reports described here.

Furthermore, it is important to note that, in the comparative methodology adopted here, care was taken that both the experimental and control groups were equalized and received the same content with respect to the developed robotic activity, except for moments that allowed the comparisons that were exposed. Later applications of robotics, still in the context of STEM education, will naturally contemplate the construction of robots based on other types of animals.

References

Ausubel, D. (2000). Aquisição e Retenção de Conhecimentos: Uma Perspectiva Cognitiva [The Acquisition and Retention of Knowledge: a Cognitive View]. *Plátano Edições Técnicas*, 35.

Bellas, F., Naya, M., Varela, G., Llamas, L., Prieto, A., Becerra, J. C., et al. (2018). The robobo project: Bringing educational robotics closer to real-world applications. In W. Lepuschitz, M. Merdan, G. Koppensteiner, R. Balogh, & D. Obdržálek (Eds.), *Robotics in education* (pp. 226–237). Cham: Springer International Publishing.

Bhagat, K. K., Chang, C., & Chang, C. (2016). The impact of the flipped classroom on mathematics concept learning in high school. *Educational Technology & Society, 19*(3), 134–142.

Bielefeldt, V., Torman, L., Coster, R., & Riboldi, J. (2012). Normalidade de variáveis: Métodos de verificação e comparação de alguns testes não-paramétricos por simulação [Normality of variables: Methods of verification and comparison of some non-parametric tests by simulation.]. *Clinical & Biomedical Research, 32*(2), 227–234.

Bussab, W. O., & Morettin, P. A. (2017). *Estatística Básica* (9a ed.) [Basic Statistics]. São Paulo: Saraiva.

Campos, F. R. (2017). Robótica Educacional no Brasil: Questões em aberto, desafios e perspectivas futuras [Robotics in Brazil: Open questions, challenges and future perspectives]. *Revista Iberoamericana de Estudos Em Educação, 12*(4), 2108–2121. https://doi.org/10.21723/riaee.v12.n4.out./dez.2017.8778.

Cramer, D., & Howitt, D. L. (2004). *The SAGE dictionary of statistics: A practical resource for students in the social sciences*. London: SAGE Publications Ltd..

Daros, C. R., da Rosa, C. T. W., & Darroz, L. M. (2016). A robótica educacional como apoio para aulas de ciências no ensino fundamental: Relato de atividade envolvendo o estudo das cores [Educational robotics as support for science classes in elementary school: Activity report involving the study of colors]., *8*(1), 57–68.

Dilshad, M., Malik, R., Tabassum, R., & Latif, M. I. (2016). Impact of computer simulation on the achievement of biology students at secondary level. *Pakistan Journal of Social Sciences, 36*(2), 1205–1214.

Gatti, B. A. (2004). Estudos quantitativos em educação [Quantitative studies in education]. *Educação e Pesquisa, 30*(1), 11–30. Retrieved from http://www.scielo.br/pdf/ep/v30n1/a02v30n1.pdf.

Godoy, A. S. (1995). Pesquisa Qualitativa: tipos fundamentais [Qualitative Research: fundamental types]. *Revista de Administração de Empresas, 35*(3), 20–29. https://doi.org/10.1017/CBO9781107415324.004.

Hohman, M., Pierce, P., & Barnett, E. (2015). Motivational interviewing: An evidence-based practice for improving student practice skills. *Journal of Social Work Education, 51*(August 2013), 287–297. https://doi.org/10.1080/10437797.2015.1012925.

Johnson, J. P., & Mighten, A. (2005). A comparison of teaching strategies: Lecture notes combined with structured group discussion versus lecture only. *Journal of Nursing Education, 44*(7), 319–322.

Karkazis, P., Balourdos, P., Pitsiakos, G., Asimakopoulos, K., Saranteas, I., Spiliou, T., & Roussou, D. (2018). Application of educational robotics on an automated water management system. *International Journal of Smart Education and Urban Society (IJSEUS), 9*(1), 25–36. https://doi.org/10.4018/IJSEUS.2018010103.

Larson, R., & Farber, B. (2015). *Estatística Aplicada* (6a ed.) [Applied statistics]. São Paulo: Pearson Education do Brasil.

Lima, W. F., Paulo, J., Ic, C., & Garcia, T. M. (2012). A robótica educacional no ensino de Química, elaboração, construção e aplicação de um robô imóvel no ensino de conceitos relacionados à tabela periódica [Educational robotics in the teaching of Chemistry, elaboration, construction and application of an immovable robot in the teaching of concepts related to the periodic table]. *XVI Encontro Nacional de Ensino de Química*.

Mataric, M. (2014). *Introdução à Robótica* [The Robotics Primer]. São Paulo: UNESP/Blucher.

Mayer, B., Braisch, U., Meule, M., Allgoewer, A., Richter, S., & Muche, R. (2018). Effect of data self-collection as an activating teaching method in a statistical software course in medical biometry – A pilot study. *GMS Journal for Medical Education, 35*(1), 1–14.

Moreira, H., & Caleffe, L. G. (2008). In Lamparina (Ed.), *Metodologia da Pesquisa para o Professor Pesquisador* (2a. Ed) [Research Methodology for the Researcher Professor]. Rio de Janeiro: Lamparina.

Moro, M., Agatolio, F., & Menegatti, E. (2018). The development of robotic enhanced curricula for the RoboESL project: Overall evaluation and expected outcomes. *International Journal of Smart Education and Urban Society, 9*(1), 48–60. https://doi.org/10.4018/IJSEUS.2018010105.

Papert, S. (1985). *LOGO: Computadores e Educação* [Mindstorms: Children, Computers, and Powerful Ideas]. São Paulo: Brasiliense.

Papert, S. (2008). *A Máquina das Crianças: repensando a escola na era da informática* [The Children's Machine: Rethinking School in the Age of the Computer]. Porto Alegre: Artmed.

Pérez-Marín, D., Hijón-Neira, R., & Santacruz, L. (2016). Active learning through collaborative knowledge building using an automatic free-text scoring system in a b-learning environment. *Behavior & Information Technology, 35*(7), 572–585.

Perrenoud, P. (2000). *Dez Novas Competências para Ensinar* [Ten New Competencies for Teaching]. Porto Alegre: Artmed.

Ramezani-monfared, N., Shahvarani, A., & Behzadi, M. H. (2015). The impact of mathematics teachers "effectiveness on students" learning in the two realms of: Knowledge and understanding. *Mathematics Education Trends and Research, 2015*(1), 35–42. https://doi.org/10.5899/2015/metr-00072.

Romero, M. (2016). La Robotique Pédagogique dans les Écoles de Québec [Pedagogical Robotics in the Quebec Schools]. Retrieved from http://www.quebecnumerique.com/robotique-dans-les-ecoles-de-quebec/.

Saleiro, M., Carmo, B., Rodrigues, J. M. F., & du Buf, J. M. H. (2013). *A low-cost classroom-oriented educational robotics system* (pp. 74–83). Springer International Publishing. https://doi.org/10.1007/978-3-319-02675-6_8.

Santos, I. (2017). *Contribuição da Robótica como Ferramenta Pedagógica no Ensino da Matemática no Terceiro Ano do Ensino Fundamental* [Contribution of Robotics as a Pedagogical Tool in Teaching Mathematics in the Third Year of Primary Education]. Curitiba-PR: Centro Universitário Internacional UNINTER. Retrieved from https://www.uninter.com/mestrado/mestrado-profissional-em-educacao-e-novas-tecnologias-dissertacoes-defendidas/.

Santos, I., & De Medeiros, L. F. (2017). Robótica com Materiais Recicláveis e a Aprendizagem Significativa no Ensino da Matemática: Estudo Experimental no Ensino Fundamental [Robotics with Recyclable Materials and Significant Learning in Mathematics Teaching: Experimental Study in Elementary Education]. In *Anais do Workshop de Informática na Escola* (pp. 275–284). WIE. https://doi.org/10.5753/cbie.wie.2017.275.

Sungur, S., & Tekkaya, C. (2006). Effects of problem-based learning and traditional instruction on. *Journal of Educational Research, 99*(5), 307–317. Retrieved from pre-test-post-test; ANCOVA.

Torcato, P. (2012). O Robô ajuda? Estudo do Impacto do uso de Robótica Educativa como Estratégia de Aprendizagem na disciplina de Aplicações Informáticas B [Does the Robot help? Study of the Impact of the use of Educational Robotics as a Learning Strategy in the discipline of Computer Applications B]. *Congresso Internacional de TIC e Educação*, 2578–2583.

Wolbers, K. A., Dostal, H. M., Graham, S., Cihak, D., Kilpatrick, J. R., & Saulsburry, R. (2018). The writing performance of elementary students receiving strategic and interactive writing instruction. *Journal of Deaf Studies and Deaf Education*, (July), 385–398. https://doi.org/10.1093/deafed/env022.

Chapter 8
Innovative Tools for Teaching Marine Robotics, IoT and Control Strategies Since the Primary School

David Scaradozzi, Lorenzo Cesaretti, Laura Screpanti, Daniele Costa, Silvia Zingaretti, and Mariantonietta Valzano

Abstract Digital skills are becoming increasingly central to the educational policy of every country and educational systems are facing new challenges in the era of the Fourth Industrial Revolution. All pupils should gain the knowledge, skills and competences they need, as they determine an individual's chance to succeed in the future labour market and to have an active role in the future society. In the last decade, a lot of projects showed how educational robotics (ER) can be a powerful tool for teaching basic skills and STEAM (science, technology, engineering, art and mathematics) subjects.

This chapter will present OpenFISH.science, a project developed to teach robotics, STEM and Internet of things (IoT). Moreover, by directly involving people in themes about the marine environment, it will raise awareness and provide knowledge

---

David Scaradozzi, Lorenzo Cesaretti and Laura Screpanti contributed equally with all other contributors.

D. Scaradozzi (✉) · L. Screpanti · S. Zingaretti
Dipartimento di Ingegneria dell'Informazione (DII), Università Politecnica delle Marche, Ancona, Italy

LSIS – umr CNRS 6168, Laboratoire des Sciences de l'Information et des Systèmes, Equipe I&M (ESIL), Case 925 - 163, Marseille Cedex 9, France
e-mail: d.scaradozzi@univpm.it

L. Cesaretti
Dipartimento di Ingegneria dell'Informazione (DII), Università Politecnica delle Marche, Ancona, Italy

LSIS – umr CNRS 6168, Laboratoire des Sciences de l'Information et des Systèmes, Equipe I&M (ESIL), Case 925 - 163, Marseille Cedex 9, France

TALENT srl, Osimo, Ancona, Italy

D. Costa
Dipartimento di Ingegneria Industriale e Scienze Matematiche (DIISM), Università Politecnica delle Marche, Ancona, Italy

M. Valzano
ICS Largo Cocconi, Largo G. Cocconi, 10, Rome, Italy

L. Daniela (ed.), *Smart Learning with Educational Robotics*,
https://doi.org/10.1007/978-3-030-19913-5_8

on RoboEthics, blue careers and ocean literacy. OpenFISH.science could empower students since the primary school to build and create distributed control systems and experiences by means of lessons and its wireless electronic building blocks and software.

Presenting curricula based on ER and constructionism that will use specific toolkits to engage (mainly) primary school kids (ages 6–12), this chapter is intended as a contribution for researchers, policymakers and everyday teachers that endeavour to help students in primary education to develop their mind and personality, as they will be their best resources to participate actively, responsibly and successfully throughout their lives in society.

Keywords Educational robotics · Curricular robotics · Primary school · Secondary school · Science · Technology · Engineering · Math · STEM · Marine robotics · IoT · Internet of things

Technical Terms and Abbreviations

ER	Educational robotics
eSTrEM	environment, Science, Technology, robotics, Engineering and Math
ISCED	International Standard Classification of Education
IoT	Internet of things
OECD	Organisation for Economic Co-operation and Development
STEM	Science, technology, engineering and mathematics
STEAM	Science, technology, engineering, art and mathematics
UNESCO	United Nations Educational, Scientific and Cultural Organization

Introduction

Digital skills are becoming increasingly central to the educational policy of every country. Educational systems are facing new challenges in the era of the Fourth Industrial Revolution. All pupils should gain knowledge, skills and competences in order to successfully enter the future labour market and to have an active role in the future society. Key competences like literacy, numeracy and basic science knowledge are essential skills for a successful professional and civic life.

In the last decade, a lot of projects showed how educational robotics (ER) can be a powerful tool for teaching STEAM (science, technology, engineering, art and mathematics) subjects. ER promotes and develops children's creative thinking, teamwork, problem-solving skills and motivation. The activity of programming and building robots is an ideal way to introduce technology and engineering skills: it can help children to develop computational thinking or the ability to design products and solutions. Moreover, ER can be a powerful tool for teaching fundamental skills and to introduce children to other important application fields: environmental awareness, human robot interaction, elderly people assistance, agricultural, craftsmanship autonomous cleaning, flying robots for surveillance and more.

This chapter will present OpenFISH.science, a project developed to teach robotics, STEM and Internet of things (IoT). Moreover, by directly involving people in themes about the marine environment, it will raise awareness and provide knowledge on RoboEthics, blue careers and ocean literacy. OpenFISH.science could empower students since the primary school to build and create distributed control systems and experiences by means of lessons and its wireless electronic building blocks and software. The chapter will present how to create a curriculum based on ER through a constructionist educational approach (Papert and Harel 1991) and the design of a toolkit to engage (mainly) primary school kids (ages 6–12). The main objective is to enhance the outcomes in STEAM subjects in primary school education through the support of academic institutions. It is necessary to help primary school students in the acquisition of knowledge about science and environment, in participating actively, responsibly and successfully throughout their lives in society. Marine environment topic is an example of how ethical considerations could be inserted in an ER curriculum. As introduced, many other topics could be relevant too. Authors, with OpenFISH.science, started with environmental awareness (marine field mainly) because it is very relevant for children and because they have competences on this field for kit developments and validation. Readers could easily reimplement the curricula adapting them to other supplementary topics besides the main robotics path and to enrich the educational programme.

By constructing and programming robots, pupils will be encouraged to use their own creative ideas and solutions in their work, thus developing transversal skills like rational thinking, creativity and innovativeness. Moreover, the chapter, thanks to the presented toolkit and strategies, will provide a detailed description of how to teach various aspects of the marine environment, thus raising awareness about the sea and all the connected problematics.

On the first section of this chapter, authors will present the relevance of introducing the ER and environment education to support future citizens in the development of a more positive attitude towards science and to help science in meeting citizens' needs. The second section will report a full curriculum of robotics within primary school education. It will provide insights of learning aims and activities on three specializations within the robotics curriculum: environmental specialization, IoT specialization and controls and automation specialization. The last section will provide the specific description of tools to implement such curriculum, with a special focus on IoT and marine environmental tools as a good application example for students in the last years.

The Relevance of Introducing Educational Robotics and (Marine) Environmental Education Since Primary Education

The vast amount of experiences in literature reporting successful implementation of robotics into STEM education is based upon the pioneering work of Seymour Papert, who saw the use of computers as an aid to learning, meaning that technology is not the centre of learning, but it can be a support in the construction of knowledge, which takes place when students build, make and publicly share meaningful objects. Papert followed Piaget's intuition that children playing is not only time pleasantly spent but also a way of learning. Following this, children use technology to build projects and teachers act as facilitators of the process (Ackermann 2001; Karim et al. 2015; Papert 1980, 1993). Even though so many activities have been carried out since Papert first introduced Logo and the robotic turtle (Karim et al. 2015; Screpanti et al. 2018a, b), robotics has not fully found its place into school's curricula yet. In Alimisis and Moro (2016), difficulties in incorporating robotics activities in school curricula are outlined and related to some features. Flexibility and openness to novelty and creativity are missing in traditional schools. Moreover, in many cases the activities of educational robotics include only building a robot providing step-by-step instruction rather than engaging students in a construction activity, which can give them the possibility of developing twenty-first-century skills. Lastly, the lack of proper design of educational activities and the lack of the means for their proper assessment is another reason for the usual absence of a strong robotics' curricula in schools. Moreover, each robotic project or course should have a clear design and assessment of the learning goals to engage children effectively in science and technology (Benitti 2012; Kandlhofer et al. 2012; Scaradozzi et al. in press; Scaradozzi et al. 2016; Scaradozzi et al. 2018).

The authors, presenting their OpenFISH.science project, intend to propose an innovative syllabus and new technological devices to teach eSTrEM (environment, Science, Technology, robotics, Engineering and Math) at the primary school and thus addressing creativity, problem-solving, teamwork and innovation capacity skills. Robotics is an approach to realize a real integrated STEM education, and in time it has developed its own methodologies and tools to effectively engage students in the construction of knowledge. The connection of the acronym STEM and the word RoboEthics is fundamental for the scientific community and society at large, to acknowledge two important areas of study and societal issue. Just like those two acknowledgements that were fundamental to focus on those issues, it is now necessary to reconnect our culture to the environment that surrounds us. To do so, we can link STEM education with a new way to design robots, which can be fully integrated with the nature and human being (Veruggio et al. 2016). In the last years, with the continuous advances of technology, different computer-assisted applications have been developed in the subject area of environmental education. For example, "computer-aided environmental education" prepared by the North American Association for Environmental Education (NAAEE) involves the problems and promises of environmental hypermedia, computer simulation/modelling interactive software (Rohwedder 1990). In other studies, by the same author, the usage of multimedia and online education in environmental education will be very useful (Rohwedder and Alm 1994; Rohwedder 1999). The factors affecting students' environmental knowledge, attitude, awareness and behaviours are investigated by scientists beginning from kindergarten to all levels of education. If school curricula are taken into account, the Next Generation Science Standards (NGSS) framework (NGSS 2013) was developed in an effort to produce K–12 science standards. In this framework engineering is fundamental because it requires students to model, plan and conduct investigations, analyse data, interpret data and hypothesize explanations across different science disciplines. This will ultimately demonstrate their understanding of core scientific ideas and taking this further to predicting future developments, interventions and research (NGSS Lead States 2013).

There are few examples of marine robotics in STEM education as a means to explore the environment and to teach students the underwater physics and dynamics. Current available underwater robots are SeaGlide from MIT (Small-scale underwater glider, n.d.), LEGO Waterbotics (Stambaugh 2015), MarineTech Project (Verma et al. 2011) and the US nationwide project known as SeaPerch (Nelson et al. 2015). These examples of successful implementation of marine robotics education in K–12 and the potential of ER into K–12 experiences stem the idea that a whole curriculum on robotics and marine environmental subjects may provide children with skills and competences that are going to be very useful in their future life as citizens, jobseekers or job creators. However, this kind of training needs a specific curriculum not only in an early stage of education. In fact, specific tools for a specific age level should be specified along with a validated methodology for teaching and evaluating results of the implementation of such curriculum (Scaradozzi et al. in press, 2018).

Introduction of Robotics, IoT and Control Strategies as Curricular Subjects in School Programmes

Many researchers and teachers agree that the inclusion of STEM subjects in early education provides a strong motivation and a great improvement in learning speed. Most curricula in primary schools include several concepts that cover science and math, but less effort is applied in teaching problem-solving, computer science, technology and robotics. The use of robotic systems and the introduction of robotics as a curricular subject can bring the possibility of introducing the basics of technology to children, giving them other kinds of human and organizational values. This work presents a new scheme that could be introduced in primary schools, designed from the Italian regular curriculum and tested for 8 years in the primary school thanks to the collaboration between National Instruments, Università Politecnica delle Marche, Istituto Comprensivo "Largo Cocconi" – Roma (Municipio V) and the start-up Talent. In the proposed curriculum, the subjects robotics, RoboEthics, IoT and control strategies become part of the primary school curricula for all the 5 years of education. This kind of subjects might seem too complicated for this stage of education, but the authors consider them fundamental for the development of the twenty-first-century citizen; other researchers proposed a similar vision, considering the relevance of introducing modelling and simulation of dynamic systems for K–12 education (Forrester 1994; Zuckerman and Resnick 2003).

The validation of the programme has been divided into two experimental stages.

For the first stage of validation, only robotics became part of the primary school curriculum for all the 5 years of cycle: this approach allowed the tuning of the teachers' training and the improvement of the proposed activities.

The first stage demonstrated various aspects (Scaradozzi et al. 2015):

- The methodology to introduce robotics into the primary school curriculum is sustainable for school and students.
- Children improve learning abilities, not only in mere technological aspects but also in cooperation and teamwork (in the following sections, results will be presented).
- At the fifth year, more advanced concepts could be introduced, preparing students for the secondary school.

In the second stage, a partially different curriculum was designed. The curriculum followed the same path designed for the first stage during the first part of the primary cycle (about 3 years), but during the last 2 years, new concepts like IoT and control strategies were introduced. The second stage has already started and some examinations have been already carried out reporting promising results. Figure 8.1 shows the curriculum distribution in the 5 years of primary school for the first and the second stage of validation.

The syllabus was developed to present a standard path for robotics with the STrEM *characterizing* learning aims to achieve, competences to develop and activities to carry out. Within the proposed syllabus, other *free choice* learning aims to achieve, competences to develop and activities to carry out have been suggested, with the goal of enriching the educational proposal presenting concepts of RoboEthics and marine environment awareness (eSTrEM), IoT and control and automation. Marine environment topic is an example of how ethical considerations could be inserted in parallel on an ER curriculum. ER can be a powerful tool for teaching fundamental skills and to introduce children to other important application fields: environmental awareness, human–robot interaction, elderly people assistance, agriculture, craftsmanship, autonomous cleaning, flying robots for surveillance and more. Authors, with OpenFISH.science, started with environmental awareness (marine field mainly) because it is very relevant for children to talk about oceans and their habitants and because authors have competences on this field for kit developments and validation. Readers could be easily reimplement the curricula adapting them to other supplementary topics besides the main robotics path and to enrich the educational programme.

First stage of validation: school years from 2010/2011 to 2014/2015*				
Primary school				
Year 1	Year 2	Year 3	Year 4	Year 5
STrEM curriculum				
* STrEM curriculum: 5-year programme of curricular Robotics with characterizing learning aims and activities				

Second stage of validation: school years from 2015/2016 to 2019/2020**				
Primary school				
Year 1	Year 2	Year 3	Year 4	Year 5
STrEM curriculum				
		IoT		
		eSTrEM		
				Control & Automation
** 5-year programme of curricular Robotics with optional learning aims and activities on IoT, Environmental education and Control & Automation				

Fig. 8.1 Curricula applied during verification and validation phases

The syllabus started from the basic knowledge and expertise and integrates them with more specialized knowledge relative both to robotics and to environmental education. The whole course was organized in different learning aims with the following scheme:

- *Mandatory* learning aims and competences from the regular Italian curriculum
- *Characterizing* learning aims and competences for introducing the robotics subject (for a STrEM curriculum)
- *Free choice* learning aims and competences characterizing the RoboEthics/environmental awareness subject (for an eSTrEM curriculum)
- *Free choice* learning aims and competences to introduce IoT competences
- *Free choice* learning aims and competences to introduce control and automation competences

A Suggested Syllabus for an Innovative ER Curriculum (Based on the "National Italian Indications" Primary School Syllabus)

The first and main objective of presented syllabus concerns the introduction of robotics at the primary school as a standard subject in the primary schools' curricula, besides being proposed as a lateral extracurricular activity to be performed out of the official school time. The project wants to increase children's capabilities, teaching them to program a machine and to consider robotics as an ordinary method of working rather than an exceptional way of operating. With robotics, the students can have a different opportunity for developing their logical ability and their creativity, features at the base of reasoning and critical thinking. The first experimental work done in the last 8 years has covered a complete primary school cycle (5 years) and the first 3 years of a new cycle; it has been performed with the priority of introducing the subject robotics as a curriculum component, improving the usual ministerial educational offer. This approach led to a change in the lesson plans, with a new teaching discipline introduced during the school year, with an impact in the weekly timetable and in the regular learning evaluation methods (robotics was added in the final report card).

The regular Italian scholastic syllabus for the primary school is divided in 5 years, so in the following syllabus, the same scheme is proposed, by authors, adding new learning aims to achieve with the respective competences to develop and activities to do in the classroom.

First Class: First Year

1.1 *Lessons' Learning Aims and Competences (from the Italian regular curriculum)*

- 1.1.1. Using the number to order and to define sets of objects
- 1.1.2. Characterizing and communicating the position of objects in a physical space, with respect to a reference or to other objects/humans
- 1.1.3. Understanding and executing instructions
- 1.1.4. Communicating own experiences in a clear way
- 1.1.5. Listening and understanding read or listened texts
- 1.1.6. Recognizing problem-solving situations
- 1.1.7. Attributing a value of truth to logical assumptions
- 1.1.8. Observing, comparing and correlating elements of the surrounding environment
- 1.1.9. Moving in the space recognizing precise references
- 1.1.10. Characterizing and applying physical measures
- 1.1.11. Doing a map localization
- 1.1.12. Collecting data and elaborating a functional diagram
- 1.1.13. Identifying a criterion to order objects
- 1.1.14. Interacting in a conversation formulating questions and giving pertinent answers presenting direct experiences
- 1.1.15. Participating actively to the games, collaborating with the others, accepting the defeat, respecting the rules and accepting the differences
- 1.1.16. Being able to get the main features of the materials
- 1.1.17. Recognizing plane and solid figures

1.1.18. Recognizing the parts working in an object to observe and to manipulate them, finding their characteristics and the materials from which they are made
1.1.19. Using knowledge associating objects and scopes
1.1.20. Knowing computers and their components
1.1.21. Being able to use computer programs, graphic programs and word processing programs
1.1.22. Being able to associate purpose to an object

1.1 *Lessons' Learning Aims and Competences* (*specific for robotics*)

1.1.23. Understanding the RoboEthics concepts and Asimov's three laws of robotics
1.1.24. Understanding the single mechanical elements: differences among shapes, materials, colours and functionalities of the elements presented on the market
1.1.25. Identifying professions involved on robot designing
1.1.26. Understanding the bases of verification concepts

1.2 *Recommendation Aims for Practical Activities* (*specific for robotics*)

1.2.1. Obtaining knowledge of the single mechanical elements through simplified ordering and planning tasks: the differences among shapes, materials, colours and functionalities of the market available elements
1.2.2. Designing a simple static "robot" using LEGO WeDo system or recycled materials
1.2.3. Measuring and mapping the positions of an entity in a real space
1.2.4. Understanding the bases of verification concepts

Second/Third Classes: Second/Third Years

2.1. *Lessons' Learning Aims and Competences* (*from the regular curriculum*)

2.1.1. Reading, understanding and interpreting critical situations
2.1.2. Understanding the four operations algorithms
2.1.3. Collecting personal data and world data
2.1.4. Representing data by means of diagrams
2.1.5. Carrying out direct measurements using various measurement units
2.1.6. Representing problem-solving situations and searching coherent solving strategies
2.1.7. Observing the space and describing it graphically
2.1.8. Moving in space using defined paths
2.1.9. Fixing criteria according to order objects using various types of graphical methods
2.1.10. Observing phenomena and formulating coherent hypotheses

2.1.11. Comparing own hypotheses with those of the classmates and debate the differences
2.1.12. Using conventional instruments to measure the time and to identify temporal cycles
2.1.13. Collecting and analysing data
2.1.14. Identifying object characteristics by means of suitable tests and comparison
2.1.15. Identifying the "structure–function" relationship of an object
2.1.16. Taking part of a debate in an adequate and pertinent way
2.1.17. Producing short texts
2.1.18. Identifying the relationship between facts, data and terms
2.1.19. Identifying temporal changes
2.1.20. Organizing a coherent and logical temporal sequence
2.1.21. Using knowledge to build objects
2.1.22. Designing objects estimating type of materials and scope
2.1.23. Being able to use computer programs, graphic programs and word processing programs in finalized contexts

2.1. *Lessons' Learning Aims and Competences (specific for robotics)*

Second Class: Second Year

2.1.24. Understanding the concept and definition of "robot" like a machine that must complete a specific task in an autonomous way
2.1.25. Designing robots, estimating type of materials and scope
2.1.26. Understanding concept and definitions of sensors and actuators through the comparison with human body system
2.1.27. Building a simple robot using LEGO WeDo system
2.1.28. Understanding verification strategies

Third Class: Third Year

2.1.29. Understanding the concept of embedded devices
2.1.30. Building a simple robot:
A free choice from:

- Using LEGO NXT/EV3 system (*characterizing STrEM curriculum*)
- Using SAM Labs system (*characterizing STrEM curriculum with IoT*)

2.1.31. Understanding the bases of procedural programming in a visual programming IDE
2.1.32. Designing with the RoboEthics concepts and understanding the operational environment concept (*free choice for eSTrEM curriculum*)
2.1.33. Understanding the bases of verification and validation concepts

2.2. *Recommendation Aims for Practical Activities (specific for robotics)*

Second Class: Second Year

2.2.1. Gaining knowledge of the concept and definition of "robot" like a machine that must complete a specific task in an autonomous way
2.2.2. Gaining knowledge of the concept and definitions of sensors and actuators through the comparison with human body system
2.2.3. Building a simple robot using LEGO WeDo system
2.2.4. Gaining knowledge of the concept of procedural programming in a visual programming IDE
2.2.5. Designing a simple robot using LEGO WeDo system
2.2.6. Understanding verification concept for a robot in an operative environment

Third Class: Third Year

2.2.7. Gaining knowledge of programming embedded devices
2.2.8. A free choice from:

- Building a simple robot using LEGO EV3 / NXT system (*characterizing STrEM curriculum*)
- Building a simple robot using SAM Labs system (*characterizing STrEM curriculum with IoT*)

2.2.9. A free choice from:

- Design a simple robot using LEGO EV3 / NXT system (*characterizing STrEM curriculum*)
- Design a simple robot using SAM Labs system (*characterizing STrEM curriculum with IoT*)

2.2.10. Gaining knowledge of the concept and definitions of sensors and actuators for a specific environment (i.e. marine environment) using LEGO EV3/ NXT or SAM Labs system (*free choice for eSTrEM curriculum*)
2.2.11. Understanding verification and validation concepts for a robot in an operative environment

Fourth/Fifth Classes: Fourth/Fifth Years

3.1. *Lessons' Learning Aims and Competences (from the regular curriculum)*

3.1.1. Understanding definition and usage of natural, relative and decimal numbers
3.1.2. Understanding how to use fractions in real cases

3.1.3. Understanding how to use four operations algorithms with integers and decimals
3.1.4. Identifying the right graphical method and strategies to solve problems
3.1.5. Identifying problems with more than one solution
3.1.6. Designing problems and hypothetical situations with or without predefined instructions
3.1.7. Understanding how to compare physical dimensions applying measurement units correctly
3.1.8. Reading and understanding texts
3.1.9. Writing different types of text
3.1.10. Writing a text with predefined scheme and representing it with a flow chart
3.1.11. Observing and characterizing a space from a point of view or a reference point
3.1.12. Estimating and classifying an object following predefined characteristics
3.1.13. Designing an object following predefined characteristics
3.1.14. Reengineering an object
3.1.15. Classifying objects in a space according to their characteristics
3.1.16. Being able to use hypertexts and graphic, word processing and presentation manager programs in finalized contexts
3.1.17. Understanding the command line, file and data type concepts
3.1.18. Understanding the computer operative system concept
3.1.19. Identifying the right object for a predefined aim

3.1. *Lessons' Learning Aims and Competences (specific for robotics)*

Fourth Class: Fourth Year

3.1.20. Building an autonomous robot that is able to communicate and react using:

- LEGO NXT/EV3 system (*characterizing STrEM curriculum*)
- SAM Labs system (*characterizing STrEM curriculum with IoT*)

3.1.21. Understanding the main program constructs: Start, Stop, Sequences, Selection and Repetition
3.1.22. Designing with the RoboEthics concepts and understanding the robot operational environment (i.e. marine environment) (*free choice for eSTrEM curriculum*)
3.1.23. Implementing verification and validation concepts

Fifth Class: Fifth Year

3.1.24. Understanding the concept of encapsulated coding
3.1.25. Understanding the concept of the sensory network
(characterizing STrEM curriculum with IoT)
3.1.26. Understanding the concept of distributed actuation
(characterizing STrEM curriculum with IoT)
3.1.27. Understanding the concept of control system and controller
3.1.28. Understanding the concept of centralized/distributed control system
(characterizing STrEM curriculum with control and automation competences)
3.1.29. Building a robot:
A free choice from:

- Using LEGO NXT/EV3 system *(characterizing STrEM curriculum)*
- Using SAM Labs system *(characterizing STrEM curriculum with IoT)*

3.1.30. Building a distributed system:
A free choice from:

- Using LEGO NXT/EV3 system
(characterizing STrEM curriculum with control and automation competences)
- Using SAM Labs system
(characterizing STrEM curriculum with IoT with control and automation competences)

3.1.31. Designing with the RoboEthics concepts and understanding the robot operational environment (i.e. marine environment) *(free choice for eSTrEM curriculum)*:
A free choice from:

- Using OpenFISH.science kit – EV3 Version
(characterizing eSTrEM curriculum)
- Using OpenFISH.science kit – SAM Labs Version
(characterizing eSTrEM curriculum with IoT)

3.1.32. Applying of verification and validation concepts

3.2. *Recommendation Aims for Practical Activities (specific for robotics)*

Fourth Class: Fourth Year

3.2.1. Building an autonomous robot able to communicate and react using:

- LEGO NXT/EV3 system *(characterizing STrEM curriculum)*
- SAM Labs system *(characterizing STrEM curriculum with IoT)*

3.2.2. Identifying the main program constructs (Start, Stop, Sequences, Selection and Repetition) on some examples
3.2.3. Designing with the RoboEthics concepts and understanding the robot operational environment (i.e. marine environment) (*free choice for eSTrEM curriculum*)
3.2.4. Implementing verification and validation concepts

Fifth Class: Fifth Year

3.2.5. Understanding the concept of re-usable coding and libraries
3.2.6. Building a sensory network
(*characterizing STrEM curriculum with IoT*)
3.2.7. Building an example of distributed actuation
(*characterizing STrEM curriculum with IoT*)
3.2.8. Identifying the control system and the controller in a robot
3.2.9. Building an example of centralized and a distributed control system
(*characterizing STrEM curriculum with control and automation competences*)
3.2.10. Building a robot:
A free choice from:

- Using LEGO NXT/EV3 system (*characterizing STrEM curriculum*)
- Using SAM Labs system (*characterizing STrEM curriculum with IoT*)

3.2.11. Building a distributed system:
A free choice from

- Using LEGO NXT/EV3 system
(*characterizing STrEM curriculum with control and automation competences*)
- Using SAM Labs system
(*characterizing STrEM curriculum with IoT with control and automation competences*)

3.2.12. Designing with the RoboEthics concepts and understanding the robot operational environment (i.e. marine environment) (*free choice for eSTrEM curriculum*):
A free choice from:

- Using OpenFISH.science kit – EV3 Version
(*characterizing eSTrEM curriculum*)
- Using OpenFISH.science kit – SAM Labs Version
(*characterizing eSTrEM curriculum with IoT*)

3.2.13. Applying of verification and validation concepts

Table 8.1 summarizes how the syllabus mandatory, characterizing and free choice learning aims to achieve, competences to transfer and activities to do have been distributed.

Table 8.1 Syllabus mandatory, characterizing and free choice learning aims to achieve, competences to transfer and activities to do distribution

<table>
<tr><th></th><th></th><th>Mandatory from regular syllabus</th><th>Subjects characterizing STrEM curriculum</th><th>Free choice subjects for IoT competences</th><th>Free choice subjects for Control and Automation competences</th><th>Free choice subjects for eSTrEM curriculum</th></tr>
<tr><td rowspan="2">1st CLASS
1st YEAR</td><td>Learning aims and competences</td><td>22</td><td>4</td><td>/</td><td>/</td><td>/</td></tr>
<tr><td>Activities concerning Robotics</td><td>\</td><td>4</td><td>/</td><td>/</td><td>/</td></tr>
<tr><td rowspan="2">2nd CLASS
2nd YEAR</td><td>Learning aims and competences</td><td rowspan="3">23</td><td>5</td><td>/</td><td>/</td><td>/</td></tr>
<tr><td>Activities concerning Robotics</td><td>6</td><td>/</td><td>/</td><td>/</td></tr>
<tr><td rowspan="2">3rd CLASS
3rd YEAR</td><td>Learning aims and competences</td><td>4</td><td>1</td><td>/</td><td>1</td></tr>
<tr><td>Activities concerning Robotics</td><td>\</td><td>4</td><td>2</td><td>/</td><td>1</td></tr>
<tr><td rowspan="2">4th CLASS
4th YEAR</td><td>Learning aims and competences</td><td rowspan="3">19</td><td>3</td><td>1</td><td>/</td><td>1</td></tr>
<tr><td>Activities concerning Robotics</td><td>3</td><td>1</td><td>/</td><td>1</td></tr>
<tr><td rowspan="2">5th CLASS
5th YEAR</td><td>Learning aims and competences</td><td>4</td><td>6</td><td>2</td><td>1</td></tr>
<tr><td>Activities concerning</td><td>\</td><td>4</td><td>5</td><td>2</td><td>1</td></tr>
</table>

Examples of Educational Activities and Timing

The OpenFISH.science project's syllabus, reported in the previous paragraph, has been designed on the basis of the regular Italian scholastic syllabus for the primary school, which is divided in 5 years. Authors planned new learning aims to achieve, competences to develop and activities to do in the classroom. In this paragraph, an example of educational activities and timing is reported for future developments in the classroom. During the quinquennial of the educational training, the main objectives have to be accomplished through different activities. The activities could be scheduled in so-called didactic units (or learning objectives), different for each school year and class. Each didactic unit could consist in specific aims and skills connected to various activities, increasing and pursuing children competences (i.e. the following scheme was adopted during the first experimentation, mentioned in the previous section).

The first activity is aimed mainly at involving children in collecting changes in the surrounding environment about the technology development. This activity has to underline the relevance of the hands-on approach, trying to increase curiosity, creativity and logic in the children. The evaluation of the accomplished knowledge is checked asking children to realize a document filled with images about different robot duties and aims useful for human being.

The second activity aims at the approach to the RoboEthics, guiding them to analyse the necessity of the three laws of robotics and their connection with laws in society. The purposes are to educate children to social values and to have respect for others; moreover, it becomes important to underline the necessity of establishing rules that save and increase the well-being of all people. Another aim is to perceive technological progress as a positive aspect in life, not a distressful one, and to increase collaboration with other subjects giving own contribution to the group. Practically, teachers help children to learn and to apply the three robotics laws working with pictures.

The third activity aims at planning and building a robot made of structured and not structured materials, using LEGO WeDo for the first 2 years, LEGO NXT/EV3 or SAM Labs for the third and the fourth classes and custom hardware with components off-the-shelf (i.e. OpenFISH.science kit) with an advanced programming language (i.e. LabVIEW) for the last year. The objective is to increase logic by discriminating and classifying materials, coherently increasing creativity in handling different materials, and trying to build objects using the acquired skills.

At the end of each activity, children are individually asked to pass two final tests:

1. Each student must correctly classify different robot pieces.
2. The student must order parts following some directives.

Other evaluation activity/activities is/are carried out in groups: each group must assemble pieces and build a robot.

The fourth activity is about structuring a software to govern robot behaviours with the aim to present theory and advices about programming using a simple and intuitive visual framework, and to discover other ways to consider computers and their possibilities. It is important to stimulate children to modify the given robot program. The skills developed with this work are the comprehension and the execution of instructions for understanding and communicating experiences in a clear way, being able to use computer and graphic programs and lastly to attribute purpose to an object. This work helps understanding the single blocks functions. The teacher could support the children when needed, but the aim of this activity relates to peer tutoring, so it would be advisable to let children be the tutors of their own classmates.

The fifth activity aims at building and programming a robot, according to specific tasks. Activities are related to getting students confident with materials manipulation and effects stimulating curiosity and dealing with new challenges: to identify robot's skills, to classify parts and functions and to apply verification and validation concepts.

The final test of this second block of activities must concentrate on how the students create the robot according to the assignment. They are required to produce a text where the process is explained. The above composition can be done individually or collectively by the working group.

Other general aims of activities are understanding how to solve a problem or mistake in what they did and to find alternative solutions and understanding the necessity of respecting ethics and environment in building and programming robots. In these phases, it is important to understand and feel the relevance of working in a group towards a common goal.

After these compulsory activities, a final activity could be proposed asking the class to elaborate a fantasy text in which the protagonists are the constructed robots. The aims are collaborating with classmates, bringing positive contributions to the group, debating other people's ideas, respecting differences, understanding the necessity of rules and, finally, increasing creativity and fantasy through the production of a coherent text. The enhanced skills are the following: understanding and executing assignments and instructions, communicating personal experiences in a clear way, interacting in a conversation through questions and narrating direct experiences and observing and comparing.

During the entire quinquennial period, it is mandatory to measure how students increase their self-confidence regarding the understanding of the technologies and how the competences are developed by students. This could be done offline by means of static instruments (questionnaires, rubrics, evaluation grids, etc.) administered before, during (to monitor the follow-up) and after the course (Scaradozzi et al. 2018) or online using data loggers of activities and expert systems (Berland et al. 2013; Blikstein et al. 2014; Jormanainen and Sutinen 2012; Scaradozzi et al. in press; Wang et al. 2017). In general, the proposed educational activities will emphasize the importance of having prefixed goals and they will stimulate logic and the analysis capacity. The requested continuous learning by doing will promote curiosity in specific cognitive instruments, will reinforce the attention and concentration capacity and will highlight the necessity to perform experiments using the scientific method. During the first experimentation in Italy, the time established for these activities is placed within the "hours for the optional disciplines" established by the Ministry of Education, University and Research in the regular weekly scheduling. The tested curriculum has been enriched with outside-the-classroom experiences, for example, meetings with robotics experts and visit to science museums, to an engineering faculty and to other exhibitions. During the closed cycle, two more experiences have been conducted (not curricular but in a formal environment) involving the general public and using students as tutors: the competition children/parents and the competition children/grandparents. These events gave the students the opportunity to strengthen their beliefs and positive attitude towards robotics by sharing meaningful experiences with people that are their first role model and affections.

Tools for STEM, STrEM and eSTrEM Curricula with a Special Focus on Marine Environment

The authors, presenting their OpenFISH.science project, intend to propose an innovative syllabus and new technological devices to teach eSTrEM (environment, Science, Technology, robotics, Engineering and Math) at the primary school and thus addressing creativity, problem-solving, teamwork and innovation capacity skills. Robotics is a demonstrated approach to realize a real integrated educational programme within a lot of subjects important for the new generations. Authors, during the verification phase and for the ongoing validation phase, identified and designed a number of tools to effectively engage students in the construction of knowledge.

Tools for Early-Stage Learning (First/Second Classes – First/Second Years)

During the first 2 years of primary school, it is necessary to work with simple tools. Solutions for early-stage learning are needed, and teachers have to use a playful approach to open up the world of math, science and language skills. It is needed to foster the love for discovery and investigation in young students and to help them develop social and emotional skills so that they will be prepared for a lifetime of successful learning.

There are several market-ready products promising to be the best solution for primary school, providing the engaging, hands-on experiences that students need to explore core STEM concepts and to link them to real-life phenomenon. For most of them, there is an integration with programming tools and a support lesson plans for teachers (see Table 2 in Scaradozzi et al. in press). For this age group, it is very important to propose a robot that must be assembled by students instead of robots already assembled. The market has a lot of robots already assembled and combined with proprietary software, but with this kind of product, it is not possible to present general

Fig. 8.2 LEGO WeDo system used for first and second classes

mechatronic components, their functionality and structure. Few products allow to leave students free to use their fantasy to explore simple cause/effect actions composition. The authors, for the experimentation, chose LEGO WeDo system (Fig. 8.2) because it triggers students' natural curiosity, helping them to develop essential communication, creativity, collaboration and critical thinking skills in a funny and exciting way. Students created original robotic artefacts solving problems and discovering how science, technology, engineering and math affect their everyday life.

It was demonstrated by the authors' experiences that, during the class activities, it is better having students divided in teams made up of four members. The activities usually started with the "brainstorming" modality: the teacher asks children to identify objects on the "robotic boxes", trying to imagine all their possible uses. The brainstorming phase, guided by teachers, has to take place after the theoretical lessons and before the technological lessons with the aim to collect ideas from the students. It's important to initially present to the students' teams the final objective of the lesson, with the aim of making them metacognitively aware: usually each activity of the syllabus presented on the previous section has one final objective. Firstly, the four roles of robotics designed by the authors are presented to the students: the designer (responsible for the project and coordinator of the team, his/her task is to communicate to the others members of the group the building instructions of the robot), the warehouse worker (responsible for the robotic kit, his/her task is to look for the LEGO pieces inside the box), the technical assembler (responsible for the robot assembling, his/her task is to build the robot receiving instructions from the designer and LEGO pieces from the technical assembler) and the validator (responsible for checking the robot assembly, observing the instructions on the computer). During the first period, students are asked to recognize single mechanical elements through some simplified tasks of ordering and planning: pupils can learn the differences among shapes, materials, colours and functionalities of the elements present in the kit. During the first year, the activities are characterized by the designing of a simple static "robot" using LEGO WeDo system or recycled materials, measuring and mapping the positions of an entity in a real space and understanding the bases of verification concepts. In general, if the teacher wants to introduce environmental consideration, the alligator robot built with LEGO WeDo kit could be a suitable choice.

During the yearly activity, each student has to experience different teams and tasks, as roles have to be switched during the year.

The activities should have two levels of learning share: a first level between students in the classroom and a second level between students, teachers, citizens and institutions. The last 15 min of each lesson, in fact, should be dedicated to debriefing: students have to express their feelings about activities and teamwork, proposing sometimes personal or collective suggestions for improvement. They also have to explain their creations and ideas to other pupils, listening to advice and constructive critiques during the discussion. Moreover, it could be useful to have some experiences outside the classroom during the scholastic year.

Tools for Secondary-Level Learning (Third/Fourth Classes – Third/Fourth Years)

Authors experimented that after 2 years of activities, students are ready to understand more complex concepts from robotics.

The third and the fourth classes are the most important for the STrEM curriculum because students could read and write, giving teachers the possibility to facilitate the design of activities in which pupils can develop competences of coding and mechatronics. In these years, the activities should promote the enhancement of students' competences, for example, building autonomous robots that are able to communicate and react thanks to their structured programs. A classic ER syllabus (without IoT and (marine) environment activities) could be well implemented with LEGO NXT/EV3 systems. The LEGO kits and their supporting IDE make building, programming and commanding robot smarter and faster. The tool also has the possibility to rapid prototyping new software blocks thanks to its connection with LabVIEW (National Instruments). This last characteristic gives researchers the possibility to assist teachers in creating personal educational kit.

Market presents other products, but authors choose LEGO EV3 (Fig. 8.3) for its completeness (see Table 2 in Scaradozzi et al. in press).

Fig. 8.3 LEGO EV3 system used for third, fourth and fifth classes during the experimentation

Tools for IoT and Control Strategies (Fourth/Fifth Classes – Fourth/Fifth Years)

A variety of data sources and services are increasing every day and the availability, type and reliability of information services are constantly changing. Therefore, information is becoming increasingly difficult for a person or machine system to collect, filter, evaluate and use in problem-solving. The problem of locating information sources (sensors), accessing (communication devices), filtering (actuators) and integrating information in support of decision-making, as well as coordinating information retrieval and problem-solving efforts has become one of the big challenges of the future society. The notion of intelligent software agents has been proposed to address this challenge (Sycara et al. 1996). Although a precise definition of an intelligent agent is still forthcoming, the current working notion is that intelligent software agents are programs that act on behalf of their human users in order to perform laborious information gathering tasks, such as locating and accessing information from various online information sources, resolving inconsistencies in the retrieved information, filtering away irrelevant or unwanted information, integrating information from heterogeneous information sources and adapting over time to their human users' information needs and the shape of the Infosphere. In the last decade, a lot of researches and industry modified the world introducing simple and low-cost machines that are able to react and reason in distributed intelligence. Nowadays, our way to retrieve information (Albayrak et al. 2005), to drive in a place (Dresner and Stone 2005) and to live our houses (Morganti et al. 2009) is assisted by agents that act on an IoT world. These considerations move authors to justify the introduction, since the primary school, of some competences related to the technologies and definitions of distributed sensitivity, actuation and computation. In order to introduce IoT and the base of multi-agent system description, authors explored the SAM Labs system (Fig. 8.4).

SAM Labs is an *edtech* company that empowers teachers with the most engaging STEAM solution including lesson plans, apps and electronics. They provide everything to deliver the most engaging STEAM learning experience with a different

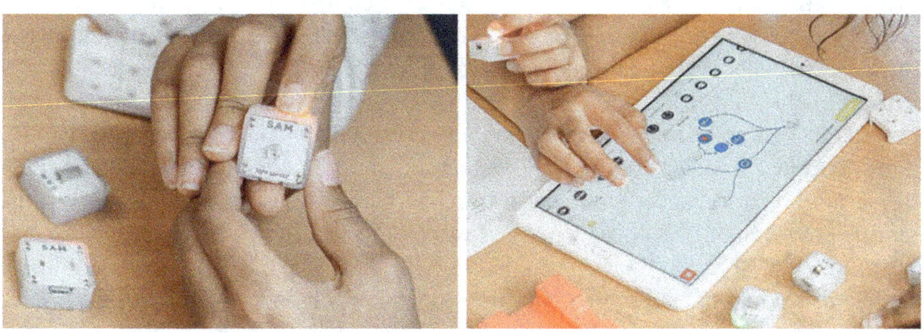

Fig. 8.4 SAM Labs system as example of IoT learning tool

point of view. SAM Labs kits are bursting with wireless electronic blocks that each have a personality of their own. From lights to motors to sliders to buzzers, every Bluetooth-enabled block can connect to the others via an app to do something different. With the app, students can code the behaviours of blocks or of the devices' network enabling them to make anything from simple reactions spread on the real world to complex creations in minutes. Students can use SAM Labs to design, write and debug programs, applying sequencing, selection and iteration. They will use logical reasoning to write algorithms, incorporating variables, inputs and outputs. Using the blocks and app, they will learn to control and simulate physical systems. The kits could be used instead of LEGO since the third class introduced both STrEM and IoT curriculum. Thanks to their way to wake up, connect in a network and act in a distributed way, SAM Labs are the right instrument to teach concepts like sensory network, distributed actuation, and centralized and distributed control systems.

Tools for eSTrEM, Marine Robotics and RoboEthics (Fourth/Fifth Classes – Fourth/Fifth Years)

After 4 years of activities, it is a good time to expand opportunities for science learning, in formal, non-formal and informal settings. It is necessary to help students in primary education to construct the required knowledge about science and environment, to participate actively, responsibly and successfully throughout their lives in the society. Society, including learners at different educational levels, should be more involved in collaborative activities because collaboration is the key to success in today's world and the collaboration skills need to be assessed and evaluated. More generally, social skills are a target themselves in the learning process. These skills are a prerequisite for other activities as they can improve science learning and ensure sustainability of open science. Considering the need to reform science education as outlined by EU, in order to support European citizens in the development of a more positive attitude towards science and environment and to help science in meeting citizens' needs, the authors started creating an educational platform for incorporating marine robotics, RoboEthics and environment awareness since the primary schools in the OpenFISH.science project. The OpenFISH.science kit has a modular architecture, composed by several blocks, devised to resemble the anatomy of a real fish. It incorporates different hardware and software components, custom-tailored to kids and capable to automatically collect data from children activities for real-time measurement. Sensors and actuators could be added to the robot mechatronic architecture by the students in order to better "sense" and "interact" in the marine environment in a broader way. Figure 8.5 shows the structure of an assembled model.

The fish could be assembled with the students or given to them partially assembled in four main blocks: Head (1); Body with the Lateral Fin_Body block (2.1), the Dislocating Mass block (2.2) and the Cerebellum block (2.3); Tail (3) and Brain (4).

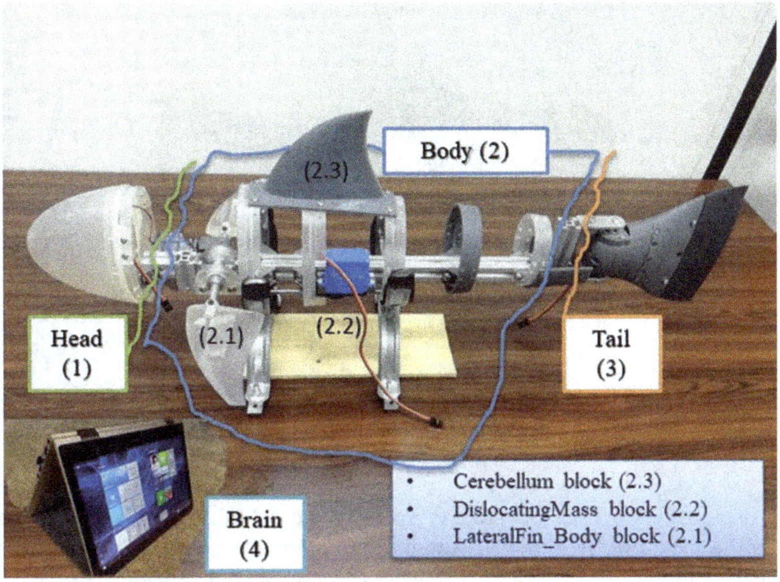

Fig. 8.5 OpenFISH.science structure

Authors decided to deeply describe here the structure in order to give an idea about how the kit could be adopted. If the reader wants to introduce different animals (i.e. for presenting different environments awareness), the same scheme could be applied. The fishbone skeleton is composed by aluminium profiles, cut and assembled by means of angular connectors, custom bolts and screws. This solution offers a wide range of combinations and thus allows to tailor the final structure to different working environments and in cooperation with other school-level students (i.e. secondary school's students involved in outside-the-classroom experiences). The fish ribs, head and lateral fins are manufactured in resin with an SLA 3D printer in order to allow students of secondary schools to participate in the fish design and upgrades. The same technology is used for the tail but with a rubber-like material. Light sensors and buttons could be installed in the head. Button has been wrapped in an insulating rubber covering the frontal part of the fish and extending the contact surface. Light sensor has been embedded in the resin, trying to keep the light-dependent resistor (LDR) as external as possible. The ideal solution would be to have more than one light sensor to create emotional effect. The central body houses a power bank, a pressure sensor used as a depth gauge, a USB hub and a servo attached to a weight by means of a transmission mechanism devised to transform the rotation of the drive in the horizontal translation of the output. By means of this solution, the position of the fish centre of mass can be artificially changed in order to create a restoring torque and consequently change the robot pitch orientation to swim upwards or downwards. The power bank serves as a power supply. The USB hub sorts the various cables that come from the central cable connected to the PC. The central cable will be passed under the belly of the fish (reducing the invol-

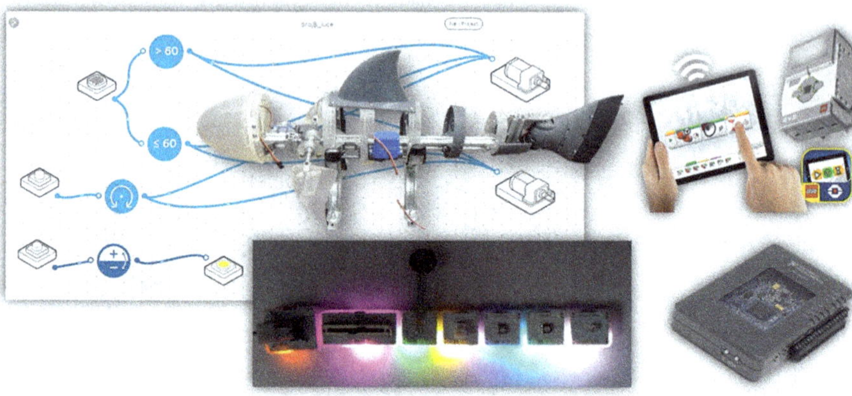

Fig. 8.6 OpenFISH.science technologies

untary oscillations with respect to a thread on the crest of the fish) near the centre of mass of the structure. With the USB hub (that could be substituted in the future with an embedded board to receive information from the sensory systems and the spinal cord and regulate motor movements like a cerebellum), the fish will be dependent on the PC-Brain where students could develop the navigation, guidance and control strategies. The idea is to use a tube, measuring either 80 × 40 or 90 × 40, as an internal compartment for the cerebellum and made the other technologies (servomotors, sensors, etc.) IP68 water resistant up to 5 m depth. A second servomotor controls the lateral fins which dynamically facilitate immersion and roll balance in water. A third servomotor and the respective power supply (or power bank) will be housed in the tail. The servomotor will have the task of activating the tail. The tail is shaped like a fin, whose size depends on the actuator velocity, meaning that the slower the motor, the larger must be the fin and vice versa.

Designers used three types of technology for creating the navigation, guidance and control strategies: National Instruments LabVIEW and MyRIO board to work with high school students, LEGO EV3 system to use the fish on eSTrEM curricula and SAM Labs systems to use the fish on eSTrEM curricula with IoT competencies (see Fig. 8.6).

With the LEGO EV3 system, the activities could be anticipated by a discussion about the environment with the third classes, allowing students to use their creativity to design a bio-inspired robot shaped like animals, not necessarily waterproof (see Fig. 8.7).

Conclusions and Final Remarks

By constructing and programming robots, pupils will be encouraged to use their own creative ideas and solutions in their work, thus developing transversal skills like rational thinking, creativity and innovativeness. These activities will help them

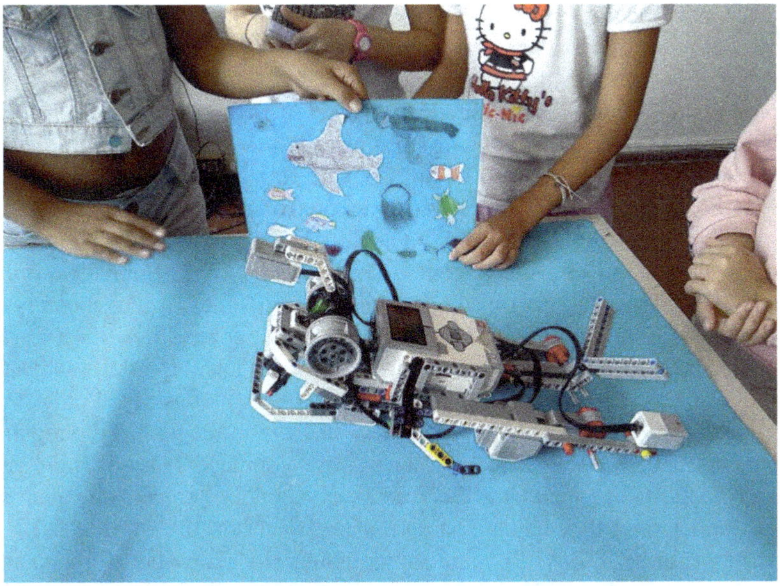

Fig. 8.7 A bio-inspired robot designed during eSTrEM classroom activities

in developing synergy of knowledge gained in different subjects. The presented syllabus and curriculum will constitute a sound opportunity both for school and connected stakeholders (local institutions, parents, project partners, target groups, etc.) to exchange practices, methodologies, approaches and tools. Introducing STrEM curricula (with IoT, eSTrEM or control and automation subjects on the syllabus) will be useful to define and strengthen the strategic skills required to better connect with the future labour market. The transnationality with more institutes that would cooperate in an experimentation could give the opportunity to reinforce and enlarge educational networks, increasing the school teachers' capacity to operate at the transnational level, sharing and confronting ideas, practices and methods within the science for citizens. Moreover, it will be a great chance to analyse and compare several "good practices" of ER intervention, especially on primary school students' curriculum, providing an insight into different teaching strategies and paying attention to potential assessment methodologies.

The proposed OpenFISH.science toolkit, allowing students to use their creativity to design a bio-inspired robot shaped like animals, will also provide a means to teach various aspects of the marine environment, in addition to increasing awareness about the sea and all the connected problematics.

The verification and validation of the programme with the schools' cohort has been divided in two experimental stages. At present, partners are at the first stage of validation: only robotics (STrEM) became part of the primary school curriculum for all the 5 years of cycle. The second stage (with the introduction of eSTrEM and more advanced subjects) has already started, and some examinations have already been carried out reporting positive results. For the first stage, it was decided to set up the educational robotics experimentation dedicating 3 h a week to make students consider this workshop and "hands-on" method normal, rather than as an extracurricular activity. Robotics has therefore become a discipline for the curriculum. Following the ordinary Italian evaluation methods for primary school, it was observed that the 5-year path of educational robotics allowed students to develop the eight key competences for lifelong learning (please see them on Recommendation of the European Parliament and of the Council 2006). The evaluation of the first stage has been accomplished, demonstrating that robotics could be introduced in a regular curriculum and the evaluation of acquired competencies was addressed with different tools directly by the teachers: self-assessment questionnaires, open and multiple-choice tests and ongoing observation. As any other regular subjects, teachers evaluate the learning level of the robotics subject contents. At the end of each year, a specific assessment of robotics was used to sum up all the skills achieved and recorded in the Evaluation Document of each student, alongside all other curriculum subjects. In order to validate the inclusion of robotics within the regular curriculum and its effect on science, Italian and math subjects learning, the results of the INVALSI tests obtained from the class involved in the project were analysed (in order to understand this Italian evaluation test, please see INVALSI 2018). The relevance of these INVALSI results are connected with the above-mentioned Recommendation of the European Parliament and Council (2006) and with the D.M. n. 13925 of 22-8-2007 of the Italian law that brings compulsory education to 10 years to give the chance to all citizens of acquiring knowledge and developing key competences of active citizenship along different axes: language, mathematics, science and technology, history and society. At the end of the fifth year, students involved in STrEM curriculum had to face some national exams among which INVALSI test of Italian and Mathematics for the fifth class of primary school. Results from the school year 2014/2015 can be found in INVALSI (n.d.). The INVALSI tests are formulated according to reliability and validity levels internationally acknowledged. Each item of the INVALSI 2015 test held a strong connection with the Italian administrative orders D.M. 31 luglio 2007 (see Indicazioni per il curricolo 2007) and the D.M. n. 254 del 16 Novembre 2012 (see Indicazioni nazionali per il curricolo 2012): they give direct indication on how curricula have to be implemented and evaluated. The INVALSI test has two parts: one test about Italian language and one test about mathematics. The Italian test is divided into three sections: one providing a text reporting of a narration and the connected questions, one providing a text explaining of something and the connected questions and one presenting ten grammar questions. The time at students' disposal for answering this test was 75 min. Regarding the mathematics test, the framework for primary school is reported in the official Italian reference scheme named "Quadro di Riferimento" (n.d.) that was based on the decree "Indicazioni nazionali per il curricolo" (2012) and took into account the comparative results from IEA-TIMSS and OCSE PISA frameworks. Mathematics test assesses two dimensions: contents dimension (knowledge of mathematics) and cognitive dimension (processes involved in answering items). The contents dimension is subdivided into four categories: numbers, space, data and predictions and relation and functions. Each category involves different processes on the basis of which items were formulated and

Table 8.2 Mathematics and Italian INVALSI results of the class experimenting STrEM curriculum

	Italian	Mathematics
STrEM	*64.1*	*59.5*
Lazio	56.3	52.9
Centre	57.4	54.7
Italy	56.6	54.6

loaded in different competences as reported in Table 3.4 in the Italian Ministry official survey "Rilevazioni nazionali degli apprendimenti 2014–2015" (n.d.). The time at students' disposal for answering this test was 75 min. Table 8.2 shows mean results in both areas of mathematics and Italian for the class experimenting STrEM curriculum (STrEM), the region it belongs (Lazio), the area to which Lazio belongs (centre) and Italy. Students in STrEM curriculum reported higher results in both areas of INVALSI test. Moreover, the socio-economic background of students involved in the STrEM curriculum was classified by the Italian Ministry of Education by means of INVALSI statisticians as low. Results reported by STrEM students were 6 points higher (Italian) and 3.9 points higher (mathematics) if compared with students with a similar socio-economic background.

These results were really encouraging, so other schools and other classes from the same school were willing to join for the second stage of experimentation. The difficulties stated by Alimisis and Moro (2016) in integrating robotics into school due to organizational issues unfortunately are still persistent in the Italian school system, but some resilient teachers and determined principals succeeded in incorporating this curriculum in their schools.

From the obtained scores in Italian and mathematics subjects, it can be seen that the average was higher than that of other schools in the Lazio region, in Italy's central area and throughout Italy, demonstrating how this approach is important not only for STEM but also for the entire children education.

References

Ackermann, E. (2001). Piaget's constructivism, Papert's constructionism: What's the difference. *Future of learning group publication, 5*(3), 438.

Albayrak, S., Wollny, S., Varone, N., Lommatzsch, A., & Milosevic, D. (2005, March). Agent technology for personalized information filtering: The PIA-system. In *Proceedings of the 2005 ACM symposium on Applied computing* (pp. 54–59). ACM.

Alimisis, D., & Moro, M. (2016). Special issue on educational robotics. *Robotics and Autonomous Systems, 77*(C), 74–75.

Benitti, F. B. V. (2012). Exploring the educational potential of robotics in schools: A systematic review. *Computers and Education, 58*(3), 978–988.

Berland, M., Martin, T., Benton, T., Petrick Smith, C., & Davis, D. (2013). Using learning analytics to understand the learning pathways of novice programmers. *Journal of the Learning Sciences, 22*(4), 564–599.

Blikstein, P., Worsley, M., Piech, C., Sahami, M., Cooper, S., & Koller, D. (2014). Programming pluralism: Using learning analytics to detect patterns in the learning of computer programming. *Journal of the Learning Sciences, 23*(4), 561–599.

Dresner, K., & Stone, P. (2005, July). Multiagent traffic management: An improved intersection control mechanism. In *Proceedings of the fourth international joint conference on Autonomous agents and multiagent systems* (pp. 471–477). ACM.
Forrester, J. W. (1994, June). Learning through system dynamics as preparation for the 21st century. In *Keynote address for systems thinking and dynamic modeling conference for K-12 education* (pp. 27–29). Concord, MA: Concord Academy.
Indicazioni nazionali per il curricolo (2012). Italian Administrative Order 16/11/2012 n. 254 "National guidelines for the kindergarten (scuola dell'infanzia) and for the primary school (scuola primaria)". Retrieved from http://www.indicazioninazionali.it/2018/08/26/indicazioni-2012/
Indicazioni per il curricolo (2007). Italian Administrative Order 31/07/2007 "National guidelines for the kindergarten (scuola dell'infanzia) and primary school (scuola primaria)". Retrieved from https://archivio.pubblica.istruzione.it/news/2007/indicazioni_nazionali.shtml
INVALSI (2018). The INVALSI test according to INVALSI. Retrieved from https://invalsi-areaprove.cineca.it/docs/2018/INVALSI_tests_according_to_INVALSI.pdf
INVALSI (n.d.). Retrieved from https://invalsi-areaprove.cineca.it/index.php?get=static&pag=precedenti_risultati
Jormanainen, I., & Sutinen, E. (2012, March). Using data mining to support teacher's intervention in a robotics class. In *2012 IEEE fourth international conference on Digital game and intelligent toy enhanced learning* (*DIGITEL*) (pp. 39–46). IEEE.
Kandlhofer, M., Steinbauer, G., Sundström, P., & Weiss, A. (2012). Evaluating the long-term impact of RoboCupJunior: A first investigation. In D. Obdrzalek (Ed.), *Proceedings of the 3rd international conference on robotics in education, RiE2012* (pp. 87–94). Prague: MatfyzPress.
Karim, M. E., Lemaignan, S., & Mondada, F. (2015, June). A review: Can robots reshape K-12 STEM education? In *2015 IEEE international workshop on Advanced robotics and its social impacts* (*ARSO*) (pp. 1–8). IEEE.
Morganti, G., Perdon, A. M., Conte, G., & Scaradozzi, D. (2009). Multi-agent system theory for modelling a home automation system. In J. Cabestany, F. Sandoval, A. Prieto, & J. M. Corchado (Eds.), *Bio-inspired systems: Computational and ambient intelligence. IWANN 2009. Lecture Notes in Computer Science* (Vol. 5517, pp. 585–593). Berlin, Heidelberg: Springer.
Nelson, S. G., Cooper, K. B., & Djapic, V. (2015, May). SeaPerch: How a start-up hands-on robotics activity grew into a national program. In *OCEANS 2015-Genova* (pp. 1–3). IEEE.
NGSS Lead States. (2013). *Next generation science standards: For states, by states*. Washington, DC: The National Academies Press.
Papert, S. (1980). *Mindstorms – Children, computers, and powerful ideas*. New York: Basic Books.
Papert, S. (1993). *The children's machine: Rethinking school in the age of the computer*. New York: Basic Book.
Papert, S., & Harel, I. (1991). Situating constructionism. *Construction, 36*(2), 1–11.
Quadro di riferimento (n.d.). Italian Ministry of Education's reference scheme for building a curriculum. Retrieved from http://www.invalsi.it/snv2012/documenti/QDR/QdR_Mat_I_ciclo.pdf
Recommendation of the European Parliament and of the Council (2006). Retrieved from http://data.europa.eu/eli/reco/2006/962/oj
Rilevazioni nazionali degli apprendimenti 2014-15 (n.d.). Italian Ministry of Education's education monitoring surveys. Retrieved from https://invalsi-areaprove.cineca.it/docs/attach/035_Rapporto_Prove_INVALSI_2015.pdf
Rohwedder, W.J. (1990). Computer-aided environmental education. Series: Monographs in environmental education and environmental studies. Grades: K 1 2 3 4 5 6 7 8 9 10 11 12 Post-Sec. ENC-013496.
Rohwedder, W. J. (1999). Environmental education goes high-tech. *Nature (Human), 4*(1), 1–2.
Rohwedder, W. J., & Alm, A. (1994). Using computers in environmental education: Interactive multimedia and on-line learning. The Environmental Education Toolbox--EE Toolbox Workshop Resource Manual.

Scaradozzi, D., Sorbi, L., Pedale, A., Valzano, M., & Vergine, C. (2015). Teaching robotics at the primary school: An innovative approach. *Procedia-Social and Behavioral Sciences, 174*, 3838–3846.

Scaradozzi, D., Screpanti, L., Cesaretti, L., Mazzieri, E., Storti, M., Brandoni, M., & Longhi, A. (2016). Rethink Loreto: We build our smart city! A stem education experience for introducing smart city concept with the educational robotics. In *9th annual international conference of education, research and innovation (ICERI 2016)*, Seville, Spain.

Scaradozzi, D., Screpanti, L., Cesaretti, L., Storti, M., & Mazzieri, E. (2018). Implementation and assessment methodologies of teachers' training courses for STEM activities. *Technology, Knowledge and Learning*, 1–21.

Scaradozzi, D., Screpanti, L., Cesaretti, L. (in press) Towards a definition of educational robotics: a classification of tools, experiences and assessments. In: Daniela, L. (Ed.),*Smart learning with educational robotics – using robots to scaffold learning outcomes*, Springer. ISBN 978-3-030-19912-8.

Screpanti, L., Cesaretti, L., Storti, M., Mazzieri, E., & Longhi, A. (2018a). Advancing K12 education through Educational Robotics to shape the citizens of the future. In *Proceeding of DIDAMATICA 2018.* AICA.

Screpanti, L., Cesaretti, L., Marchetti, L., Baione, A., Natalucci, I. N., & Scaradozzi, D. (2018b). An educational robotics activity to promote gender equality in STEM education. In *Proccedings of the eighteenth International Conference on Information, Communication Technologies in Education (ICICTE 2018)* (pp. 336–346), Chania, Crete, Greece.

Small-scale underwater glider (n.d.). Retrieved from http://www.seaglide.net/about/

Stambaugh, B. (2015, April). Waterbotics: Pooling students to stem. In *EGU General Assembly Conference Abstracts* (Vol. 17).

Sycara, K., Pannu, A., Williamson, M., Zeng, D., & Decker, K. (1996). Distributed intelligent agents. *IEEE Expert-Intelligent Systems and their Applications, 11*(6), 36–46. https://doi.org/10.1109/64.546581.

Verma, A. K., Dickerson, D., & McKinney, S. (2011). Engaging students in STEM careers with project-based learning – MarineTech project. *Technology & Engineering Teacher, 71*(1), 25–31.

Veruggio, G., Operto, F., & Bekey, G. (2016). RoboEthics: Social and ethical implications. In *Springer handbook of robotics* (pp. 2135–2160). Cham: Springer.

Wang, L., Sy, A., Liu, L., & Piech, C. (2017). Learning to represent student knowledge on programming exercises using deep learning. In *Proceedings of the 10th International conference on educational data mining* (pp. 324–329), Wuhan, China.

Zuckerman, O., & Resnick, M. (2003, April). A physical interface for system dynamics simulation. In *CHI'03 extended abstracts on Human factors in computing systems* (pp. 810–811). ACM.

Chapter 9
Robot Programming to Empower Higher Cognitive Functions in Early Childhood

Maria Chiara Di Lieto, Chiara Pecini, Emanuela Castro, Emanuela Inguaggiato, Francesca Cecchi, Paolo Dario, Giuseppina Sgandurra, and Giovanni Cioni

Abstract This chapter describes a new approach of educational robotics (ER) aimed at empowering higher cognitive functions in school. As robot programming requires mentally planning complex action sequences before the motor act, ER may promote several crucial cognitive processes underlying learning. During robot programming, the child has to first set the target, second sequentially think through the steps needed to achieve that target, then verify the goal, and eventually reset the plan. All these mental acts involve executive functions (EFs), which are complex higher cognitive processes, crucial in early development because they are the base for abstraction and logical reasoning, decision-making, sequential thinking, and maintaining and updating information in memory and problem-solving. Robot programming may empower EFs not only by improving top-down cognitive control, working memory, and inhibition skills but also by placing the child, more than other passive thought technologies, in front of "objects to think with" in a group setting that stimulates the use of EFs for social and emotional purposes. Recent studies demonstrating, through a rigorous and scientific approach, the effect of ER on EFs in typical and atypical development will be discussed.

Keywords Educational Robotics · Robot Programming · Executive Functions · Early Childhood

---

M. C. Di Lieto · C. Pecini · E. Inguaggiato
Department of Developmental Neuroscience, IRCCS Fondazione Stella Maris, Pisa, Italy
e-mail: mcdilieto@fsm.unipi.it; cpecini@fsm.unipi.it; einguaggiato@fsm.unipi.it

E. Castro · F. Cecchi · P. Dario
The BioRobotics Institute, Scuola Superiore Sant'Anna, Pisa, Italy
e-mail: emanuela.castro@santannapisa.it; francesca.cecchi@santannapisa.it; paolo.dario@santannapisa.it

G. Sgandurra (✉) · G. Cioni
Department of Developmental Neuroscience, IRCCS Fondazione Stella Maris, Pisa, Italy

Department of Clinical and Experimental Medicine, University of Pisa, Pisa, Italy
e-mail: gsgandurra@fsm.unipi.it; gcioni@fsm.unipi.it

L. Daniela (ed.), *Smart Learning with Educational Robotics*,
https://doi.org/10.1007/978-3-030-19913-5_9

Introduction

Educational robotics (ER) refers to a new learning method based on the programming, designing, and/or assembling of robots through play and hands-on activities. ER was developed at the end of the 1960s thanks to the integration of theories on pedagogical learning and cognitive development, such as constructionism of Seymour Papert and Jean Piaget (Papert 1980; Piaget and Inhelder 1966), but also on theories of the relationship between the social world and cognitive development, such as the social constructivism of Lev Semënovič Vygotskij and the social learning theory of Albert Bandura (Bandura 1986; Vygotsky 1980). Most in-school ER studies examined the impact of ER activities on "STEM" areas (science, technology, engineering, and mathematics), with particular focus on the design and assembly of robots (Barker and Ansorge 2007; Conrad et al. 2018; Hussain et al. 2006; Nugent et al. 2008; Nugent et al. 2010), while others aimed at using ER as an assistive device for motor and social-communication problems and for inclusive education (Daniela and Strods 2018; Daniela and Lytras 2018; Krebs et al. 2012; Srinivasan et al. 2016; Vanderborght et al. 2012).

However, ER activities may also promote other crucial learning processes such as robot programming, which requires mentally planning complex sequences of actions before the motor act. During robot programming, the child has to first set the target, second sequentially think through the steps needed to achieve that target, then programme actions, and finally verify behaviour. For example, if you want a robot to reach one of the balls placed on a table, you must first decide which ball is your target, after that you must choose the sequential steps needed to reach it, then give the robot the correct commands, and finally verify the results. This task involves several complex superior cognitive functions crucial in development, such as abstraction and logical reasoning, decision-making, sequential thinking, maintaining and updating information in memory, and finally problem-solving. In this framework, ER activities focused on robot programming places the child, more than any other passive thought technology, in front of "objects to think with" (Papert 1980) stimulating and empowering top-down cognitive control, and metacognition. A few studies have recently tried to identify the role of robot programming in learning processes, describing changes after ER activities in auto-monitoring and reflection on math tasks, attention, decision-making, problem-solving representations, computational thinking, visuo-spatial working memory, and inhibition skills (Di Lieto et al. 2017b; Kazakoff and Bers 2014; la Paglia et al. 2011). However, none of these studies were based on reliable designs with large samples, thus no conclusive and exhaustive findings were available to clearly define the significant ER effect on cognitive functions (Alimisis 2013; Benitti 2012). For this reason, our clinical and scientific studies have attempted to verify and describe the ER effect, based on robot programming activities, on superior cognitive processes in samples of children with typical development or neurodevelopmental disorders. For this purpose, this chapter will present executive functions (EFs), as higher cognitive processes involved in robot programming activities and developed in early childhood, an intervention proposal for cognitive empowerment through in-school ER laboratories based on our previous experiences, and possible ER adaptations in terms of goals and methods in order to promote EFs in children with various neurodevelopmental disorders and report on some of our clinical experiences.

Executive Functions: Higher Cognitive Processes Crucial for Robot Programming

Some studies have only recently described the effects of ER activities on automonitoring and reflection on math tasks, attention, decision-making, problem-solving representations, computational thinking, visuo-spatial working memory, and inhibition skills (Di Lieto et al. 2017b; Kazakoff and Bers 2014; la Paglia et al. 2011). However, most of these studies were not based on cognitive models of mental functioning and development or on reliable designs with large samples; thus, no conclusive and exhaustive findings have been provided to clearly define the significant effect of ER on cognitive control (Alimisis 2013; Benitti 2012).

Higher cognitive processes involved in robot programming, such as problem solving, cognitive control, and logical reasoning, belong to the EF cognitive domain, which consists of a group of top-down cognitive functions important for adaptive and goal-directed behaviour (Lehto et al. 2003; Miyake et al. 2000). Thanks to EFs, people may "mentally play with ideas, take the time to think before acting, meet novel, unanticipated challenges, resist temptations and stay focused" (Diamond 2013). There is agreement that EFs are made up of three main basic components (Diamond 2013; Friedman and Miyake 2017; Lehto et al. 2003; Miyake et al. 2000): (1) *Inhibitory control*: the ability to suppress an automatic response in favour of a goal-appropriate action, when there are interference stimuli or predominant mental representations, such as unwanted thoughts or memories. This ability requires us to selectively focus on what we are doing, suppressing attention to other stimuli or mental representations (e.g., a classmate talking or daydreaming while the teacher is explaining something). Inhibition control plays an important role both in "cool" EFs, the component of EFs evoked under relatively abstract, non-affective situations, as well as in "hot" EFs, elicited in settings that engender emotion, motivation, and conflict between immediate gratification and long-term rewards (Zelazo and Muller 2002; Zelazo et al. 2005). Thus, inhibition control is crucial also for the development of self-regulation capacity in situations which require cool as well as hot EF components. (2) *Working memory*: the ability to mentally retain and elaborate information (Baddeley and Hitch 1994; Smith and Jonides 1999). This is important for complex cognitive activities, interpreting written or spoken language, mentally reordering items (such as reorganizing a to-do list), translating instructions into action plans, incorporating new information into thinking or action plans (updating), considering alternatives and mentally relating information to derive a general principle, or seeing relations between items or ideas (Diamond and Lee

2011). Research has documented the role of working memory in academic achievement, such as math and reading (Holmes et al. 2009; Van de Weijer-Bergsma et al. 2015), because it requires the child to pay attention to instructions or information, to retain information and integrate that information so as to derive meaning from it. (3) *Cognitive Flexibility*: the ability to switch between two or more tasks, mental sets, or response rules. Cognitive flexibility is responsible for spatially or interpersonally changing perspectives, that is, to see something from a different point of view, inhibiting previous perspectives and loading different ones into working memory. Another aspect of this EF component involves changing how we think about something (looking outside the box), adjusting priorities, or taking advantage of sudden, unexpected opportunities.

These three EF components are strongly interrelated, and from these, high-order EFs are constructed, such as reasoning, problem-solving, and planning (Collins and Koechlin 2012; Diamond 2013; Lunt et al. 2012). These processes, that progressively mature during child development, above all in the preschool and school ages, are strongly associated with academic learning and must be continually challenged during development. Robot programming may become a new tool to empower EFs, because it requires inhibiting automatic responses in favour of goal-appropriate actions, retaining and manipulating visuo-spatial and verbal information in working memory, shifting or switching a mental set, if necessary, and reasoning and sequential programming before the motor act.

The Executive Functions Development in Early Childhood

The development of different EFs seems to partially happen in a sequential manner. More complex cognitive abilities that emerge later are placed over already present simpler fundamental ones. Studies suggest a unitary view of EFs in the preschool period, whereas the multifaceted nature of EFs appears later in development (Hughes et al. 2009).

Studies by Usai and colleagues indicate, more specifically, that the best fit for the EFs organisation in 5–7-year-old children was a two-factor model, with inhibition as a distinct dimension from a working memory-flexibility factor (Usai et al. 2014). The differentiation between working memory and shifting may emerge later (Usai et al. 2014).

EFs develop over a long period, beginning in the first year of life and continuing up to late adolescence (Garon et al. 2008). The most impressive growth in EFs takes place between the ages of 3 and 5 years (Garon et al. 2008). The significant age-related improvements in 5–7-year-old children appeared in all three EF components and mainly in complex skills that implicate the management of simpler abilities (Garon et al. 2008). According to Garon et al. (2008), there are two possible explanations for this development pattern: according to the first, changes in this period are simply quantitative for all EFs components and in the second hypothesis, non-mutually exclusive to the first, changes in an underlying factor, such as the attentional system, could affect all aspects of EFs.

Planning Education Robotic Laboratories Focused on In-School Robot Programming: An Intervention Proposal

As in the processes involved in robot programming, planning ER laboratories should follow the same steps: (1) identify and explicitly clarify the target, (2) programme sequential steps needed to reach the target in a social and collaborative setting, and (3) implement laboratories and verify their evolution. This intervention proposal derives from our previous studies and experience with the in-school ER and in clinical setting (Bargagna et al. 2018; Di Lieto et al. 2017b; Di Lieto et al., submitted).

Regarding the first step, as in evidence-based rehabilitation setting, it is important to choose and to explain the aims of planned activities. However, the need to programme academic activities and the variability of cognitive, linguistic, and social skills of children in sometimes large classrooms make the setting of targets difficult and often inappropriate. Indeed, in several previous studies and scholastic programmes employing ER laboratories the focus of planning was mainly oriented around activities or the robot device rather than on clearly established hypotheses or targets. ER aims may be short or long term and concern cognitive or robot programming processes. Subdividing long-term goals into different short-term ones may be useful in organizing activities and obtaining clear feedback about the achievement of the first short-term goal before introducing the second one. ER aims may be distinguished according to the type of processes involved, such as specific cognitive functions (e.g., visual imagery, inhibition, working memory) or robot programming capacities (e.g., robot familiarization, complex visuo-spatial planning with the robot).

The second step, concerning the organization of activities needed to reach the target in a social and collaborative setting, is also crucial.

In accordance with recent literature on EFs training, some critical and methodological aspects have been highlighted in order to impose significant changes and positive effects. The first of these changes was to create incremental challenging activities based on adaptive and intensive paradigms (Klingberg et al. 2005; Thorell et al. 2009) and on the concept introduced by Lev Vygotsky (1987) of "zone of proximal development", that is, "the distance between the actual developmental level as determined by independent problem solving and the level of potential development as determined through problem solving under adult guidance, or in collaboration with more capable peers". Other critical aspects consist of the following points: (i) administration of longer training phases suitable for very young subjects and actively monitoring participation levels during training (Diamond and Ling 2016; Wass 2015); (ii) the need to continually challenge EFs in order to induce improvements (Diamond and Ling 2016) through different and heterogeneous activities with similar goals (Rueda et al. 2005; Wass et al. 2011); (iii) planning of more enjoyable and social activities since benefits will be greater if emotional, social, and physical needs are also addressed (Diamond and Ling 2016).

With these aspects in mind and through an approach based on the "errorless learning" method (Warmington et al. 2013) and on an adaptive paradigm in a social and collaborative setting, incremental more difficult ER activities should be planned, so that children can gradually achieve more complex competencies in terms of cognitive and robot programming goals.

Finally, the last step for planning ER laboratories focused on robot programming should include implementation and assessment of classroom activities. Assessment could be provided through concrete or verbal feedback to children in order to sustain the gradual development of auto-monitoring and self-control, in addition to a careful reflection about the established goals in order to evaluate the need for changes or modifications. Moreover, at the end of the session it may be useful to verify the attention and motivation of the children and their collaboration to see if any other changes are necessary.

Experiences of In-School Educational Robotics to Empower Higher Cognitive Functions

Two experiences of in-school ER are reported: a pilot study with 12 children attending the last year of kindergarten (Di Lieto et al. 2017b) and a randomized study with a sample of 187 children attending Grade 1 in several primary schools in Pisa, Italy (Di Lieto et al., submitted). These research projects were approved by the Paediatric Ethics Committee of the Tuscany Region and parents gave written consent for the participation of their children and publication of results. Both studies aimed at responding to the lack of quantitative evidence on how ER can increase learning in students by providing intensive ER training to improve EFs in preschool and school-aged children, a crucial evolutionary window for the maturation of these functions and for the acquisition of academic competences (Diamond 2013; Usai et al. 2014). To pursue this aim, ER laboratories were organized with intensive, challenging, enjoyable, and incrementally more difficult activities, with a child-friendly bee-shaped robot, called Bee-bot® (Campus Store), frequently used in lower primary school-aged children (Janka 2008), and different colourful mats characterized by a grid 15 × 15 cm to guide robot programming (Fig. 9.1). The Bee-bot consists of seven colourful buttons positioned on its back: four orange buttons which move the robot either forwards or backwards (15 cm), right or left (90°rotation); a central green GO button which launches the programmed sequence; a blue CLEAR or X button to erase memory; and another blue PAUSE or II button to program a short interruption in the robot motion. Children control the Bee-bot by giving it a sequence of instructions, using the seven commands, for a maximum of 40 potential instructions in one programmed sequence.

Fig. 9.1 Bee-bot in one of its mats

ER laboratories were conducted twice a week for 6 weeks in the pilot study and for 10 weeks in the randomized study; each session lasted about 60–75 min. Incrementally more difficult activities were planned by the experimenters and proposed to the children, allowing them to gradually achieve more complex competencies in terms of cognitive and robot programming goals. Some additional and optional activities, directed at consolidating learned abilities, were also included. A metacognitive approach was encouraged, stimulating sequential reasoning, encouraging formulation of hypotheses and solutions, and favouring learning based on collaboration and feedback among peers. Proposed ER activities were mainly focused on visuo-spatial planning, response inhibition, interference control, working memory, and cognitive flexibility. An example of activities promoting working memory was "Bee is hungry!", in which the Bee-bot has to reach some flowers represented by geometric shapes of different colours, shapes, and sizes on the mat to pick up pollen. The child has to follow incremental challenging instructions given by a teacher. For example, a simple instruction is "the best pollen is in red flowers" while a more complex one would be "the best pollen is in big yellow flowers and in little red ones" (see Di Lieto et al., submitted for a detailed description of weekly goals and activities).

To pursue the aims of both studies, a set of neuropsychological tests were administered to each child before and after training to verify differences within groups (Di Lieto et al. 2017b) and with respect to a control group (Di Lieto et al., submitted). In particular, in the Randomized Controlled Study (RCT) study, according to the design, children were randomly split into two groups (Custer A and Cluster B) and sequentially participated in rollout training. The children were assessed by a neuropsychological evaluation before the start of the training period. The evaluation was made by standardized neuropsychological tests in order to assess visuo-spatial memory, EFs such as working memory, response inhibition and interference control, and cognitive flexibility. After the first evaluation, Cluster A immediately began ER laboratories, while Cluster B continued their normal scholastic program. After 10 weeks, a second evaluation was performed. Then, Cluster B started ER laboratories and Cluster A continued their normal scholastic program. At the end of the activities, a third follow-up and post-training assessment evaluation of all the children was made for both clusters. Parametric analyses ($p < 0.05$) comparing the two clusters demonstrated that intensive, challenging, and enjoyable ER laboratories, with incrementally more difficult activities both in terms of cognitive and robot programming goals, improve visuo-spatial working memory and inhibition processes in 5–7-year-old children who performed ER laboratories with respect to control group. Moreover, as emerged by repeated measure analyses within Cluster A, the positive effects of training were still present 3 months later, as demonstrated by the absence of differences in neuropsychological test scores at the three evaluation time points of the group of children that immediately begin training. Thus, this study documented for the first time, in a quantitative manner, the positive and long-term effects of ER activities focused on robot programming, working memory, and inhibition EF components at school. This suggests the possibility of including this approach for improving EFs in the early school years, if based on rigorous methodological aspects and evidence-based methods. Moreover, the possibility of carrying out these laboratories in school could really be important in order to directly empower scholastic performance and support children with EFs deficit or impairment in a more ecological and social setting.

Promote Executive Functions in Neurodevelopmental Disabilities with Educational Robotics

Neurodevelopmental disabilities are a group of disorders with an onset in the developmental period characterized by alterations of the central nervous system due to early brain damage, genetic/chromosomal abnormalities, epilepsy or environmental conditions which can give rise to intellectual disability, neuropsychological deficits, specific learning deficits, movement and posture disorders, such as cerebral palsy, and psychiatric disturbances. Other neurodevelopmental disorders frequently co-occur and determine impairments in personal, social, academic, and vocational functioning (American Psychiatric Association 2013). Neurodevelopmental disabilities are currently studied on different description levels, that is, behavioural, cognitive, an neural, within a cognitive neuroscience framework, very different from the adult domain-specific modularization approach. The modern cognitive neuroscience framework, fuelled by recent breakthroughs in the fields of brain connectivity and genetics, posits that brain-behaviour relationships in childhood are best understood as being subsumed by progressive specialization and localization of function within the complex two-way interaction between genes and various environments. Within this theoretical framework, it is crucial to investigate neurodevelopmental disorders through different levels of description correlating, in the same patient, neuropsychological, neurofunctional, and psychiatric evidence. Increasing research on EFs for different neurodevelopmental disorders highlights the transversal role of EFs in these different description levels, identifying their role in academic achievement (Jenks et al. 2009), quality of life (Sharfi and Rosenblum 2016), psychopathological problems (Schoemaker et al. 2010; Shimoni et al. 2012), and health.

Although EF deficits are typically found in children with attention deficits and hyperactivity disorder (De La Fuente et al. 2013) or traumatic brain injury (Beauchamp and Anderson 2013), several studies suggest that EFs impairment may be part of several neurodevelopmental disorders, such as specific learning disabilities (Kudo et al. 2015), specific language impairment (Kapa and Plante 2015), intellectual disabilities (Bexkens et al. 2014; Costanzo et al. 2013), autism spectrum disorders (Chen et al. 2016), and cerebral palsy, including spastic diplegia (Bodimeade et al. 2013; Bottcher et al. 2010). Based on these findings, the interest in studying the effect of EFs training in several neurodevelopmental disorders has increased in the last years (Green et al. 2012; Grunewaldt et al. 2013; Klingberg et al. 2005; Klingberg et al. 2002; Mak et al. 2017; Piovesana et al. 2017) Di Lieto et al., submitted). Consistent with this new trend of studies, our recent clinical and research experience has been oriented around adapting ER laboratories in terms of goals and methods to some clinical developmental populations that need to improve EFs: in particular, children with specific congenital spastic diplegia, Down syndrome (DS) and specific learning disorders. Moreover, in order to promote the integration of children with neurodevelopmental disabilities at school, some adaptation and changes in ER laboratories were conducted in the classroom during the randomized study (called e-Rob project) as shown below.

Experiences of Educational Robotics in Different Developmental Disorders: Aims and Methodological Adaptations

Congenital Spastic Diplegia

Congenital spastic diplegia is a form of cerebral palsy which involves both sides of the body with a predominance in lower limbs (Rosenbaum et al. 2007). It commonly occurs in preterm children (preterm spastic diplegia, pSD) and it is generally due to periventricular leukomalacia, a form of white matter brain injury typically affecting neural pathways lying close to the lateral ventricles, such as the corticospinal tract and optic radiations (Cioni 2000; Fazzi et al. 2009; Guzzetta et al. 2010; Jacobson and Button 2000; Pavlova and Krägeloh-Mann 2013). Children with congenital spastic diplegia are consistently impaired in non-verbal intelligence and visuo-perceptual and visuo-spatial abilities, while verbal abilities, as measured by verbal IQ tests, are generally spared (Fazzi et al. 2009; Ito et al. 1996; Pavlova and Krägeloh-Mann 2013; Sigurdardottir et al. 2008). Beyond visuo-spatial deficits and impaired non-verbal intelligence, weak EFs have been also reported, such as the inability to quickly process, maintain, update, and inhibit information (Di Lieto et al. 2017a; Pirila et al. 2011; Pueyo et al. 2009; Schatz et al. 2001; White and Christ 2005) and a multilevel organization of the neuropsychological profile has been suggested beyond the common core visuo-spatial and sensory-motor deficits. When congenital spastic diplegia is associated with thinning of the anterior/middle corpus callosum, impairments in attention and EFs seemed to act as additional factors in further affecting visuo-spatial, sensory-motor, and social skills (Di Lieto et al. 2017a).

Due to the presence of common core visuo-spatial and sensory-motor deficits and impairment in working memory and inhibition components of EFs in the spastic diplegia population, ER activities have been used not only to promote robot programming but also to improve impaired processes, such as visuo-motor integration (VMI), visuo-spatial processing, visuo-spatial working memory, inhibition, and speed of processing.

A group of three preschool children (case G, case F, case E) diagnosed with congenital spastic diplegia were selected from a larger group of cerebral palsy children, classified as spastic diplegia according to ICD-9 based register in 2014–2015 of the Department of Developmental Neuroscience of IRCCS Stella Maris. The children presented a neuroradiological diagnosis of periventricular leukomalacia documented at brain MRI, mild to moderate upper limb functional impairment (from level I to III) at the Manual Ability Classification System (MACS; Eliasson et al. 2006), and a verbal intelligence level in the average range. Written consent was obtained from participants' parents to analyse research results. The ER laboratories were conducted once a week for 12 weeks for about 75 min. Incremental and multidisciplinary activities were planned by neuropsychologists and psychomotor therapists, allowing the children to gradually achieve more complex competencies in terms of psychomotor, cognitive, and robot programming goals. Differently from the method used with typically developing children, some methodological adaptations were imposed, such as the need to repeat the same activities during ER laboratories to strengthen competencies, to simplify sequences of robot programming due to deficits in visuo-spatial processing, to suggest verbal strategies to compensate for visuo-spatial difficulties, to employ mats with fewer visual elements and targets, and to give a Bee-bot to each child because of frustration of waiting (Fig. 9.2).

A funny story, "*The adventure of Bee-Bot: a special bee born with the wheel and not with a wing*", created ad hoc was told at the beginning of each ER laboratory to motivate and engage children. A metacognitive approach was encouraged, favouring learning based on collaboration and feedback among peers.

Fig. 9.2 ER Laboratory

Despite the initial adaptation difficulties due to anxiety separation from parents, children progressively participated in the ER laboratories with motivation, interest, and amusement, reciprocally helping each other during the activities.

At the end of the 3 months of ER laboratories, based on clinical needs of participants, two children (case G and case F) continued to exercise EFs through a new home-based software training programme, specially created to promote working memory, inhibition, and cognitive flexibility (MemoRan® Anastasis), and one child (case E) started a commercial home-based software training programme named Run the RAN ® (RANt, Anastasis; http://www.ridinet.it) to increase rapid automatizated naming due to significant difficulties in visuo-verbal integration and verbal fluency.

For both programmes, the training period lasted about 3 months with three to five daily sessions lasting 10–15 min per week.

In order to verify and describe the effect of the programmes, a short test protocol was administered before the ER laboratories and at the end of home-based training (6 months later). This short test protocol assessed verbal and visuo-spatial working memory (Digit and Corsi span, BVN 5-11), cognitive flexibility (Card Sort, FE-PS 2-6), cognitive inhibition with visuo-verbal stimuli (Inhibition subtest, NEPSY-II), inhibition of motor responses (Little fishes play, FE-PS 2-6), and VMI.

Qualitative comparative analysis between pre- and post-assessments for each child showed improvement performances in all three children in the speed of processing both in cognitive and motor inhibition tasks and in the ability to spontaneously change their responses based on visual targets. Two children (case G and case F), who performed home-based EFs training after ER laboratories, showed better performances in verbal working memory at the end of the training period, while case G also showed increased abilities in visuo-spatial working memory task. Nevertheless, case F and case E presented higher performances in VMI test at the end of the training period with respect to the beginning.

Although conducted on only a few children, this clinical experience clearly indicates the practicality of including ER, if based on a rigorous methodology and clinical needs, to promote EF components also in preschool children with congenital spastic diplegia.

Down Syndrome

An ER laboratory was ad hoc adapted for children with a diagnosis of DS. Eight children were involved in the laboratory, selected from a larger group of DS children referred to the Department of Developmental Neuroscience of IRCCS Fondazione Stella Maris. Children were selected according to rigorous inclusion criteria (Bargagna et al. 2018). The laboratory consisted of 45 min weekly sessions for 8 weeks, using Bee-bot. The children were evaluated by a neuropsychological assessment at the beginning and at the end of the ER laboratory. The adaptations of the ER laboratory were multiple in order to respond to the cognitive, behavioural, and attentional characteristics of the children. A highly structured format of time and activities was created to maintain a simple narrative setting in order to facilitate the engagement and attractiveness of activities. In addition, an easier narrative setting with respect to the ones utilized for typical developing children was used and the laboratories were organized in small groups or individual sessions. Because of the limited attention span of the children involved in the laboratory, periodic pauses in the activities were planned. Moreover, in order to focalize the attention of the children, possible sources of distraction in the environment were eliminated. Finally, a personalized positive reinforcement at the end of each session was organized in order to maintain motivation.

This experience has enabled us to draw some conclusions concerning the applicability of ER in children with a diagnosis of DS. Bee-bot turned out to be a very attractive device, able to promote interest and capture the attention of the children and favour relationships with adults and peers, acting as a mediator. Nevertheless, the robotic kit was not always sufficient to obtain full compliance. In these situations, adaptations of activities resulted decisive to maintain the efficacy of laboratories. In this sense, the possibility of organizing small work groups was fundamental in order to maintain the attention of the children and promote imitation learning. A critical aspect of ER laboratories is the duration: one weekly session per 8 weeks resulted insufficient to consolidate learning. Longer periods of familiarization with the Bee-bot may promote an easier access to the proposed activities. Nevertheless, because all the children were able to perform ER activities, this first experience suggested practicability in a DS population. However, because of the small sample size and significant heterogeneity in terms of recorded results (e.g., not all the children were able to perform all the evaluations due to opposition behaviour), it is difficult to draw general conclusions. Because of this, two representative case studies are reported, in order to qualitatively describe two different experiences of children involved in the ER laboratories: S and F.

S is a 7-year-old DS female with a mild/moderate intellectual disability. During her experience in ER laboratories, behavioural and relational difficulties caused her difficulties in participating in the activities. For example, S had problems in a peer-group setting, so individual sessions were frequently proposed. The goal of the activities was shifted from the cognitive domain to the relational one. Bee-bot was utilized as a mediator for relationships with adults and peers and to promote attention and motivation. Once S felt comfortable and once she was positively inserted in a group setting, basic functions of the Bee-bot were proposed. At the end of the ER laboratory programme, S was able to accept a play-group setting and was able to perform a one-step forward movement with the Bee-bot. At the beginning of the study, the abovementioned behavioural and relational problems made it impossible for her to perform most of the neuropsychological evaluations. At the end of the training, a greater collaboration was observed during assessment which indicates a small enhancement in passive visuo-spatial span and spatial orientation on maps.

The second case study is represented by F, a 12-year-old DS male with a mild/moderate intellectual disability. He was very passionate and infatuated by the ER activities, up to the point of being unable to share the robot with peers. Therefore, it was necessary to focalize the training goals on respect of turn and to promote a more reflexive approach to the use of the robot. During the ER laboratories, F was easily able to plan complex routes (e.g., forward and backward movements and left and right turns) using his hand on the mat as a concrete support to guide robot programming. During the training, F achieved adequate behaviour during play, collaborating with peers and acting as a tutor if necessary. Pre- and post-assessments showed an enhancement in the abilities of the passive visuo-spatial span. The boy achieved the highest scores in most of the other tests in pre-evaluation, producing a ceiling effect.

Specific Learning Disorders

By definition, children are reported to be affected by a specific learning disorder if they show developmental language or literacy impairments but normal non-verbal intelligence in the absence of other neurological or clinical conditions. For many years, this was seen as a consequence of a deficit either in perceptual processing or in underlying language representations, depending on one's theoretical persuasion. More recently, however, there has been interest in the idea that the learning process itself might be impaired and that language abilities are not "encapsulated" modules, acquired and used without conscious effort, but rather that they interplay with more controlled and deliberate operations probed by EFs (Bishop et al. 2014). Several studies have documented EF deficits in specific learning disorders (Gooch et al. 2011; de Jong 2006; Varvara et al. 2014) and the need to plan EF programmes with the hypothesis that a high cognitive control may facilitate language learning across modalities.

Within this framework, learning to plan a robot may represent a motivating "extra-deficit" setting where children with language or literacy disorders exercise those EFs that are usually impaired in these populations, such as sequential planning, updating in working memory, and inhibition of automatic responses. Accordingly, we recently set up a three-month weekly programme where preschool children with oral language deficit worked in small groups with the Bee-bot. Each hourly session consisted of different activities in agreement with a multifactorial risk model of learning disorder aimed at integrating cognitive control language production or comprehension with visual elaboration and motor action. For example, in order to improve motor-language integration, children had to imitate the movement of the robot pre-planned by the operator, while they were mentally engaged in counting each step. During this activity, children decreased the tendency to anticipate robot movement and improved the ability to complete a language-motor dual task.

In another exercise, a pre-programmed robot would move on a floor covered with animal shapes and children were instructed to pick up, as quickly as possible, all the shapes different from those touched by the robot. For example, if the robot touched a dog and an elephant, children would have to pick up all the other animal shapes except those representing a dog or an elephant. Children were instructed to do this quickly, before the robot had touched all the types of animals. During these activities, children learned to rapidly access semantic representations while using VMI. A third exercise required a child to program the robot to move across different colour squares while another child, next to the former, had to pronounce a word belonging to a certain semantic category or a word starting with a certain sound when the robot touched a specific colour or word from another category or a word with a different sound for another colour. This exercise worked mainly on verbal fluency and visual control. For each session, activities from about 20 different exercises were preplanned in order to train different cognitive abilities. In each session, two psychologists recorded the performance online by means of medical notes and chose the next exercise in order to vary the activities and training for the different cognitive abilities. Although quantitative results are not available for this programme, qualitative analysis suggests that it is suitable to be used in parallel with speech therapy in order to reinforce what can be achieved by a specific-domain programme.

Educational Robotic Adaptations in Classroom

After these feasibility studies with specific populations, we aimed at verifying the feasibility of ER laboratories in different special educational needs (SEN) children. To pursue this aim, ER laboratories were organized at school and specific adaptations were made. SEN children attending the first grade of primary school were enrolled. The other children in the classes participated in the ER activities too. SEN children participating in the study were divided into the following categories:

- Motor or visual problems
- Linguistics problems
- Cognitive problems
- Relational problems
- Attentional problems

To choose the most appropriate robot for our research purposes, a search was conducted individuating two models:

- Bee-bot (Campus Store, TTS Group), a bee-shaped robot with seven buttons on its back for planning
- Pro-bot (Campus Store, TTS Group), a car-shaped robot with seven buttons on its back for planning

After careful analysis of the two possibilities, the Bee-bot robot was selected for the following conceptual and technical reasons:

- Bee-bot is one of the most utilized robots in the age-range considered (Janka 2008).
- Bee-bot is considered one of the most impressive hardware devices for kindergarten and lower primary school children in educational technology (Janka 2008).
- Pro-Bot resulted more difficult to use for a SEN child in the first grade of the primary school.

To carry out these activities, the children were divided into small groups in order to favour imitation learning, collaboration, and involvement among peers. To stimulate attention and motivation, the proposed activities were made part of a narrative setting. All the ER laboratory activities were organized with an increasing complexity, both in terms of cognitive goals and in terms of robot programming ability. A metacognitive approach to the activities was encouraged, stimulating reflection about what was happening, encouraging the formulation of hypothesis and solutions, favouring an anxiety-free and fear-of-error-free learning environment, and promoting collaboration and feedback among peers.

For SEN children, ad hoc adaptations of both the robots and activities were proposed.

Bee-bot itself was adapted in order to respond to the needs of children with motor or visual disabilities, for which a simple change in activities would have resulted insufficient.

Regarding the adaptation to motor disabilities, it was necessary to make the robot accessible to children who would have had difficulties in programming the Bee-bot with the standard buttons. For this reason, a modification of the interface was carried out using some specific sensors in place of the original ones. The sensors are 65 mm diameter on/off buttons (Jelly Bean, see Fig. 9.3).

These sensors allow for an easier and customizable access because they can be freely positioned next to the child or inserted in a case that resembles the back of the Bee-bot in order to facilitate motor planning. Moreover, the sensors can temporarily be put offline, in order to limit the choices of planning, simplifying the activities. In Fig. 9.4, it is possible to see the final adapted prototype.

Fig. 9.3 Jelly bean

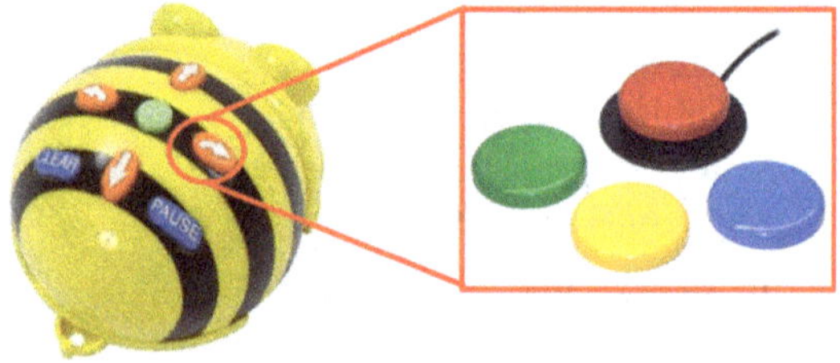

Fig. 9.4 The adapted Bee-bots

Regarding the adaptation of activities, the general indications to perform the activities with SEN children were as follows:

- To place the child in a small group.
- SEN children were positioned in a place with few distractions to foster working with a partner and stimulate attention/motivation towards the activity.

Here are some examples of specific adaptations of the activities for SEN children. For children with motor or visual problems, the use of the adapted Bee-bot was proposed. For children with linguistic/cognitive problems, some instruction cards were used to help the child understand the correct command. These cards represented the different buttons present on the Bee-bot. Gestural commands associated with the oral ones were also encouraged. For children with attention problems, frequent breaks and utilizing a token economy strategy helped to progressively increase their attention span. Finally, for children with relational problems, particular attention was paid to promote imitation learning, collaboration, and involvement among peers through relational reinforcements.

All the children were able to perform ER activities and therefore the results of the study were promising in terms of feasibility. By inserting ER into a scholastic system, it is possible to facilitate the inclusion of SEN children in learning programs, exploiting the adaptability of the robots. Our aim was to familiarize the students to new technologies, valorising the role of robotics not only in education but also as a support for situations of discomfort, as in SEN children.

Conclusions

Besides the historical use of ER to improve school engagement and STEM learning, in recent years interest in investigating ER effects on cognitive processes underlying learning has increased exponentially. This approach appears relevant not only for teachers and pedagogues to enhance cognitive operations that are engaged and strengthened by robot programming, but also for clinicians who would like to set up new programmes in order to ameliorate cognitive control in several developmental disorders. Indeed, within specific methodological characteristics, ER has the potential of becoming a tool that, by incorporating complexity, novelty, and diversity, is highly suitable to improve EFs in children with congenital or specific disorders.

Nevertheless, so far, few studies have used a rigorous and scientific approach and further research is needed in order to gather data on larger samples of children with typical development or neurodevelopmental disorders. Furthermore, available evidence has suggested that robot programming could be a powerful tool for improving EFs. However, to be effective, it must be used by embedding EF exercises within the cognitive area of major development for a certain age or within the domain that is dysfunctional for a certain disorder. Indeed, for children unable to act properly in the real word or who lack high order cognitive tools and thus struggle to access complex elaborations, robots represent the chance to facilitate action, representation, and thinking. However, achieving this potential requires strong multidisciplinary collaboration. While developmental psychology and neuropsychiatry address developmental trajectories of several typical and atypical cognitive processes, biotechnologies can adapt old and new robots to overcome sensory, motor, and cognitive limitations that characterize many children with developmental disorders.

Acknowledgements We would like to thank the children and parents who participated in this study and the TIM Foundation for its support of the "e-ROB Project" (aimed at creating a platform for ER through e-learning). We also thank Vincent Corsentino for English reviewing.

References

Alimisis, D. (2013). Educational robotics: Open questions and new challenges. *Themes in Science and Technology Education, 6*(1), 63–71. Retrieved from http://earthlab.uoi.gr/theste/index.php/theste/article/view/119.

American Psychiatric Association. (2013). Diagnostic and statistical manual of mental disorders (DSM-5®).

Baddeley, A., & Hitch, G. J. (1994). Development in the concept of working memory. *Neuropsychology, 8*, 485–493. Retrieved from http://psycnet.apa.org/journals/neu/8/4/485.html?uid=1995-04539-001.

Bandura, A. (1986). *Social foundations of thought and action : a social cognitive theory / Albert Bandura*. Englewood Cliffs, N.J: Prentice-Hall, 1986. xiii, 617 pp.

Bargagna, S., Castro, E., Cecchi, F., Cioni, G., Dario, P., Dell'Omo, M., et al. (2018). Educational robotics in down syndrome: A feasibility study. *Technology, Knowledge and Learning*. https://doi.org/10.1007/s10758-018-9366-z.

Barker, B. S., & Ansorge, J. (2007). Robotics as means to increase achievement scores in an informal learning environment. *Journal of Research on Technology in Education, 39*(3), 229–243. https://doi.org/10.1080/15391523.2007.10782481.

Beauchamp, M. H., & Anderson, V. (2013). Cognitive and psychopathological sequelae of pediatric traumatic brain injury. In Handbook of clinical neurology (Vol. 112, pp. 913-920). Elsevier.

Benitti, F. B. V. (2012). Exploring the educational potential of robotics in schools: A systematic review. *Computers & Education, 58*(3), 978–988. Retrieved from http://www.sciencedirect.com/science/article/pii/S0360131511002508 LK - link%7C http://www.sciencedirect.com/science/article/pii/S0360131511002508 SRC - BaiduScholar FG - 0.

Bexkens, A., Ruzzano, L., Collot d' Escury-Koenigs, A. M. L., Van der Molen, M. W., & Huizenga, H. M. (2014). Inhibition deficits in individuals with intellectual disability: A meta-regression analysis. *Journal of Intellectual Disability Research*. https://doi.org/10.1111/jir.12068.

Bishop, D., Nation, K., & Patterson, K. (2014). When words fail us: insights into language processing from developmental and acquired disorders. Retrieved from http://rstb.royalsocietypublishing.org/content/369/1634/20120403.short

Bodimeade, H. L., Whittingham, K., Lloyd, O., & Boyd, R. N. (2013). Executive function in children and adolescents with unilateral cerebral palsy. *Developmental Medicine & Child Neurology, 55*(10), 926–933. https://doi.org/10.1111/dmcn.12195.

Bottcher, L., Flachs, E. M., & Uldall, P. (2010). Attentional and executive impairments in children with spastic cerebral palsy. *Developmental Medicine and Child Neurology, 52*, e42–e47. https://doi.org/10.1111/j.1469-8749.2009.03533.x.

Chen, S. F., Chien, Y. L., Wu, C. T., Shang, C. Y., Wu, Y. Y., & Gau, S. S. (2016). Deficits in executive functions among youths with autism spectrum disorders: An age-stratified analysis. *Psychological Medicine, 46*(8), 1625–1638. https://doi.org/10.1017/S0033291715002238.

Cioni, G. (2000). Correlation between visual function, neurodevelopmental outcome, and magnetic resonance imaging findings in infants with periventricular leucomalacia. *Archives of Disease in Childhood - Fetal and Neonatal Edition, 82*(2), 134F–140. https://doi.org/10.1136/fn.82.2.F134.

Collins, A., & Koechlin, E. (2012). Reasoning, learning, and creativity: Frontal lobe function and human decision-making. *PLoS Biology, 10*(3), e1001293. https://doi.org/10.1371/journal.pbio.1001293.

Conrad, J., Polly, D., Binns, I., & Algozzine, B. (2018). Student perceptions of a summer robotics camp experience. *The Clearing House: A Journal of Educational Strategies, Issues and Ideals, 91*(3), 131–139. https://doi.org/10.1080/00098655.2018.1436819.

Costanzo, F., Varuzza, C., Menghini, D., Addona, F., Gianesini, T., & Vicari, S. (2013). Executive functions in intellectual disabilities: A comparison between Williams syndrome and Down syndrome. *Research in Developmental Disabilities, 34*(5), 1770–1780. https://doi.org/10.1016/j.ridd.2013.01.024.

Daniela, L., & Lytras, M. D. (2018). Educational robotics for inclusive education. *Technology, Knowledge and Learning*. https://doi.org/10.1007/s10758-018-9397-5.

Daniela, L., & Strods, R. (2018). Robot as agent in reducing risks of early school leaving. *Innovations, Technologies and Research in Education*, 140–158. Retrieved from https://books.google.it/books?hl=it&lr=&id=UJ9fDwAAQBAJ&oi=fnd&pg=PA138&dq=Daniela,+%26+Strods,+2018+robot&ots=tTctvNcwWq&sig=9Bbgv4npUuUN_Bvu2xI523gy6bU.

de Jong, P. F. (2006). Understanding normal and impaired reading development: A working memory perspective. In *Working memory and education* (pp. 33–60). Retrieved from https://books.google.it/books?hl=it&lr=&id=J48mU2acHoYC&oi=fnd&pg=PA33&dq=Understanding+normal+and+impaired+reading+development:+A+working+memory+perspective+PF+De+Jong&ots=4AX88Bfg8T&sig=5x9PsQkiU_AzEJqSwtbOnaYJwvM

De La Fuente, A., Xia, S., Branch, C., & Li, X. (2013). A review of attention-deficit/hyperactivity disorder from the perspective of brain networks. *Frontiers in Human Neuroscience, 7*. https://doi.org/10.3389/fnhum.2013.00192.

Di Lieto, M. C., Brovedani, P., Pecini, C., Chilosi, A. M., Belmonti, V., Fabbro, F., et al. (2017a). Spastic diplegia in preterm-born children: Executive function impairment and neuroanatomical correlates. *Research in Developmental Disabilities, 61*, 116–126. https://doi.org/10.1016/j.ridd.2016.12.006.

Di Lieto, M. C., Inguaggiato, E., Castro, E., Cecchi, F., Cioni, G., Dell'Omo, M., et al. (2017b). Educational robotics intervention on executive functions in preschool children: A pilot study. *Computers in Human Behavior, 71*, 16–23. https://doi.org/10.1016/j.chb.2017.01.018.

Di Lieto, M. C., Pecini, C., Castro, E., Inguaggiato, E., Cecchi, F., Dario, P., Cioni, G., & Sgandurra, G. (submitted). Empowering executive functions in 5–6-year-old typically developing children by educational robotics: An RCT study.

Diamond, A. (2013). Executive functions. *Annual Review of Psychology, 64*(1), 135–168. https://doi.org/10.1146/annurev-psych-113011-143750.

Diamond, A., & Lee, K. (2011). Interventions shown to aid executive function development in children 4 to 12 years old. *Science*. https://doi.org/10.1126/science.1204529.

Diamond, A., & Ling, D. S. (2016). Conclusions about interventions, programs, and approaches for improving executive functions that appear justified and those that, despite much hype, do not. *Developmental Cognitive Neuroscience, 18*, 34–48. https://doi.org/10.1016/j.dcn.2015.11.005.

Eliasson, A. C., Krumlinde-Sundholm, L., Rösblad, B., Beckung, E., Arner, M., Öhrvall, A. M., & Rosenbaum, P. (2006). The Manual Ability Classification System (MACS) for children with cerebral palsy: scale development and evidence of validity and reliability. Developmental medicine and child neurology, 48(7), 549–554.

Fazzi, E., Bova, S., Giovenzana, A., Signorini, S., Uggetti, C., & Bianchi, P. (2009). Cognitive visual dysfunctions in preterm children with periventricular leukomalacia. *Developmental Medicine & Child Neurology, 51*(12), 974–981. https://doi.org/10.1111/j.1469-8749.2009.03272.x.

Friedman, N. P., & Miyake, A. (2017). Unity and diversity of executive functions: Individual differences as a window on cognitive structure. *Cortex*. https://doi.org/10.1016/j.cortex.2016.04.023.

Garon, N., Bryson, S., & Smith, I. (2008). Executive function in preschoolers; a review using an integrative framework. *Psychological Bulletin, 134*, 31–60. Retrieved from http://psycnet.apa.org/record/2007-19419-002.

Gooch, D., Snowling, M., & Hulme, C. (2011). Time perception, phonological skills and executive function in children with dyslexia and/or ADHD symptoms. *Journal of Child Psychology and Psychiatry and Allied Disciplines, 52*(2), 195–203. https://doi.org/10.1111/j.1469-7610.2010.02312.x.

Green, C. T., Long, D. L., Green, D., Iosif, A.-M., Dixon, J. F., Miller, M. R., et al. (2012). Will working memory training generalize to improve off-task behavior in children with attention-deficit/hyperactivity disorder? *Neurotherapeutics, 9*(3), 639–648. https://doi.org/10.1007/s13311-012-0124-y.

Grunewaldt, K. H., Løhaugen, G. C. C., Austeng, D., Brubakk, A.-M., & Skranes, J. (2013). Working memory training improves cognitive function in VLBW preschoolers. *Pediatrics*. https://doi.org/10.1542/peds.2012-1965.

Guzzetta, A., Tinelli, F., Bancale, A., & Cioni, G. (2010). Visual and oculomotor disorders. In *The spastic forms of cerebral palsy* (pp. 115–142). Milano: Springer Milan. https://doi.org/10.1007/978-88-470-1478-7_7.

Holmes, J., Gathercole, S. E., & Dunning, D. L. (2009). Adaptive training leads to sustained enhancement of poor working memory in children. *Developmental Science, 12*(4), F9–F15. https://doi.org/10.1111/j.1467-7687.2009.00848.x.

Hughes, C., Ensor, R., Wilson, A., & Graham, A. (2009). Tracking executive function across the transition to school: A latent variable approach. *Developmental Neuropsychology, 35*(1), 20–36. https://doi.org/10.1080/87565640903325691.

Hussain, S., Lindh, J., & Shukur, G. (2006). The effect of LEGO training on pupils' school performance in mathematics, problem solving ability and attitude: Swedish data. In *Educational Technology and Society* (Vol. 9, pp. 182–194). https://doi.org/10.2307/jeductechsoci.9.3.182.

Ito, J., Araki, A., Tanaka, H., Tasaki, T., Cho, K., & Yamazaki, R. (1996). Muscle histopathology in spastic cerebral palsy. *Brain and Development*. https://doi.org/10.1016/0387-7604(96)00006-X.

Jacobson, L. K., & Button, G. N. (2000). Periventricular leukomalacia: An important cause of visual and ocular motility dysfunction in children. *Survey of Ophthalmology*. https://doi.org/10.1016/S0039-6257(00)00134-X.

Janka, P. (2008). Using a programmable toy at preschool age: why and how. ... , *International Conference on Simulation, Modelling and ...*, (pp. 112–121). Retrieved from http://www.dei.unipd.it/~emg/downloads/SIMPAR08-WorkshopProceedings/TeachingWithRobotics/pekarova.pdf

Jenks, K. M., De Moor, J., & Van Lieshout, E. C. D. M. (2009). Arithmetic difficulties in children with cerebral palsy are related to executive function and working memory. *Journal of Child Psychology and Psychiatry and Allied Disciplines, 50*(7), 824–833. https://doi.org/10.1111/j.1469-7610.2008.02031.x.

Kapa, L. L., & Plante, E. (2015). Executive function in SLI: Recent advances and future directions. *Current Developmental Disorders Reports, 2*(3), 245–252. https://doi.org/10.1007/s40474-015-0050-x.

Kazakoff, E. R., & Bers, M. U. (2014). Put your robot in, put your robot out: Sequencing through programming robots in early childhood. *Journal of Educational Computing Research, 50*(4), 553–573. https://doi.org/10.2190/EC.50.4.f.

Klingberg, T., Forssberg, H., & Westerberg, H. (2002). Increased brain activity in frontal and parietal cortex underlies the development of visuospatial working memory capacity during childhood. *Journal of Cognitive Neuroscience, 14*(1), 1–10. https://doi.org/10.1162/089892902317205276.

Klingberg, T., Fernell, E., Olesen, P. J., Johnson, M., Gustafsson, P., Dahlström, K., et al. (2005). Computerized training of working memory in children with ADHD--a randomized, controlled trial. *Journal of the American Academy of Child and Adolescent Psychiatry., 44*(2), 177–186. https://doi.org/10.1097/00004583-200502000-00010.

Krebs, H. I., Fasoli, S. E., Dipietro, L., Fragala-Pinkham, M., Hughes, R., Stein, J., & Hogan, N. (2012). Motor learning characterizes habilitation of children with hemiplegic cerebral palsy. *Neurorehabilitation and Neural Repair*. https://doi.org/10.1177/1545968311433427.

Kudo, M. F., Lussier, C. M., & Swanson, H. L. (2015). Reading disabilities in children: A selective meta-analysis of the cognitive literature. *Research in Developmental Disabilities*. https://doi.org/10.1016/j.ridd.2015.01.002.

la Paglia, F., Rizzo, R., & la Barbera, D. (2011). Use of robotics kits for the enhancement of metacognitive skills of mathematics: A possible approach. *Annual Review of CyberTherapy and Telemedicine, 9*(1), 22–25. https://doi.org/10.3233/978-1-60750-766-6-26.

Lehto, J. E., Juujärvi, P., Kooistra, L., & Pulkkinen, L. (2003). Dimensions of executive functioning: Evidence from children. *British Journal of Developmental Psychology, 21*(1), 59–80. https://doi.org/10.1348/026151003321164627.

Lunt, L., Bramham, J., Morris, R. G., Bullock, P. R., Selway, R. P., Xenitidis, K., & David, A. S. (2012). Prefrontal cortex dysfunction and "Jumping to Conclusions": Bias or deficit? *Journal of Neuropsychology, 6*(1), 65–78. https://doi.org/10.1111/j.1748-6653.2011.02005.x.

Mak, C., Whittingham, K., Cunnington, R., & Boyd, R. N. (2017). Efficacy of mindfulness-based interventions for attention and executive function in children and adolescents—A systematic review. *Mindfulness*. https://doi.org/10.1007/s12671-017-0770-6.

Miyake, A., Friedman, N. P., Emerson, M. J., Witzki, A. H., Howerter, A., & Wager, T. D. (2000). The unity and diversity of executive functions and their contributions to complex "frontal lobe" tasks: A latent variable analysis. *Cognitive Psychology*. https://doi.org/10.1006/cogp.1999.0734.

Nugent, G., Barker, B., & Grandgenett, N. (2008). The effect of 4-H robotics and geospatial technologies on science, technology, engineering, and mathematics learning and attitudes. In *World conference on educational multimedia, hypermedia and telecommunications 2008* (Vol. 2008, pp. 447–452). Retrieved from https://www.learntechlib.org/p/28433/

Nugent, G., Barker, B., Grandgenett, N., & Adamchuk, V. I. (2010). Impact of robotics and geospatial technology interventions on youth STEM learning and attitudes. *Journal of Research on Technology in Education, 42*(4), 391–408. https://doi.org/10.1080/15391523.2010.10782557.

Papert, S. (1980). Mindstorms: Children, computers, and powerful ideas. *Basic Books, 12*(4), 285–286. Retrieved from http://dl.acm.org/citation.cfm?id=1095592%22 LK - link%7Chttp://dl.acm.org/citation.cfm?id=1095592%22 SRC - BaiduScholar FG - 0.

Pavlova, M. A., & Krägeloh-Mann, I. (2013). Limitations on the developing preterm brain: Impact of periventricular white matter lesions on brain connectivity and cognition. *Brain*. https://doi.org/10.1093/brain/aws334.

Piaget, J., & Inhelder, B. (1966). *Child., L'image mentale chez l'enfant./the mental image in the*. Presses Universitaires de France.

Piovesana, A., Ross, S., Lloyd, O., Whittingham, K., Ziviani, J., Ware, R. S., et al. (2017). A randomised controlled trial of a web-based multi-modal therapy program to improve executive functioning in children and adolescents with acquired brain injury. *Clinical Rehabilitation, 31*(10), 1351–1363. https://doi.org/10.1177/0269215517695373.

Pirila, S., Van Der Meere, J. J., Rantanen, K., Jokiluoma, M., & Eriksson, K. (2011). Executive functions in youth with spastic cerebral palsy. *Journal of Child Neurology, 26*(7), 817–821. https://doi.org/10.1177/0883073810392584.

Pueyo, R., Junqué, C., Vendrell, P., Narberhaus, A., & Segarra, D. (2009). Neuropsychologic impairment in bilateral cerebral palsy. *Pediatric Neurology*. https://doi.org/10.1016/j.pediatrneurol.2008.08.003.

Rosenbaum, P., Paneth, N., Leviton, A., Goldstein, M., Bax, M., Damiano, D., et al. (2007). A report: The definition and classification of cerebral palsy April 2006. *Developmental Medicine and Child Neurology, 49*(Suppl. 109), 8–14. https://doi.org/10.1111/j.1469-8749.2007.tb12610.x.

Rueda, M. R., Rothbart, M. K., McCandliss, B. D., Saccomanno, L., & Posner, M. I. (2005). Training, maturation, and genetic influences on the development of executive attention. *The National Academy of Sciences, 102*(2), 2–7. https://doi.org/10.1073/pnas.0506897102.

Schatz, J., Craft, S., White, D., Park, T. S., & Figiel, G. S. (2001). Inhibition of return in children with perinatal brain injury. *Journal of the International Neuropsychological Society: JINS, 7*, 275–284. Retrieved from https://www.cambridge.org/core/journals/journal-of-the-international-neuropsychological-society/article/inhibition-of-return-in-children-with-perinatal-brain-injury/D5595103DE1CB9F18B43056823B5F0F3.

Schoemaker, K., Bunte, T., Wiebe, S. A., Espy, K. A., Dekovic, M., & Matthys, W. (2010). Executive functions in preschool children with ADHD and disruptive behavior disorders. *European Child & Adolescent Psychiatry, 19*, S64–S65. Retrieved from https://www.tandfonline.com/doi/abs/10.1080/87565641.2014.911875?casa_token=g5kZaOF_MZQAAAAA:TmFHNpIzgK8rnoIowMfY7rtPNTpk-Z3Z31FdaGO27wry3bL9BQzkba1CJCff4jRtiwia1xlBjWSe3A.

Sharfi, K., & Rosenblum, S. (2016). Executive functions, time organization and quality of life among adults with learning disabilities. *PLoS One, 11*(12). https://doi.org/10.1371/journal.pone.0166939.

Shimoni, M., Engel-Yeger, B., & Tirosh, E. (2012). Executive dysfunctions among boys with Attention Deficit Hyperactivity Disorder (ADHD): Performance-based test and parents report. *Research in Developmental Disabilities, 33*(3), 858–865. https://doi.org/10.1016/j.ridd.2011.12.014.

Sigurdardottir, S., Eiriksdottir, A., Gunnarsdottir, E., Meintema, M., Arnadottir, U., & Vik, T. (2008). Cognitive profile in young Icelandic children with cerebral palsy. *Developmental Medicine & Child Neurology, 50*(5), 357–362. Retrieved from https://onlinelibrary.wiley.com/doi/abs/10.1111/j.1469-8749.2008.02046.x.

Smith, E. E., & Jonides, J. (1999). Neuroscience – Storage and executive processes in the frontal lobes. *Science, 283*(5408), 1657–1661. https://doi.org/10.1126/science.283.5408.1657.

Srinivasan, S. M., Eigsti, I. M., Gifford, T., & Bhat, A. N. (2016). The effects of embodied rhythm and robotic interventions on the spontaneous and responsive verbal communication skills of children with Autism Spectrum Disorder (ASD): A further outcome of a pilot randomized controlled trial. *Research in Autism Spectrum Disorders*. https://doi.org/10.1016/j.rasd.2016.04.001.

Thorell, L. B., Lindqvist, S., Bergman Nutley, S., Bohlin, G., & Klingberg, T. (2009). Training and transfer effects of executive functions in preschool children. *Developmental Science, 12*(1), 106–113. https://doi.org/10.1111/j.1467-7687.2008.00745.x.

Usai, M. C., Viterbori, P., Traverso, L., & De Franchis, V. (2014). Latent structure of executive function in five- and six-year-old children: A longitudinal study. *European Journal of Developmental Psychology, 11*(4), 447–462. https://doi.org/10.1080/17405629.2013.840578.

Van de Weijer-Bergsma, E., Kroesbergen, E. H., & Van Luit, J. E. H. (2015). Verbal and visual-spatial working memory and mathematical ability in different domains throughout primary school. *Memory and Cognition*. https://doi.org/10.3758/s13421-014-0480-4.

Vanderborght, B., Simut, R., Saldien, J., Pop, C., Rusu, A. S., Pintea, S., et al. (2012). Using the social robot probo as a social story telling agent for children with ASD. *Interaction Studies*. https://doi.org/10.1075/is.13.3.02van.

Varvara, P., Varuzza, C., Sorrentino, A. C. P., Vicari, S., & Menghini, D. (2014). Executive functions in developmental dyslexia. *Frontiers in Human Neuroscience, 8*. https://doi.org/10.3389/fnhum.2014.00120.

Vygotsky, L. S. (1980). *Mind in society: The development of higher psychological processes*. Cambridge, MA: Harvard University Press. https://doi.org/10.1007/978-3-540-92784-6.

Vygotsky, L. (1987). Zone of proximal development. *Mind in Society The Development of Higher Psychological Processes, 157, 5291 SRC.*

Warmington, M., Hitch, G. J., & Gathercole, S. E. (2013). Improving word learning in children using an errorless technique. *Journal of Experimental Child Psychology, 114*(3), 456–465. https://doi.org/10.1016/j.jecp.2012.10.007.

Wass, S. V. (2015). Applying cognitive training to target executive functions during early development. *Child Neuropsychology*. https://doi.org/10.1080/09297049.2014.882888.

Wass, S., Porayska-Pomsta, K., & Johnson, M. H. (2011). Training attentional control in infancy. *Current Biology, 21*(18), 1543–1547. https://doi.org/10.1016/j.cub.2011.08.004.

White, D., & Christ, S. E. (2005). Executive control of learning and memory in children with bilateral spastic cerebral palsy. *Journal of the International Neuropsychological Society, 11*(07), 920–924. https://doi.org/10.1017/S1355617705051064.

Zelazo, P. D., & Muller, U. (2002). Executive function in typical and atypical development. In *Blackwell handbook of childhood cognitive development* (pp. 445–469). Retrieved from https://onlinelibrary.wiley.com/doi/abs/10.1002/9780470996652.ch20

Zelazo, P. D., Qu, L., & Müller, U. (2005). Hot and cold aspects of executive functions: Relations in early development. *Young Children's Cognitive Development: Interrelationships Among Executive Function, Working Memory, Verbal Ability, and Theory of Mind*, 71–93. Retrieved from https://books.google.it/books?hl=it&lr=&id=Ob95AgAAQBAJ&oi=fnd&pg=PA71&dq=Hot+and+cool+aspects+of+executive+function:+Relations+in+early+development&ots=pN8ovlHvkf&sig=iOeiEJtBzQTubqgaZULN3xgPN4E.

Chapter 10
Activities with Educational Robotics: Research Model and Tools for Evaluation of Progress

Linda Daniela, Raimonds Strods, and Ilze France

Abstract The use of robots in the learning process has been popular since S. Papert developed his LOGO Turtle idea and argued that students can construct their own knowledge, test their constructive solutions and be motivated to learn if they use robotics in the learning process. Today, the idea of using elements of robotics in the learning process is no longer new and innovative but there are still elements that can be developed and issues that should be discussed. In this chapter, the authors provide the research model and five research tools (structured observation protocol, evaluation of the possible risks of early school leaving to be filled in by teachers before and after activities, students' questionnaires to be filled in before and after activities) for evaluating the outcomes of organized after-school robotics activities. The research model and tools were tested and approbated with students who are at risk of early school leaving and students who participate in robotics activities to develop computational thinking.

Keywords Educational robotics · Educational robotics curriculum · Early school leaving · Computational thinking · After-school classes · Evaluation tools

Introduction

The use of robots in the world is no longer a novelty. There are some who think that the origins of robotics came with the work of the Czech author Karel Capek's RUR, or *Rossum's Universal Robots*, 1921, where the word 'robot' is first mentioned, which means the term describing devices that do the work (Niku 2011). Others say that robotics is beginning its success with Papert's ideas that children need to learn

L. Daniela (✉)
Faculty of Education, Psychology, & Art, University of Latvia, Rīga, Latvia
e-mail: linda.daniela@lu.lv

R. Strods · I. France
University of Latvia, Rīga, Latvia
e-mail: Raimonds.strods@lu.lv; ilze.france@lu.lv

L. Daniela (ed.), *Smart Learning with Educational Robotics*,
https://doi.org/10.1007/978-3-030-19913-5_10

computational thinking while working with computers and later developing the LOGO turtle (1983). Three major directions in robotics could be distinguished at this time:

- Industrial robots
- Educational robots
- Assistive robots

This chapter will analyse the use of educational robotics (ER) and the possible evaluation of results. In general, it is clear that robots have become a part of our daily lives, but to prepare the new generation for robots, to create new innovations, to create new software and to be prepared for the challenges posed by robotization, we need to develop computational thinking, but also to raise awareness of the side effects of robotization of various processes, such as interactions between robotics and inclusive education (Bargagna et al. 2018; Catlin and Blamires 2018; Daniela and Lytras 2018; Jung et al. 2019; Ronsivalle et al. 2018), the possibilities of using robotics to reduce early school leaving risks (Alimisis 2014; Daniela and Strods 2018; Karampinis 2018; Karkazis et al. 2018; Moro et al. 2018) and meeting the risks of robotization, which are mentioned in the European Civil Law Rules in Robotics (Nevejans 2017).

Theoretical Background

The use of ER in the learning process has been popular since the time Papert came up with the idea that children themselves should be allowed to work with robots to promote their computational thinking and defined this direction as constructionism (1984). Currently, the ER learning process is widely used at various levels of education – starting from preschool (Bers 2008; Cejka et al. 2006; Kazakoff and Bers 2012, 2014) to higher education (Danahy et al. 2014; Sünderhauf et al. 2018; Sünderhauf et al. 2016) – in various fields and in various dimensions: science, technology, engineering and mathematics (STEM) (Williams et al. 2007), engineering (Ariza et al. 2017; Zaldivar et al. 2013) and other branches of STEM (Eck et al. 2014; Witherspoon et al. 2018). The relationship between science and engineering practice (Bell et al. 2012; Li et al. 2016) is described, but the relationship between them should be clearly seen by the students as well and can be purposefully developed by supporting the development of computational thinking (Bocconi et al. 2016) by developing such competencies as asking questions and defining problems; developing and using models; planning and carrying out investigations; analysing and interpreting data; using mathematics and algorithmical thinking; constructing explanations and designing solutions; engaging in argument from evidence; and obtaining, evaluating, and communicating information.

The issues of the use of ER have been actively thought about and studied over the past decades, with a particular focus on mastering STEM as a whole. Many researchers are discussing what we understand today by integrated STEM education, which dates back to 1990 in the USA. However, there is still a lot of focus on improving science and mathematics as isolated disciplines (Kelley and Knowles 2016), with less emphasis on teaching technology and engineering (Castro et al. 2018). Nowadays, with the advancement of technology, it is essential to promote the development of both computational thinking (Bocconi et al. 2016) and thinking about the possibilities of various activities with technology, to foster mutual cooperation between students in order to strengthen the formation of social links that are essential to making the individual feel accepted in society.

Despite the recent emergence of more and more initiatives to use robotics creatively, to use different robots to develop certain competencies, to promote the development of an inclusive society and to strengthen the development of STEM competencies, there are authors who point out that in the early stages, children are offered too few activities to help them develop their STEM competencies in their broader sense; more activities are focused on natural sciences through analysing plants, animals, natural conditions and so on. Sullivan and Bers state that these aspects are important in fostering children's comprehensive knowledge, point out the importance of preparing children for the world as it is today as well as preparing for the progress that humans have created and therefore emphasize the need to promote the development of the competencies required for technology collaboration (2016). There are many examples where the acquisition of STEM is associated with robotics activities initially organized as after-school activities (Smyrnova-Trybulska et al. 2017), but which later try to include them in the compulsory curriculum (Ntemngwa and Oliver 2018).

The Context

The research model and tools that are offered to the reader are originally designed to meet the needs of the project 'Robotics-based learning interventions for preventing school failure and early school leaving', where Italy, Latvia and Greece jointly implemented a project whose main objective was to reduce early school leaving (ESL) risks using ER. The focus of this RoboESL project was more on the involvement of pupils at risk of ESL, so it is important to remember that they are pupils who often have low learning motivation, have lost the desire to actively engage in the compulsory learning process because their learning needs are not being satisfied and have a relatively negative attitude towards teachers. As a result of the project, evidence was provided that purposefully organized robotics activities reduce the risks of ESL, and original lesson descriptions for working with LEGO Mindstorm robots were developed. The pupils involved in the project activities were selected according to the methodology developed for the project (Daniela 2016), where teachers had to assess the risks of social exclusion of pupils, and those students who were at the highest risk of social exclusion and whose parents agreed that their children would engage in such activities took part in the project. Robotic activities were organized as after-school activities after a compulsory school day once a week.

Teachers working with pupils represented a variety of subjects – ICT teachers, mathematics, physics, English, philosophy, home economics and so on. The participation of teachers in project activities was voluntary, where their benefit was the opportunity to participate in an international project, participate in training organized by the project team, and use innovative teaching methods in their work.

The evaluation tools were re-examined in the context of organized after-school activities with LEGO Mindstorm educational robotics. The curriculum for after-school activities was developed by integrating elements of robotics, science and maths. Students aged 11–13 years old were offered the chance to work with the developed curriculum and were motivated by teachers to develop different LEGO robotics constructions by themselves. This kind of activity is like an intermediate step between the compulsory process in school, where students have to complete certain tasks for which they receive assessment, and leisure time activities, where children have full freedom in choosing the activities they want to do at a given moment.

Students worked on an integrated curriculum that included science, mathematics and programming. The total number of classes in the programme during the school year is 56 lessons, each 120 min long. The programme can be used as a deepening and extension of the school curriculum. In these classes, students engage voluntarily, but the learning process is structured and designed to achieve specific learning goals according to the programme developed.

The course included elements of engineering design process:

1. Identify the need or problem.
2. Find the information to create a robot.
3. Design the idea and plan the activities.
4. Prototype the program.
5. Test and evaluate the program.
6. Provide feedback. Feedback can be asked for and/or given at any point during engineering design.
7. Redesign the product to improve it.

In these robotics lessons, pupils were mostly given a problem, to which they need to find a solution, except for the final lesson block – creating their own project.

Research Model and Tools for Evaluation of Educational Robotics Activities

In this section the authors offer a research model, tools and methodology on how student activities and the outcomes reached can be evaluated during and after educational robotics activities. Developed tools are designed to evaluate students' attitude, motivation, collaboration, problem-solving skills and learning activity, so they can also be used in other contexts.

Before the research is started, data are collected and further analysed and it is imperative to have the parents' permission for the researchers to obtain and analyse the results of the study. It is also important to observe the principles of data protection in data collection and use the obtained results only to improve the learning process and reduce problems that can cause risks in order to promote the development of certain competencies as well as to allow students to express their views on organized activities.

The first tool provides (see Appendix 1) the evaluation instrument, which consists of several parts. In part one there are a few questions which can be used to detect the learning subjects that are problematic for particular students and the learning subjects in which these students achieve their best results. This information can be used to evaluate intervention activities to find out whether they are effective or whether additional intervention activities should be added to support the student. Next, information is asked regarding the number of lessons that students are skipping to find out whether there is a problem with truancy, and these data can be analysed after specific activities are provided to conclude whether interventional activities are successful or not. Moreover, there is a section where teachers assess students' difficulties that can influence the learning process. This part of the evaluation tool can be used separately to detect risks of ESL in order to plan the activities that are aimed at reducing such risks. The subsequent parts of the questionnaire consist of statements where teachers have to evaluate students' attitude, motivation, behaviour and problem-solving skills on a Likert scale of 1–5, where 1 = never, 2 = rarely, 3 = sometimes, 4 = often and 5 = always.

The second tool (see Appendix 2) is the questionnaire for students, which should be filled in before they start robotics activities. This tool is divided into several parts: first, students give the information about themselves; second, they are asked to evaluate the statements about learning and achievements on a scale of 1–5, where 1 = completely disagree, 2 = rarely agree, 3 = sometimes agree, 4 = mostly agree and 5 = completely agree. This is followed by the part where students are asked to evaluate the statements about missed lessons on a scale of 1–5, where 1 = always, 2 = often, 3 = sometimes, 4 = rarely and 5 = never. These questions are aimed at finding out the reasons why they are skipping lessons and provide the opportunity to focus on problems that are emerging for the students in order to find the best possible solution. The last part of this questionnaire is to find out the students' opinion about the learning subjects that they like or don't like. This information, together with the information given by the teachers about the evaluation grades in particular subjects, can help to provide an understanding of the linkage between students' attitude and learning outcomes.

The third tool (see Appendix 3) offers a structured observation protocol, which is used by teachers to evaluate students' outcomes during ER activities according to the criteria, which have to be evaluated on a Likert scale as 0 = can't be observed, 1 = low level, 2 = can be observed almost in all situations and 3 = does more than expected. This tool can be used quite frequently to see the changes in students' outcomes.

The fourth tool is a questionnaire that can be used after the activities (see Appendix 4), whereby teachers fill in the information about changes in students' attitude, motivation and problem-solving skills. These statements should be evaluated on a scale of 1–5, where 1 = no changes at all, 2 = some signs of improvement observed occasionally/rarely, 3 = some signs of improvement observed sometimes, 4 = signs of improvement observed in most situations and 5 = strong improvement observed in all situations. Statements about the student's behaviour should be evaluated on a scale of 1–5, where 1 = no changes at all, 2 = some signs of positive improvement observed occasionally/rarely, 3 = some signs of positive improvement observed sometimes, 4 = signs of positive improvement observed in most situations and 5 = strong positive improvement observed in all situations. This tool can be used after a period of intervention activities to find out whether there are improvements, but the authors suggest that the first minor changes can be observed after at least 3 months of activities because the risks of ESL develop over a longer period of time and reducing them is not a fast process.

The fifth tool (see Appendix 5) is the questionnaire for students which should be filled in after participation in robotics activities to find out whether there are any changes in students' attitude to learning. The tool is organized into several parts where the first part is to get the information about the students; the second part is to get to know the students' opinions about robotics activities to give teachers the opportunity to understand whether there are any changes needed in organizing such activities. The next part is organized as substatements about 'learning with robots' and the statement 'Activities with robots helped me to improve my: . .' where stu-

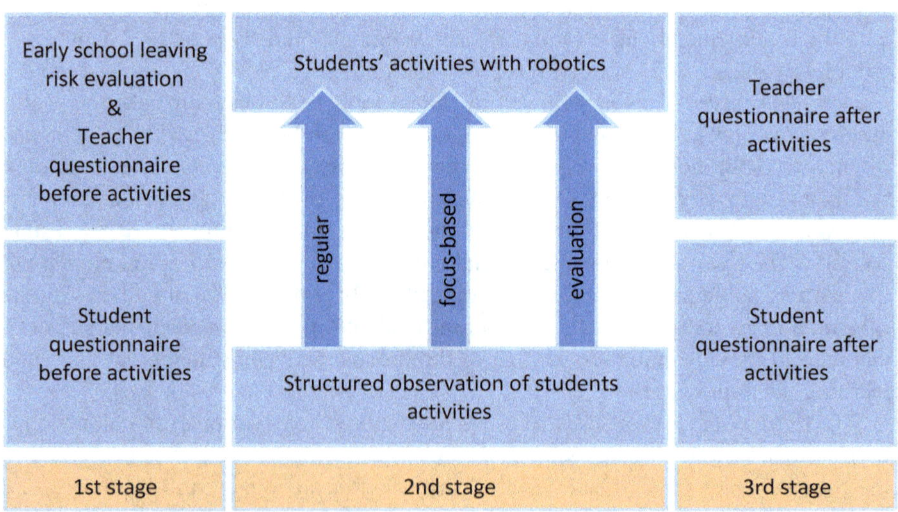

Fig. 10.1 Research model

dents had to evaluate these substatements in a scale of 1–5, where 1 = completely disagree, 2 = rarely agree, 3 = sometimes agree, 4 = mostly agree and 5 = completely agree. The last question in this questionnaire included the part where students were asked to name three learning subjects where their learning outcomes improved after participation in robotics activities.

The research model is summarized in Fig.10.1.

Discussion

The proposed research model and all suggested evaluation tools have been tested in several contexts in Latvia, Italy and Greece. The results are summarized in several articles that have already been published (Daniela 2016; Daniela and Strods 2016; Daniela et al. 2017), and one chapter, 'Educational Robotics for Reducing Early School Leaving from the Perspective of Sustainable Education', is included in this book where results are analysed from the perspective of sustainable development – by Daniela and Strods. All these results confirmed that the research model and tools can be used to work with students and evaluate their progress.

These tools have also been tested in another context where 11–13-year-old students participated in organized after-school robotics activities to develop their computational thinking.

The proposed research model can be replicated, and the developed tools can be used in different contexts to evaluate outcomes such as learning motivation, improved attitude to learning and improved behaviour and problem-solving skills. To extend the usability of this set of tools, authors continue to work on the development of specific tools to evaluate knowledge improvement during robotics activities.

Appendix 1

EVALUATION INSTRUMENT for detecting students who are at risk of ESL

EVALUATION INSTRUMENT

General information about student

Student ______________________________ (code of the student)

____ (age)

Subject/s you teach for particular student ______________________

☐ subject teacher ☐ class coordinator/class teacher ☐ robotics teacher

Average evaluation grade for the 1st term for all subjects _______

Three learning subjects with LOWEST grade for the 1st term (starting from the lowest):	**average mark**	**Three learning subjects with HIGHEST grade for the 1st term (starting from the highest):**	**average grade**
1.		1.	
2.		2.	
3.		3.	

Missed lessons 3 months before starting participation in activities	**Number of missed lessons**
1. Excused by doctor	
2. Excused for other reasons (parents etc.)	
3. Without an excuse	

Difficulties which influence learning Mark with 'X' if the statement characterizes the student	
Has learning difficulties connected with reading	
Has learning difficulties connected with calculation	
Has difficulties in understanding graphs and schemes	
Has other different learning difficulties	
Has other special needs	
Has attention concentration problems	
Attitude to learning Please evaluate these statements about the student on a scale of 1–5, where 1 = never, 2 = rarely, 3 = sometimes, 4 = often, 5 = always	
Completes homework	
Cooperates with teachers in a positive way	
Cooperates with classmates during lessons in a positive way	
Is ready for work in lessons	
Understands the connection between learning and achievements	
Is ready to do extra assignments to improve achievements	
Obeys behavioural rules in classroom	
Is ready to join out-of-class/school activities together with other classmates	
Is involved in sport/art activities not connected with learning at school	
Knows that the learning is important for him/her	
Problem-solving skills Please evaluate these statements about the student on a scale of 1–5, where 1 = never, 2 = rarely, 3 = sometimes, 4 = often, 5 = always	
Solves the learning problems by himself/herself	
Asks for help from teachers	
Solves conflicts in a calm way	
Motivation Please evaluate these statements about the student on a scale of 1–5, where 1 = never, 2 = rarely, 3 = sometimes, 4 = often, 5 = always	
Is motivated to learn the subject you teach	
Is motivated to understand his/her mistakes to correct them	

Is motivated to improve achievements	
Is motivated to overcome difficulties in learning	
Has an aim and works to achieve it	
Observed problems Please evaluate these statements about the student on a scale of 1–5, where 1 = never, 2 = rarely, 3 = sometimes, 4 = often, 5 = always	
Is late for the beginning of lessons	
Has problematic behaviour during recess (break)	
Is aggressive to other students	
Is aggressive to teachers	
Uses rude language with classmates	
Uses rude language with teachers	
Refuses doing assignments during the lessons	
In situation of conflict reacts aggressively	

Appendix 2

Hello! You are going to learn how to work with Robots! Congratulations! It is fun! Before starting learning, can you answer some questions? Your responses are anonymous!

1. I am:

   ☐ boy ☐ girl

   I am ______years old (your age)

2. **Learning and achievements** Please evaluate these statements on a scale of 1–5, where 1 = completely disagree, 2 = rarely agree, 3 = sometimes agree, 4 = mostly agree, 5 = completely agree	
3.1. Learning is fun	
3.2. My achievements depend on my learning	
3.3. I do all the homework	
3.4. I like to cooperate with my classmates in lessons	
3.5. I like to work individually to do assignments	
3.6. I like to do extra assignments	
3.7. I like it when there are different activities in lessons	
3.8. I like it when I can do something active in lessons	
3.9. I like to solve learning problems by myself	
3.10. I like to look for extra information needed for learning	
3. **If you miss lessons it happens because** Please evaluate these statements on a scale of 1–5, where 1 = always, 2 = often, 3 = sometimes, 4 = rarely, 5 = never	
4.1. I was sick or had an appointment with the doctor	

4.2. I had to participate in another activity – sports, art, etc.	
4.3. I had to help my parents	
4.4. It was an unpredictable situation	
4.5. I had to work	
4.6. I had to babysit my sister/brother	
4.7. I did not want to go to school	
4.8. I overslept	
4.9. I did not want to meet my classmates	
4.10. I do not like learning	

5. **Please write here the three learning subjects you like the most**

   5.1. ______________________________
   5.2. ______________________________
   5.3. ______________________________

6. **Please write here the three learning subjects you like the least**

   6.1. ______________________________
   6.2. ______________________________
   6.3. ______________________________

Appendix 3

Protocol of observation

Teacher______________________

Legend: 0 = can't be observed, 1 = low level, 2 = can be observed almost in all situations, 3 = can be observed during all projects, 4 = does more than expected

Criteria / Code of the student									
Cooperates with other members of the group during activities in a positive way									
Is ready to do extra assignments to improve achievements									
Obeys behavioural rules during the projects									
Knows the aim of the task and how to achieve it									
Solves the faced problems by himself/herself									
Asks for help from teachers									
Solves problem situations in a calm way									
Is motivated to overcome difficulties in doing tasks									
Is motivated to understand mistakes to correct them									
Does assignments during the robotics classes									
Participates in group work									
Helps peers in problem solving									

Appendix 4

Dear Teacher,

Please assess the changes in the attitude of the student who has participated in the activities. Use the same code for the student as was used before the activities. This survey is very important. It will take approximately 5 min to fill in the questionnaire. Thank you in advance for your time.

Student ____ (code)

Gender_____

Subject/s you teach ____

Attitude to learning (statements are the same as in the first questionnaire, but evaluation is based on changes in the student's attitude) Please evaluate these statements about the student's attitude on a scale of 1–5, where: 1 = no changes at all 2 = some signs of improvement observed occasionally/rarely 3 = some signs of improvement observed sometimes 4 = signs of improvement observed in most situations 5 = strong improvement observed in all situations	
Preparation of homework	
Cooperation with teachers in a positive way	
Cooperation with classmates during lessons in a positive way	
Readiness for work in lessons	
Understanding of the connection between learning and achievements	
Readiness to do extra assignments to improve achievements	
Following of the behavioural rules in the classroom	
Readiness to join out-of-class/school activities together with other classmates	
Readiness to join activities led by other classmates	
Readiness to reach learning aims	
Motivation (statements are the same as in the first questionnaire, but evaluation is based on changes in the student's motivation) Please evaluate these statements about the student's motivation on a scale of 1–5, where: 1 = no changes at all 2 = some signs of improvement observed occasionally/rarely 3 = some signs of improvement observed sometimes 4 = signs of improvement observed in most situations 5 = strong improvement observed in all situations	
Motivation to learn the subject you teach	
Motivation to understand his/her mistakes to correct them	
Motivation to improve achievements	
Motivation to overcome difficulties in learning	
Readiness to work hard to achieve the aim	

Observed problems (statements are the same as in the first questionnaire, but evaluation is based on changes in the student's behaviour) Please evaluate these statements about the student's behaviour on a scale of 1-5, where: 1 = no changes at all 2 = some signs of positive improvement observed occasionally/rarely 3 = some signs of positive improvement observed sometimes 4 = signs of positive improvement observed in most situations 5 = strong positive improvement observed in all situations	
Being late for the beginning of lessons	
Problematic behaviour during recess (break)	
Aggressiveness to other students	
Aggressiveness to teachers	
Using rude language with classmates	
Using rude language with teachers	
Refuses to do assignments during the lessons	
Aggressive reaction in situations of conflict	
Problem-solving skills Please evaluate these statements about the student on a scale of 1–5, where 1 = never, 2 = rarely, 3 = sometimes, 4 = often, 5 = always	
Solves the learning problems by himself/herself	
Asks for help from teachers	
Solves the conflicts in a calm way	

Thank you!

Appendix 5

You had a wonderful opportunity to learn how to work with robotics. We hope you enjoyed that! Can you answer some questions about your experience? Your responses are anonymous!

1. I am:
   ☐ boy ☐ girl
2. I am ______years old (your age)
3. Which robotics activities did you like most? Please name at least three of them

   ____

   ____

   ____

   ____

4. Which robotics activities were most challenging? Please name them

____

____

____

____

5. **Learning with robots** Please evaluate these statements on a scale of 1–5, where 1 = completely disagree, 2 = rarely agree, 3 = sometimes agree, 4 = mostly agree, 5 = completely agree	
5.1. Learning by using robots was fun	
5.2. I have learned how to program robots	
5.3. I liked to work in groups to do assignments with robots	
5.4. I liked to make calculations for programming	
5.5. I can use this knowledge in other activities	
5.6. I liked to solve problems with programming by myself	
5.7. I liked that others helped me to solve problems with programming	
5.8. I liked to look for extra information needed for using robots	
5.9. Other outcome (please name it)	
6. **Activities with robots helped me to improve my:** Please evaluate these statements on a scale of 1–5, where 1 = completely disagree, 2 = rarely agree, 3 = somehow agree, 4 = mostly agree, 5 = completely agree	
6.1. understanding of maths	
6.2. understanding of physics	
6.3. understanding of informatics and technologies	
6.4 attitude to learning	
6.5. cooperation skills with my classmates	
6.6. cooperation skills with teachers	
6.7. other outcome (please name it)	

7. **Please write here three learning subjects where your learning outcomes improved**

7.1. ______________________________________

7.2. ______________________________________

7.3. ______________________________________

References

Alimisis, D. (2014). Educational robotics in teacher education: An innovative tool for promoting quality education. In L. Daniela, I. Lūka, L. Rutka, I. Žogla (Eds.), *Teacher of the 21st century: Quality education for quality teaching* (pp. 28–39). Newcastle upon Tyne: Cambridge Scholar Publishing. ISBN (13): 978-1-4438-5612-6.

Ariza, D. V., Palacio, A. M., Aragón, I. P., Pulido, C. M., Logreira, E. A., & McKinley, J. R. (2017). Application of color sensor programming with LEGO-Mindstorms NXT 2.0 to recreate a simplistic plague detection scenario. *Scientia et Technica, 22*(3), 268–272.

Bargagna, S., Castro, E., Cecchi, F., Cioni, G., Dario, P., Dell'Omo, M., et al. (2018). Educational robotics in Down Syndrome: A feasibility study. *Technology, knowledge and learning*, 1–9. https://doi.org/10.1007/s10758-018-9366-z.

Bell, P., Bricker, L., Tzou, C., Lee, T., & Van Horne, K. (2012). Exploring the science framework: Engaging learners in science practices related to obtaining, evaluating, and communicating information. *Science Scope, 36*(3), 18–22.

Bers, M. (2008). *Blocks to robots: Learning with technology in the early childhood classroom*. New York, NY: Teacher's College Press.

Bocconi, S., Chioccariello, A., Dettori, D., Ferari, A., & Engelhardt, K. (2016). Developing computational thinking in compulsory education. Retrieved from http://publications.jrc.ec.europa.eu/repository/bitstream/

Castro, E., Cecchi, F., Valente, M., Buselli, E., Salvini, P., & Dario, P. (2018). Can educational robotics introduce young children to robotics and how can we measure it? *Journal of Computer Assisted Learning., 34*, 970–977. https://doi.org/10.1111/jcal.12304.

Catlin, D., & Blamires, M. (2018) Designing robots for special needs education. *Technology, knowledge and learning*, 1–23. Retrieved from https://doi.org/10.1007/s10758-018-9378-8.

Cejka, E., Rogers, C., & Portsmore, M. (2006). Kindergarten robotics: Using robotics to motivate math, science, and engineering literacy. *International Journal of Engineering Education, 22*(4), 711–722.

Danahy, E., Wang, E., Brockman, J., Carberry, A., Shapiro, B., & Rogers, C. B. (2014). LEGO-based robotics in higher education: 15 years of student creativity. *International Journal of Advanced Robotic Systems*. https://doi.org/10.5772/58249.

Daniela, L. (2016) Preliminary results of the project 'Robotics-based learning interventions for preventing school failure and early school leaving. ICERI 9th Annual International Conference of Education, Research and Innovation Proceedings (pp. 4336–4344). ISBN: 978-084-0617-05895-01. https://doi.org/10.21125/iceri.2016.

Daniela, L., & Lytras, M. D. (2018). Educational robotics for inclusive education. *Technology, Knowledge and Learning*, 1–7. https://doi.org/10.1007/s10758-018-9397-5.

Daniela, L., & Strods, R. (2016). The role of robotics in promoting the learning motivation to decrease the early school leaving risks. References: p.13. ROBOESL Conference, November 26, 2016, Athens, Greece: Proceedings Athens, 2016, pp. 7–13. Retrieved from http://roboesl.eu/conference/?page_id=508

Daniela, L., & Strods, R. (2018). Robot as agent in reducing risks of early school leaving. In L. Daniela (Ed.), *Innovations, technologies and research in education* (pp. 140–158). Newcastle upon Tyne: Cambridge Scholars Publishing. ISBN (10): 1-5275-0622.

Daniela, L., Strods, R., Alimisis, D. (2017) Analysis of robotics-based learning interventions for preventing school failure and early school leaving in gender context // 9th International Conference on *Education and New Learning Technologies (Edulearn17)*, 3rd-5th July, 2017, Barcelona, Spain: Conference Proceedings Barcelona, pp.810–818. ISBN 9788469737774. ISSN 2340-1117.

Eck, J., Hirschmugl-Gaisch, S., Kandlhofer, M., & Steinbauer, G. (2014). A cross-generational robotics project day: Pre-school children, pupils and grandparents learn together. *Journal of Automation, Mobile Robotics and Intelligent Systems, 8*(1), 12–19. https://doi.org/10.14313/JAMRIS_1-2014/2.

Jung, S. E., Lee, K., Cherniak, S., & Cho, E. (2019). Non-sequential learning in a robotics class: Insights from the engagement of a child with Autism spectrum disorder in technology, knowledge and learning, pp. 1–19. Retrieved from https://doi.org/10.1007/s10758-018-9394-8.

Karampinis, T. (2018). Activities and experiences through RoboESL project opportunities. *International Journal of Smart Education and Urban Society, 9*(1), 13–24. https://doi.org/10.4018/IJSEUS.

Karkazis, P., Balourdos, P., Pitsiakos, G., Asimakopoulos, K., Saranteas, I., Spiliou, T., & Roussou, D. (2018). To water or not to water. The Arduino approach for the irrigation of a field. *International Journal of Smart Education and Urban Society, 9*(1), 25–36. https://doi.org/10.4018/IJSEUS.

Kazakoff, E., & Bers, M. (2012). Programming in a robotics context in the kindergarten classroom: The impact on sequencing skills. *Journal of Educational Multimedia and Hypermedia, 21*(4), 371–391.

Kazakoff, E. R., & Bers, M. U. (2014). Put your robot in, put your robot out: Sequencing through programming robots in early childhood. *Journal of Educational Computing Research, 50*(4), 553–573. https://doi.org/10.2190/EC.50.4.f.

Kelley, T. R., & Knowles, J. G. (2016). A conceptual framework for integrated STEM education. *International Journal of STEM Education, 3*(1). Retrieved from https://stemeducationjournal.springeropen.com/articles/10.1186/s40594-016-0046-z.

Li, Y., Huang, Z., Jiang, M., & Chang, T. W. (2016). The effect on pupils' science performance and problem-solving ability through Lego: An engineering design-based modeling approach. *Educational Technology & Society, 19*(3), 143–156.

Moro, M., Agatolio, F., & Menegatti, E. (2018). The development of robotic enhanced curricula for the RoboESL project: Overall evaluation and expected outcomes. *International Journal of Smart Education and Urban Society (IJSEUS), 9*(1), 48–60. https://doi.org/10.4018/IJSEUS.

Nevejans, N. (2017). European civil law rules in robotics, European Commission. Retrieved from http://www.europarl.europa.eu/committees/fr/supporting-analyses-search.html

Niku, S. B. (2011). *Introduction to robotics: Analysis, control, applications*. Hoboken, NJ: Wiley.

Ntemngwa, C., & Oliver, J. S. (2018). The implementation of integrated science technology, engineering and mathematics (STEM) instruction using robotics in the middle school science classroom. *International Journal of Education in Mathematics, Science and Technology (IJEMST), 6*(1), 12–40. https://doi.org/10.18404/ijemst.380617.

Papert, S. (1983). Talking Turtle. Horizon, Part 1 – 02:48 to 03:10. BBC and the Open University. Retrieved from http://research.roamer-educational-robot.com/1983/12/14/talking-turtle/

Papert, S. (1984). New theories for new learnings. *School Psychology Review, 13*(4), 422–428.

Ronsivalle, G. B., Boldi, A., Gusella, V., Inama, C., & Carta, S. (2018). *How to implement educational robotics' programs in Italian schools: A brief guideline according to an instructional design point of view in Technology, knowledge and learning* (pp. 1–19). https://doi.org/10.1007/s10758-018-9389-5.

Smyrnova-Trybulska, E., Morze, N., Kommers, P., Zuziak, W., & Gladun, M. (2017). Selected aspects and conditions of the use of robots in STEM education for young learners as viewed by teachers and students. *Interactive Technology and Smart Education, 14*(4), 296–312. https://doi.org/10.1108/ITSE-04-2017-0024.

Sullivan, A., & Bers, M. U. (2016). Robotics in the early childhood classroom: Learning outcomes from an 8-week robotics curriculum in pre-kindergarten through second grade. *International Journal of Technology and Design Education, 26*(1), 3–20.

Sünderhauf, N., Leitner, J., & Milford, M. (2016). Robotics: Science and systems (RSS) workshop. Are the sceptics right? Limits and potentials of deep learning in robotics. Retrieved from http://juxi.net/workshop/deep-learning-rss-2016/

Sünderhauf, N., Brock, O., Scheirer, W., Hadsell, R., Fox, D., Leitner, J., et al. (2018). The limits and potentials of deep learning for robotics. *The International Journal of Robotics Research, 37*(4–5, 405), –420. https://doi.org/10.1177/0278364918770733.

Williams, D. C., Yuxin, M., Prejean, L., Ford, M. J., & Lai, G. (2007). Acquisition of physics content knowledge and scientific inquiry skills in a robotics summer camp. *Journal of Research on Technology in Education, 40*(2), 201–216.

Witherspoon, E. B., Schunn, C. D., Higashi, R. M., & Shoop, R. (2018). Attending to structural programming features predicts differences in learning and motivation. *Journal of Computer Assisted Learning, 34*(2), 115–128. https://doi.org/10.1111/jcal.12219.

Zaldivar, D., Cuevas, E., Pérez-Cisneros, M. A., Sossa, J. H., Rodríguez, J. G., & Palafox, E. O. (2013). An educational fuzzy-based control platform using LEGO robots. *International Journal of Electrical Engineering Education, 50*(2), 157–171. https://doi.org/10.7227/IJEEE.50.2.5.

Chapter 11
The Use of Robotics for STEM Education in Primary Schools: Teachers' Perceptions

Ahmad Khanlari

Abstract This study aims to better understand elementary teachers' perspective on the use of robotics for STEM education. Through an 8-h workshop, the participants engaged in hands-on activities and filled out a pre-survey and a post-survey. The results indicated that the participants' perceptions significantly changed as a result of participating in the workshop, learning about robotics, and being involved in hands-on robotics activities.

Keywords Robotics · STEM education · Elementary schools · Professional development

Introduction

STEM education aims to increase STEM literacy which includes "the knowledge and understanding of scientific and mathematical concepts and processes required for personal decision making, participation in civic and cultural affairs, and economic productivity for all students" (National Research Science 2011, p.5). Another goal of STEM education is to persuade students to explore degrees and careers in STEM-related fields. A review of the literature shows that "[c]hildren undergo many developmental changes between the ages of 6 and 12, particularly in terms of their cognitive development" (Canadian Child Care Federation 2010, p. 6). Therefore, STEM education is more effective if it starts in early education. As a result, it is recommended to lay the foundations of science and technology and mathematics education as early as the elementary grades (Marulcu 2010). Early engagement in STEM education facilitates students' understanding of subject matter (Marulcu 2010), reduces barriers for entering jobs related to STEM fields (Madill et al. 2007),

---

A. Khanlari (✉)
University of Toronto, Toronto, ON, Canada
e-mail: a.khanlari@mail.utoronto.ca

L. Daniela (ed.), *Smart Learning with Educational Robotics*,
https://doi.org/10.1007/978-3-030-19913-5_11

and diminishes the gender-based stereotypes about STEM careers (Metz 2007). Although early STEM education is very important, educators pay little attention to STEM education in early childhood classrooms (Bers 2008; Marulcu 2010). Therefore, I have chosen to focus on STEM education in elementary schools to fill this gap. I specifically focus on robotics, since it is a "gateway to STEM because it integrates all these different disciplines in an applied way" (Kazakoff et al. 2013, p. 246) and has "the potential to significantly impact the nature of engineering and science education at all levels, from K–12 to graduate school" (Mataric 2004, p. 1). Furthermore, Rogers and Portsmore (2004) found that one of the best ways for improving students' performance in mathematics and science is conducting simple hands-on activities in elementary schools. Robotics offers students' hands-on experience in a wide range of subjects, improves student knowledge of STEM-related topics, and provides an alternative teaching method to traditional lecture-style classes (Gura 2012).

Moreover, I have chosen to focus on teachers' perceptions, because, while their perceptions are very crucial and may encourage or discourage them from using robotics in their class activities, there is limited research about teachers' perceptions of using robotics (Scaradozzi et al. 2018; Kim et al. 2015), especially in the elementary grades. Most of the studies have been conducted in other grades rather than elementary grades, and some studies (e.g., Scaradozzi et al. 2018) had participants from a variety of grades. Other researchers who focused on elementary schools (e.g., Kim et al. 2015) mainly focused on teachers' engagements and learning, not their perceptions. Thus, more exploration is needed to fill this gap in the existing literature and explore what elementary teachers perceive about the use of robotics for STEM education. Knowing the elementary teachers' perceptions would help the researchers, educators, and policymakers to take the required actions to encourage them to integrate robotics into their teaching activities for STEM education.

Studies (e.g., Gürcan-Namlu and Ceyhan 2003; Hallam 2008; Thorpe and Brosnan 2008) show that teachers' prior experience in using technologies decreases their level of anxiety and motivates them to integrate technologies into their teaching activities. Therefore, providing an opportunity for teachers to know more about technologies like robotics and engaging them in hands-on activities may affect teachers' perceptions and may encourage them to integrate robotics into their lessons. The present study is the second phase of a study (Khanlari 2016a) conducted to gauge elementary teachers' perceptions of the use of robotics for STEM education. In the first phase, the teachers were provided with an opportunity to explore some materials about robotics and learn more about robotics, without being asked to do hands-on activities. The teachers were asked to fill out a survey indicating the kinds of needs and support they need in order to be able to integrate robotics into their teaching activities. The participants indicated that one of the main challenges is the lack of access to materials, including educational robotics kits and handbooks (Khanlari 2016a). Also, the majority of participants believed that inadequate technical support and their lack of knowledge in making connections between robotics and the subject matter are other major obstacles. The majority of the participants expressed their needs in attending in-service professional development in order to be able to use robotics in their classes. The present study aims to provide teachers with the support they indicated they need through running a one-day robotics workshop. This workshop aims to provide an opportunity for teachers to engage in hands-on robotics activities and explore their perceptions of the use of robotics for STEM education. The research questions to be addressed in this study include:

1. To what extent do elementary teachers believe that robotics can help students to learn STEM subjects?
2. To what extent do elementary teachers believe that using robotics in the classroom will foster positive attitudes toward STEM disciplines in students, resulting in encouraging them to pursue their education and career in these fields?

State of Art

Along with robotics technology development, researchers, and educators in many countries, including Canada, Japan, South Korea, Taiwan, and the United States, have employed robots to support education (Han 2012). Educational robots, as a new type of manipulative learning, engages students in hands-on activities: students learn concepts while they are doing some activities and projects. Hands-on nature of robotics creates an active learning environment and increases conceptual understanding of subject matter (Adolphson 2005; Brosterman 1997). Robotics might be used as a learning object or as a learning tool (Alimisis and Kynigos 2009). In the first category (i.e., learning object), robotics on its own is studied as a subject, while in the second category (i.e., learning tool), robotics is used as a tool for teaching and learning other school subjects such as mathematics and science. Several studies (e.g., Attard 2012; Bauerle and Gallagher 2003; Druin and Hendler 2000; Jeschke et al. 2008; Khanlari 2016b) have shown that hands-on robotics is engaging, creates authentic learning environments that are suitable for a better understanding of STEM disciplines, has positive long-term effects such as attracting students to technological and scientific studies, and leads students to a love of STEM subjects. A review of the literature shows that robotics can help students to learn many subjects, including mathematics, physics, science, mechanics, electronics, computer engineering, geography, art, and biology (Eguchi 2007; Kolberg and Orlev 2001; Kazakoff et al. 2013; Marulcu 2010; Oppliger 2002; Sklar et al. 2002; Sklar et al. 2003). Additional goals and objectives of STEM education in Canada include developing positive attitudes in students about STEM fields, promoting students' interests toward STEM disciplines, and encouraging students to pursue education and careers in STEM-related fields (Canadian Association of Science Centre 2010; STEM n.d.).

Method

The participants of this study included 58 elementary teachers (43 female and 15 male) who are participating in an in-service professional development program. As parts of their professional development, the participants were provided with an opportunity to participate in a one-week workshop and learn about integrating technologies into their teaching activities. One day of this one-week workshop was allocated to robotics and STEM education. During four sessions, each of which lasted two hours, the participants were engaged in hands-on robotics activities. The first session included a pre-survey about using robotics for teaching STEM-related fields, a lecture on robotics, and some videos about the importance of robotics in industry and education. Then, the educational robotics kits and the coding environment were introduced to the participants. The educational package used for this workshop was LEGO® MINDSTORMS® EV3, which is a well-known robotics kit in education settings. The instructors provided lectures on LEGO EV3 and its different parts, as well as the programming environment and the way the EV3 can be programmed. Due to the time limitations, the workshop providers had pre-built the LEGO bases, and they were ready to be used by the participants. In the second session, the participants started doing some simple coding activities. Also, the instructors provided the participants with a set of activities about exploring Mars using their robots, which needs to be done by the participants. The workshop instructors had mocked up the Mars rocky surface, and ask the participants to use appropriate sensors and program their robots to do some activities, involving math and science problems. Some of the activities included:

1. Write a code and download into the robot to move around and look for hills as quickly as possible. While doing so, the participants should consider different design considerations, such as the types of wheels required to move on the surface, the size of the wheels, the suitable sensors, etc.
2. Calculate the time the robot spent to find the hills, measure the distance the robots traveled, calculate the area the robot covered, and come up with some ideas to improve their program so that the robot would spend less time and consume lower energy.

Participants spent the sessions 2, 3, and 4 to do the activities. After 8 h of engagement in learning about robotics (through lectures, handouts, and videos) and learning with robotics (through hands-on robotics activities), the participants were asked to fill out an online post-survey (Fig. 11.1).

Data Presentation

The first few questions aimed to gauge teachers' experience and knowledge in using technologies in general and robotics in particular. According to the data collected as the first question of the survey, the majority of the participants had no prior knowledge/experience in using robotics in their teaching activities (Fig. 11.2).

Fig. 11.1 Mars rocky surface and the explorer robot

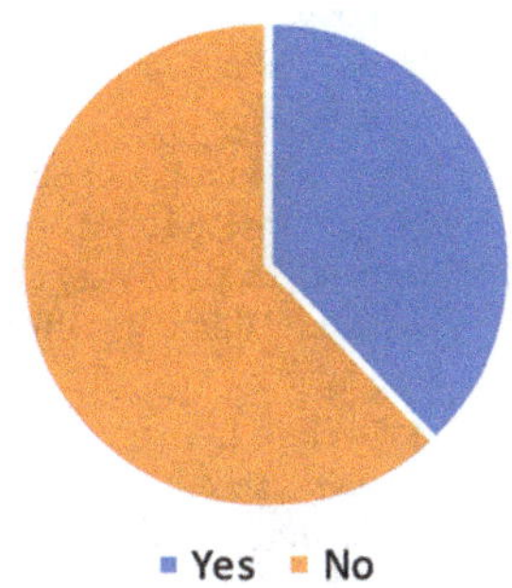

Fig. 11.2 Participants' responses about their prior experience/knowledge in integrating robotics into teaching activities

The survey also included 11 Likert scale questions, including two questions about robotics and learning math, three questions about robotics and science and technology literacy, two questions about robotics and interest toward STEM disciplines, and four general questions about the effects of robotics. Table 11.1 shows the Likert scale questions.

The results of the pre-survey and post-survey are presented in Figs. 11.3 and 11.4.

Table 11.1 The Likert scale questions

Likert scale questions	Strongly disagree	Disagree	Neutral	Agree	Strongly agree
Q1. Robotics has the potential to facilitate learning of mathematics.					
Q2. Using robotics in mathematics can help students to improve their mathematical reasoning and problem-solving skills.					
Q3. Robotics has the potential to facilitate learning of science subjects.					
Q4. Using robotics in schools can help students in the process of scientific inquiry and improving their skills for initiating and planning, performing and recording, and analyzing and interpreting.					
Q5. Using robotics can develop a positive attitude about STEM disciplines.					
Q6. Using robotics can encourage students to pursue their education and career in STEM-related fields.					
Q7. Robotics has the potential to improve technology literacy in schools.					
Q8. Overall, students will be more actively involved in the lesson/unit in which robotics technology is used, compared to the lessons/units they are not involved in robotics activities.					
Q9. Overall, in robotics sessions, students' different learning styles are better accommodated than they are with comparable lessons/units that don't involve robotics technologies.					
Q10. Overall, in robotics activities, student work would show a more in-depth understanding of the content compared to the lessons that robotics is not being used.					
Q11. Overall, student work is more creative when engaging in robotics activities, compared to a class/unit in which robotics is not being used.					

Data Analysis

Robotics and Math Education

The results show that, while before the workshop 43% of the participants had positive perceptions (i.e., agree or strongly agree) and 48% had negative perceptions (i.e., disagree or strongly disagree) on the effects of robotics on learning math subjects (Q.1), after the workshop 78% of the participants had a positive perception on

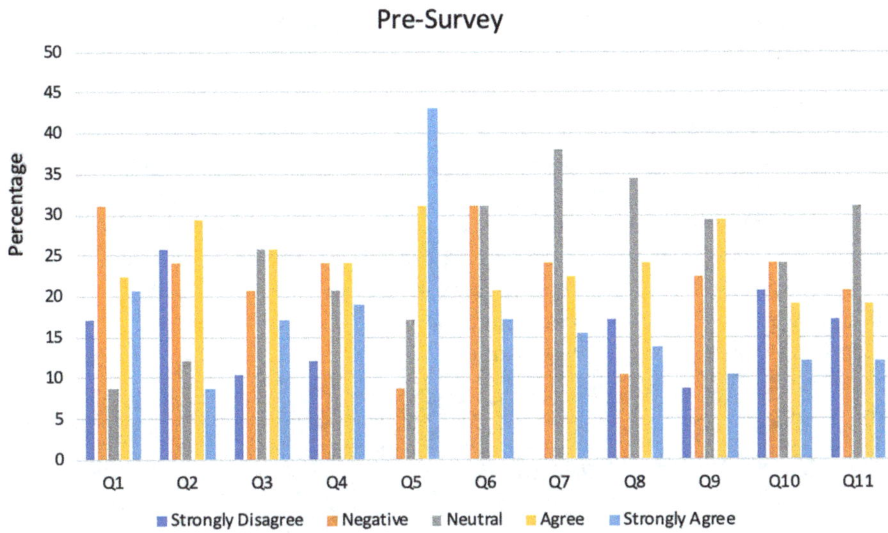

Fig. 11.3 The results of the pre-survey

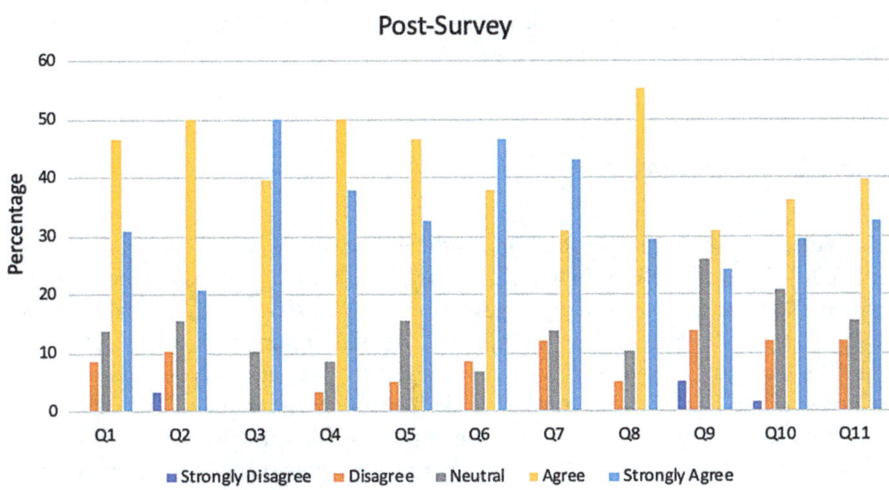

Fig. 11.4 The results of the post-survey

the positive effects of robotics for math education. Also, the perception of the participants regarding the effects of robotics on mathematical reasoning and problem-solving (Q.2) has dramatically changed. Before the workshop, about 40% of the participants had positive perceptions, and 50% of the participants had negative perceptions, but after the workshop, 71% of the participants had positive attitudes toward the effectiveness of robotics on problem-solving and mathematical reasoning.

Robotics and Science and Technology Literacy

The results of the pre-survey show that before the workshop, 43% of the participants believed robotics has positive effects on learning of science subjects (Q3) and on the process of scientific inquiry (Q.4). Also, 31% of the participants disagreed with Q3's statement, and 36% disagreed with Q4's statement. However, after the workshop, about 90% of the participants agreed or strongly agreed with these two statements. Also, before the workshop, 74% of the participants believed that robotics has the potential to improve technology literacy in schools (Q.5). After the workshop, however, 79% of the participants had a positive attitude about the positive effects of robotics on technology literacy.

Robotics and STEM Education

Before the workshop, about 38% of the participants believed robotics can develop a positive attitude on students about STEM (Q6) and can encourage them to pursue their education and career in STEM-related field (Q7). After the workshop, about 84% of the participants agreed or highly agreed with Q6, and 74% of the participants agreed or highly agreed with Q7.

Overall Perceptions

The participants' overall perceptions on the effectiveness of the robotics have significantly changed; before the workshop, 38% of the participants agreed or highly agreed that robotics can actively engage students in lessons (Q8), and 40% agreed or highly agreed that robotics can support different learning styles (Q9). However, after the workshop, these percentages changed to 84% and 55%, respectively. In response to the last two questions, only 31% of the participants used to have positive attitudes about the robotics potential to provide a more in-depth understanding of the content (Q10) and create more engaging activities (Q11). However, after the workshop, 65% and 72% of the participants believed robotics has these potentials.

The last question asked the participants whether they would like to integrate robotics into their teaching activities for STEM education. In the pre-survey, the majority of the participants (45%) indicated they are not sure, 22% of the participants answered yes, and 33% answered no. However, in the posttest, 24% indicated they are not sure, 53% answered yes, and 22% answered no.

Discussions and Conclusion

The result of the analysis first highlights the importance of learning about robotics and being engaged in hands-on activities with robotics, because the participants' perceptions of the use of robotics have dramatically changed after being engaged in the pre-service workshop.

The participants of this study, after participating in the robotics workshop, perceived that robotics has the potential to facilitate learning of mathematics and can help students to improve their mathematical reasoning and problem-solving skills. Therefore, this study concurs with the studies that indicate robotics facilitates learning of mathematics subjects (e.g., Allen 2013; Bers and Portsmore 2005). Hussain, Lindh, and Shukur (2006) in their research concluded that employing robotics in a grade 5 class resulted in a better performance in mathematics. Faisal, Kapila, and Iskander (2012) in their study examined the effects of using LEGO robotics on engaging fourth-grade students in mathematics and enhancing their visual understanding of concepts. The analysis of their pre-assessment and post-assessment tests revealed that robotics increases students' performance: "the average performance of the class increased from 36% to 92% after the activity" (p. 10). The authors also reported that robotics helps 87% of students to learn and improves their understanding of abstract concepts such as unit conversion. Mathematical skills, such as basic algebra, trigonometry, counting, measuring, estimating, and geometry, are embedded in designing and programming robots, and students can learn these subjects during robotics projects (Gura 2012; Johnson 2002; Samuels and Haapasalo 2012). A review of the literature surrounding mathematics learning shows that one of the best ways for improving students' performance in mathematics is conducting simple hands-on activities (Rogers and Portsmore 2004). The results of the present study, along with the existing literature, confirm that hands-on robotics can actively engage students in lessons and has the potential to provide an alternative teaching method to traditional lecture-style classes in order to improve students' understanding of math concepts. In other words, educational robots, as a new type of manipulative learning, have the potential to improve students' understanding of mathematical concepts.

Moreover, after attending the robotics workshop, the majority of the participants believed that robotics has the potential to facilitate learning of science subjects; can help students in the process of scientific inquiry; can improve students' skills for initiating and planning, performing and recording, and analyzing and interpreting; and has the potential to improve technology literacy in schools. Therefore, the result of this study is in agreement with the literature (e.g., Barker and Ansorge 2007; Bers and Portsmore 2005; Bers et al. 2002; Grubbs 2013; Nugent et al. 2010) regarding the positive effects of robotics on teaching and learning science and technology. For example, Carbonaro, Rex, and Chambers (2004) conducted an action research project to examine the effects of robotics on learning computer and science subjects.

The authors found that robotics provides a challenging learning environment in which "the abstract levels of concepts (programs) are directly mapped to the concrete physical level (robots) and that students themselves can observe the results of their designs at both levels" (p. 4549). The results of the present study, along with other studies that examined the effects of educational robotics on learning science and technology literacy, show that robotics has the potential to improve students' scientific conceptual understanding and technology literacy.

Furthermore, the participants of this study, after attending in the robotics workshop, perceived that using robotics can develop a positive attitude about STEM disciplines and has the potential to encourage students to pursue their education and career in STEM-related fields. Therefore, this study is in agreement with the reviewed literature surrounding the positive effects of robotics on STEM education. For example, Grubbs (2013) has shown that robotics creates an exciting and authentic environment that provides students with the opportunity to apply their knowledge that they thought is unusable; therefore, robotics encourages students to pursue a STEM field in the future and has the potential to increase the number of students entering STEM fields. Allen (2013) in a study expressed that robotics has the potential to present a strong example of STEM education, is a powerful tool for changing students' perceptions of STEM fields, and leads students to "fall in love with these subjects and all that science, technology, engineering, and mathematics make possible in our world" (p. 345). Allen also stated that robotics helps students to "see themselves as future scientists, tech specialists, engineers, and mathematicians" (p. 345) and can prepare students in all grade levels to succeed in the future that is strongly STEM-based. In fact, the results of the present study show that running a robotics workshop provided an opportunity for teachers to understand the effects of robotics on STEM education, which may encourage them to integrate robotics into their math curriculum. The implication of this study is that if the elementary teachers are provided with the support they need (in this study, in-service professional development), they would think of integrating robotics into their teaching activities, which would be beneficial for students. Also, providing an opportunity for teachers to engage in hands-on activities with technologies like robotics would diminish their anxiety and would encourage them to integrate technologies into their teaching activities. Another implication of this study is that robotics is well-suited for universal design for learning, because the majority of the participants, after attending the workshop, perceived that robotics can create more engaging activities and can support different learning styles.

One of the limitations of this study is that as the study is done in a day-long workshop about robotics, there was no opportunity to conduct an interview or focus group. Therefore, the researcher was unable to ask some follow-up questions in order to better understand the teachers' perceptions. However, some of the participants have indicated their interest to be interviewed for further research and have provided their contact information. Therefore, future work would be to contact the interested participants and conduct face-to-face interviews. Another limitation is that the majority of the participants were female teachers. This setting may have influenced the results. For the future studies, it might be better to have an equal number of male and female teachers.

References

Adolphson, K. (2005). *Robotics as a context for meaningful mathematics*. Paper presented at the annual meeting of the North American Chapter of the International Group for the Psychology of Mathematics Education, Roanoke.

Alimisis, D., & Kynigos, C. (2009). Constructionism and robotics in education. In D. Alimisis (Ed.), *Teacher education on robotics-enhanced constructivist pedagogical methods* (pp. 11–26). Athens: ASPETE.

Allen, K. C. (2013). Robots bring math-powered ideas to life. *Mathematic Teaching in the Middle School, 18*(6), 340–347.

Attard, C. (2012). Teaching with technology: Exploring the use of robotics to teach mathematics. *Australian Primary Mathematics Classroom, 17*(2), 31–32. Retrieved from http://eric.ed.gov/?id=EJ978138.

Barker, B., & Ansorge, J. (2007). Robotics as means to increase achievement scores in an informal learning environment. *Journal of Research on Technology Education, 39*, 229–243.

Bauerle, A., & Gallagher, M. (2003). Toying with technology: Bridging the gap between education and engineering. In C. Crawford et al. (Eds.), *Proceedings of society for information technology & teacher education international conference* 2003 (pp. 3538–3541). Chesapeake: AACE.

Bers, M. U. (2008). *Blocks to robots: Learning with technology in the early childhood classroom*. New York, NY: Teacher's College Press.

Bers, M., & Portsmore, M. (2005). Teaching partnerships: Early childhood and engineering students teaching math and science through robotics. *Journal of Science Education and Technology, 14*(1), 59–74.

Bers, M., Ponte, I., Juelich, K., Viera, A., & Schenker, J. (2002). Integrating robotics into early childhood education. In *Information technology in childhood education annual* (pp. 123–145).

Brosterman, N. (1997). *Inventing kindergarten*. New York: H.N. Abrams.

Canadian Association of Science Centre. (2010). *Youth STEM guidelines*. Retrieved from http://www.canadiansciencecentres.ca/?n=29-161-153&newsId=73

Canadian Child Care Federation and Canadian Language and Literacy Research Network. (2010). *Foundations for numeracy: An evidence-based toolkit for the effective mathematics teacher*. Retrieved from http://foundationsfornumeracy.cllrnet.ca/pdf/SY_NumeracyKit09_ENG.pdf

Carbonaro, M., Rex, M., & Chambers, J. (2004). Exploring middle school children's problem solving during robot construction tasks. In R. Ferdig et al. (Eds.), *Proceedings of the society of information technology and teacher education* (pp. 4546–4550). Charlottesville: AACE.

Druin, A., & Hendler, J. (2000). *Robots for kids: Exploring new technologies for learning*. San Diego: Academic Press.

Eguchi, A. (2007). Educational robotics for elementary school classroom. In R. Carlsen et al. (Eds.), *Proceedings of society for information technology & teacher education international conference* (pp. 2542–2549). Chesapeake: AACE.

Faisal, A., Kapila, V., & Iskander, M. G. (2012). Using robotics to promote learning in elementary grades. Paper presented at the 19th ASEE Annual Conference & Exposit.

Grubbs, M. (2013). Robotics intrigue middle school students and build STEM skills. *Technology and Engineering Teacher, 72*(6), 12–16. Retrieved from http://eric.ed.gov/?id=EJ1006898.

Gura, M. (2012). Lego robotics: STEM sport of the mind. *Learning and Leading with Technology, 40*(1), 12–16. Retrieved from http://eric.ed.gov/?id=EJ991224.

Gürcan-Namlu, A., & Ceyhan, E. (2003). Computer anxiety: Multidimensional analysis on teacher candidates. Educational Sciences. *Theory & Practice, 3*(2), 401–432.

Hallam, T.A. (2008). *Sociocultural influences on computer anxiety among preservice teachers: An exploratory study*. Doctoral Dissertation, University of Akron, Akron.

Han, J. (2012). Emerging technologies: Robot assisted language learning. *Language Learning & Technology, 16*(3), 1–9. Retrieved from http://llt.msu.edu/issues/october2012/emerging.pdf.

Hussain, S., Lindh, J., & Shukur, G. (2006). The effect of LEGO training on pupils' school performance in mathematics, problem solving ability and attitude: Swedish data. *Educational Technology & Society, 9*(3), 182–194.

Jeschke, S., Kato, A., & Knipping, L. (2008). The engineers of tomorrow: Teaching robotics to primary school children. *Proceedings of the SEFI 36th Annual Conference*, Aalborg, Denmark.

Kazakoff, E. R., Sullivan, A., & Bers, M. U. (2013). The effect of a classroom-based intensive robotics and programming workshop on sequencing ability in early childhood. *Journal of Early childhood Education, 41*, 245–255. https://doi.org/10.1007/s10643-012-0554-5.

Khanlari, A. (2016a). Teachers' perceptions of the benefits and the challenges of integrating educational robots into primary/elementary curricula. *European Journal of Engineering Education*, 1–11.

Khanlari, A. (2016b). Robotics integration to create an authentic learning environment in engineering education. *In the proceedings of the IEEE conference on Frontiers in Education*, 1–4.

Kim, C., Kim, D., Yuan, J., Hill, R. B., Doshi, P., & Thai, C. N. (2015). Robotics to promote elementary education pre-service teachers' STEM engagement, learning, and teaching. *Computers & Education, 91*, 14–31.

Kolberg, E., & Orlev, N. (2001). Robotics learning as a tool for integrating science-technology curriculum in K-12 schools. *31st ASEE/IEE Frontiers in Education Conference*. Reno.

Johnson, J. (2002). Children, robotics, and education. In the Proceedings of the 7th International Symposium on Artificial Life and Robotics (491 – 496).

Madill, H., Campbell, R. G., Cullen, D. M., Armour, M. A., Einsiedel, A. A., Ciccocioppo, A. L., et al. (2007). Developing career commitment in STEM-related fields: myth versus reality. In R. Burke, M. Mattis, & E. Elgar (Eds.), *Women and minorities in science, technology, engineering and mathematics: Upping the numbers* (pp. 210–244). Northhampton: Edward Elgar Publishing.

Marulcu, I. (2010). *Investigating the impact of a LEGO(TM)-based, engineering-oriented curriculum compared to an inquiry-based curriculum on fifth graders' content learning of simple machines*. Doctoral dissertation.

Mataric, M. J. (2004). Robotics education for all ages. In *Proceedings of American Association for Artificial Intelligence Spring Symposium on Accessible, Hands-on AI and Robotics Education*, Palo Alto.

Metz, S. S. (2007). Attracting the engineering of 2020 today. In R. Burke, M. Mattis, & E. Elgar (Eds.), *Women and minorities in science, technology, engineering and mathematics: Upping the numbers* (pp. 184–209). Northampton: Edward Elgar Publishing.

National Research Council. (2011). *Successful K-12 STEM education*. Washington, DC: The National Academies Press.

Nugent, G., Barker, B., Grandgenett, N., & Adamchuck, V. (2010). Impact of robotics and geospatial technology interventions on youth STEM learning and attitudes. *Journal of Research on Technology in Education, 42*(4), 391–408.

Oppliger, D. (2002). Using FIRST LEGO league to enhance engineering education and to increase the pool of future engineering students. *32nd ASEE/IEEE Frontiers in Education Conference*, Boston.

Rogers, C., & Portsmore, M. (2004). Bringing engineering to elementary school. *Journal of STEM Education, 5*, 17–28.

Samuels, P. & Haapasalo, L. (2012). Real and virtual robotics in mathematics education at the school–university transition. International Journal of Mathematical Education in Science and Technology, 43(3), 285–301.

Scaradozzi, D., Screpanti, L., Cesaretti, L., Storti, M., & Mazzieri, E. (2018). Implementation and assessment methodologies of teachers' training courses for STEM activities. *Technology, Knowledge and Learning*, 1–21. https://doi.org/10.1007/s10758-018-9356-1.

Sklar, E., Eguchi, A., & Johnson, J. (2002). Children's learning from team robotics: RoboCupJunior 2001. In *Proceedings of RoboCup-2002*. Robot Soccer World Cup VI.

Sklar, E., Eguchi, A., & Johnson, J. (2003). Scientific challenge award: RoboCupJunior – learning with educational robotics. *AI Magazine, 24*(2), 43–46.

STEM NS. (n.d.). Growing a better tomorrow with STEM. Retrieved from http://www.stemns.com/index.html

Thorpe, S. J., & Brosnan, M. (2008). Does computer anxiety reach levels which conform to DSM IV criteria for specific phobia? *Computers in Human Behavior, 23*(3), 1258–1272.

Chapter 12
Using Robots to Introduce First-Year College Students to the Field of Electrical Engineering

Mounir Ben Ghalia

Abstract Improving the retention of first-year college students has been the focus of many engineering programs. Often, students declare an engineering major without having an understanding of what that major entails or what their career goals are. This lack of understanding combined with the absence of proper mentoring results in a large percentage of students who either switch to other majors or drop out of college without receiving a degree. To lessen the attrition rate among first-year engineering majors, the common initiative that has been adopted by many engineering programs is to offer an introductory-level engineering course. The main educational objective of these courses is to help students gain a better understanding of the engineering field and what engineers do. At the University of Texas Rio Grande Valley, located in the South Central Region of USA, we developed and introduced a robot project in the curriculum of our introductory course for first-year electrical engineering majors. Using the Pololu 3pi mobile robot, the project allows students to experience a real-world engineering problem. To prepare students for the robot project, we used the experiential learning model that aimed at engaging students in hands-on learning experiences consisting of robot navigation examples. Students gained a glimpse into the different technical subjects within the electrical engineering field through their experience working with the robots. Data collected over several semesters showed that educational robots contribute to keeping first-year students in the program.

Keywords First-year college students · Educational robotics · Mobile robots · Using robots to learn engineering · Project-based learning · Student retention in engineering

M. Ben Ghalia (✉)
University of Texas Rio Grande Valley, Edinburg, TX, USA
e-mail: mounir.benghalia@utrgv.edu

L. Daniela (ed.), *Smart Learning with Educational Robotics*,
https://doi.org/10.1007/978-3-030-19913-5_12

Introduction

Educators introduce robots in entry-level engineering courses as one of the tools to improve student motivation (Aroca et al. 2016; Blanding and Meah 2014) and to allow students learn about different engineering disciplines (Roth 2002). Several engineering programs have developed introductory courses for their first-year college students (Mix and Balda 1997; Huettel et al. 2007; Mercede 2008; Blanding and Meah 2014; Behrens et al. 2010; Carley et al. 2000). These courses provide an introduction to the field of engineering and help students decide if they want to follow a degree in engineering. The most reported successful approaches to introducing students to engineering are those that involve hands-on laboratory and project-based learning. For instance, other engineering programs use robot kits such as the LEGO MINDSTORMS for first-year course projects (Behrens et al. 2010; Williams 2009; Roth 2002). Besides providing students insight into the technical field of engineering, such projects also help students gain skills in teamwork and communication (Behrens et al. 2010). Engaging first-year students in hands-on learning not only increases their motivation (Mix and Balda 1997; Huettel et al. 2007; Behrens et al. 2010) but also helps to reduce attrition (Karam and Mounsef 2011).

This chapter describes a robot project introduced in the introductory-level course offered for first-year electrical engineering students enrolled at the University of Texas Rio Grande Valley[1] (UTRGV). The chapter is organized as follows. Section "The Introductory Engineering Course" describes the introductory engineering course and provides the rationale for using robots as a tool to expose students to different technical subjects within the field of electrical engineering. Section "Preparing Students for the Robotics Project with Experimental Learning" explains the motivation behind the use of the experiential learning model to introduce students to robotics programming. The section also presents and discusses the series of active hands-on learning experienced by our students. Section "Main Robotics Project" presents the culminating main robot project. Section "Project Impact" discusses data that show the impact of the robot project on student performance. Finally, Section "Conclusion" provides the overall conclusion of this study.

The Introductory Engineering Course

Course Description

The course, ELEE 1101,[2] introduces first-year students to the field of electrical engineering, its different specializations, and career paths. Typically, this is the first engineering course that our students take in their first year with calculus and physics

[1] UTRGV is located in the South Central Region of USA.

[2] The course numbering at UTRGV consists of four letters followed by four digits. The letters

courses, which are prerequisites for advanced electrical engineering courses. The topics of the introductory course include:

- Overview of specializations within the field of electrical engineering
- Electrical engineering as a career
- Introduction to digital systems
- Basic electrical laws
- Introduction to electrical laboratory instruments
- Graphical representation of data using the MATLAB software
- Engineering design cycle
- Mobile robot project

Our electrical engineering program schedules the course as a 160-minute weekly laboratory meeting. The class meets either in a computer laboratory, the electric circuit laboratory, or the robotics and controls laboratory, all housed in the electrical engineering department. At the end of the course, students are expected to demonstrate their ability to:

- Understand the differences between the engineering majors
- Identify the qualities of a successful engineering student
- List the different types of engineering job classifications
- Understand the purpose of internships and cooperative education
- Perform engineering calculation
- Graph data using the MATLAB software
- Understand basic digital circuit concepts
- Understand and apply the engineering design process
- Program a mobile robot using the programming language C++
- Carry out a reasonably complex navigation task
- Communicate project results verbally and in written reports

We introduced improvements to the course curriculum to include a project-based learning experience that helps students gain a better insight into the different technical engineering subjects. These include electronics, controls, microprocessors, and programming.

The 3pi Mobile Robot Project

One of the critical aspects of planning the course project was to select a mobile robot that is affordable and easily programmable. After considering several alternatives, we decided to use the Pololu 3pi robot shown in Fig. 12.1 (Pololu Corporation 2014). We bought several robots to allow a team of two students to share one robot.

---

indicate the subject (e.g., ELEE indicates electrical engineering and CSCI indicates computer science). The first digit of the course number indicates the class year, and the second digit indicates the number of credit hours.

Fig. 12.1 The Pololu 3pi robot

The 3pi robot is a wheeled mobile robot which is capable of speeds up to 100 cm/s. The robot has two micro metal gearmotors, five infrared (IR) reflectance sensors, and 8x2 LCD. The microcontroller of the 3pi robot is an ATmega328P from microchip running at 20 Mhz and featuring 32 KB of a flash program, 2 KB RAM, and 1 KB of persistent EEPROM memory. To program the robot, we use the development environment Atmel Studio and the free GNU C/C++ compiler (Fig. 12.2). The robot software comes with an extensive set of libraries and a number of sample programs which demonstrate the capabilities of the robot. To transfer a compiled program on a computer to the robot, we use an in-system programmer (ISP) which connects to the computer's USB port via a USB-A to Micro-B cable. The programmer connects to the robot via a 6-pin ISP programming cable (Fig. 12.2).

The goal of introducing a challenging project in this introductory engineering course is threefold:

1. Allow students to experience working on a real engineering problem
2. Help students improve their problem-solving skills
3. Inform students which advanced coursework will teach them the technical details of the mobile robot modules

We designed several short tutorials to help students become familiar with the various modules of a mobile robot. Students who take this course in their first semester do not have a background on the fundamental or advanced electrical engineering curriculum. Also, most of our students have not received any formal training

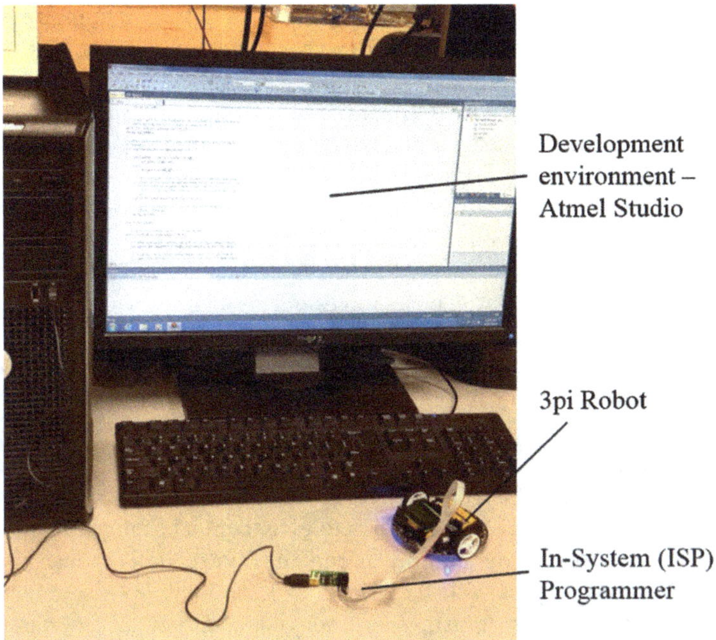

Fig. 12.2 The development environment for the 3pi robot

in computer programming during their secondary education. Therefore, it is important to design the tutorials to allow for a basic and qualitative understanding of how the different modules of the robot are designed and how they work. Most importantly, in these short tutorials, we point to specific fundamental and advanced electrical engineering courses that cover the topics needed to fully understand the robot modules. This approach provides students with a glimpse at the curriculum of the electrical engineering program. The tutorials include:

- *The Robot Drive Module Tutorial*: This tutorial explains in qualitative terms how DC motors work and how software and hardware can control motor speed. Fundamental courses that cover these topics include ELEE 2330 Digital Systems I, ELEE 2305 Electric Circuits I, and ELEE 3301 Electronics I. Advanced courses that cover these topics include ELEE 3321 Signals and Systems, ELEE 3302 Electronics II, and ELEE 4328 Solid States.
- *The Sensor Module Tutorial*: This tutorial explains how an IR reflectance sensor works and how the reflected energy is converted into a voltage signal. The tutorial also explains the rationale for using IR reflectance sensors to allow the robot to navigate along a black line. Courses that cover fundamental topics that are helpful to understand the working of IR reflectance sensors include ELEE 2305 Electric Circuits I, ELEE 3301 Electronics I, and ELEE 4328 Solid States.
- *The Robot Microcontroller Tutorial:* This tutorial provides a brief introduction to the role of a microcontroller. The course ELEE 3435 Microprocessor Systems

covers the software and hardware, architecture, and programming of microprocessors. The course ELEE 4380 Computer Architecture covers advanced topics in computer systems.

- *The Robot Navigation Module:* This tutorial provides an explanation of how the robot can navigate following a black line. Qualitative description of the role of the proportional-integral-derivative (PID) control strategy that uses the IR reflectance sensor readings to compute the appropriate speeds of both motors to ensure that the robot follows the black line is presented. Advanced courses that cover the topic on automatic control include ELEE 3321 Signals and Systems and ELEE 4321 Automatic Control. Advanced topics in control design are offered in ELEE 4323 Rapid Control Prototyping that is offered as a technical elective course. Applications to robotics are covered in ELEE 4325 Introduction to Robotics that is also offered as a technical elective course.
- *Computer Programming and the Use of the Development Environment Atmel Studio*: The introductory course does not provide formal teaching of computer programming using the C++ language. However, we guide our students through a set of samples of 3pi robot programs to give them a basic, yet relevant understanding of the robot programming as explained in section "Preparing Students for the Robotics Project with Experimental Learning". We use a step-by-step tutorial and computer screenshots to teach students how to program the robot using the development environment Atmel Studio. Students in our engineering program learn the C++ programming language in CSCI 1380 Computer Science I and the MATLAB language in ELEE 2319 Numerical Computation.

Preparing Students for the Robotics Project with Experiential Learning

Most of the first-year college students enrolled in our engineering program have not taken their first course in C/C++ programming or have not had a background in computer programming in their secondary education. Therefore, preparing students for the robotic project using a traditional lecture-based pedagogy, and within the course time constraint, presents a significant challenge. This is why we opted for an experiential learning model to actively engage students in hands-on learning that progressively immerse them in programming the robot with incremental levels of challenge. Our active experiential learning approach follows three phases. In the first phase, a series of demonstrations of code examples range from the basic black line following tasks to complex robot tasks such as robot navigation of a complicated path to solve a maze. Through this direct experience, students learn the capabilities of the robot and become better positioned to understand the commented source codes of the examples that they have just experienced. After reflecting on the experience with the provided robot application examples, students move to the second phase of active learning by working on a series of mini-challenges. In this second phase, students learn how to write short programs that allow a robot to stop at junctions on a navigation track and make left and right turns. Finally, the third phase of the project is the main challenge. In this phase, we ask students to solve the challenge by designing and writing a program that allows the robot to autonomously navigate through a mock-urban course containing traffic obstacles. In this last phase, students solve the robot challenge on their own and the instructor provides minimal support.

Several engineering programs (Conger et al. 2010; Hajshirmohammadi 2017) have promoted learning through experience following the Kolb's model (Kolb 1984). Experiential learning is useful in meeting the needs of the 21st-century industries that seek innovative and creative engineers to help them keep their competitiveness. Courses that integrate experiential learning help students transition from academia to the workplace where problems are often complex and require multiple design iterations (Regev et al. 2008). Our goal is for the first-year college students to gain new content knowledge and develop their problem-solving skills through an active experience working with robots. The experiential learning model actively involves students in their learning. Also, the experiential learning approach is likely to help students in solving complex problems (Bernik and Žnidaršič 2012).

The basic experiential learning model follows three phases:

- *Experience*: In this phase, students perform an activity.
- *Reflection*: In this phase, students reflect on the experience and develop an understanding of the result of the experience or activity.
- *Generalization*: In this phase, students apply what they learned to solve new problems.

Figure 12.3 shows the experiential learning model used in our engineering course. The learning model aims at introducing students to robot programming through a series of demos and mini-challenges. We specifically designed the tutorials and mini-challenges to prepare the student for a culminating project where we ask them to demonstrate their understanding of robotic programming and their problem-solving skills.

In the next sections, we describe the various demos and mini-challenges designed to prepare students for the culminating robotic project.

Line Following Example

The line following example project was provided by Pololu (Pololu Corporation 2014). The application shows how to program the 3pi robot to follow a black line on a white background. For students to experience this application, we built the line following course shown in Fig. 12.4. We used black electric tape measuring ¾ inch to trace the path line on a white poster board. We guided the students through multiple steps to compile and download the line following program to the robot. After experiencing the demo, students reflected on how the robot followed the black line

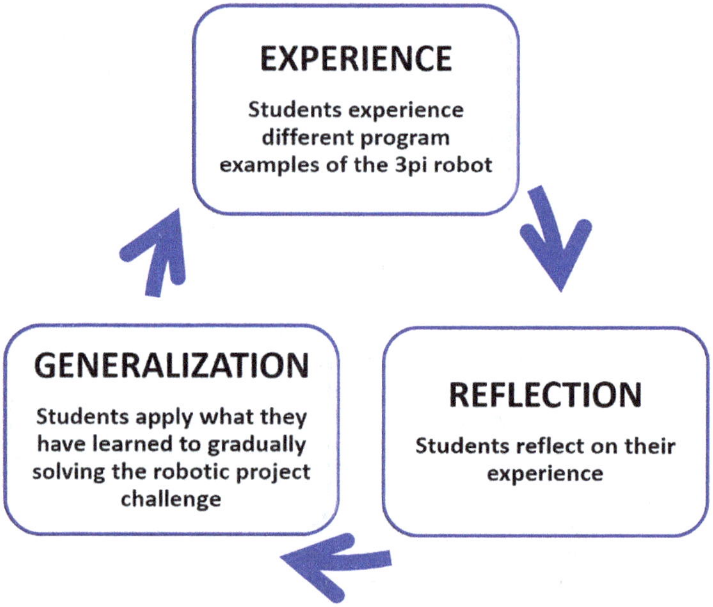

Fig. 12.3 Experiential learning process to help students solve the course robotics project

Fig. 12.4 Line following course

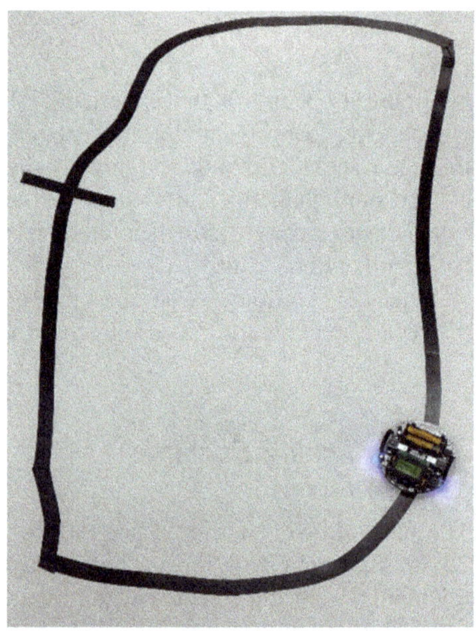

without veering off at the junction. We conducted class discussions to help students understand the role of the IR reflectance sensors further and how the PID controller uses the readings of the sensors to calculate the speeds of the motors. The students learned the controller is responsible for making the robot navigate along the black line. The line following project was easy to understand and presented a good first introduction to robot programming. We then asked students to read and study the code of the line following application and reflect on the sections that deal with IR sensor reading.

Maze Solver Example

The next example that students experienced was the maze solver project (Pololu Corporation 2014). In this project, the 3pi robot navigates complicated paths of intersecting black lines and can make sharp turns and 180-degree turns at dead ends. The task of the robot is to navigate this complicated track (the maze) from a starting location to a goal destination represented by a black circle (Fig. 12.5). The robot navigates the maze and memorizes the paths that lead to dead ends. It continues its search till it reaches the goal location. After the robot finishes exploring the maze and reaches the goal destination, it memorizes the shortest path to the destination. Students run the robot again to verify if it navigates to the goal destination following the fastest and shortest path. The maze solver project provides students with a further understanding of how IR sensors allow detecting intersections and dead ends. Students study the maze code to learn how to control the robot motors to make left, right, and 180-degree turns. This understanding is useful to tackle the next three mini-challenges and the project main challenge.

Fig. 12.5 Maze course for the 3pi robot (Pololu Corporation 2014)

Mini-Challenge 1

In this mini-challenge, we provide students the simple track shown in Fig. 12.6. We ask them to write a program that makes the robot navigate from a starting location to a dead-end destination where the robot must stop and display "Stopped." Also, the robot must detect the junction and display "Junction" on the LCD without stopping.

This mini-challenge guides students to go through all steps of writing, compiling, and downloading a C++ program to the robot microprocessor using the development environment Atmel Studio. We provide hints and important sections of the code to students to help them solve the mini-challenge. The hints provided are reported in Table 12.1. Students could recognize most of the code because of the reflective observation phase that they experienced during the line following and the maze solver examples.

Mini-Challenge 2

This mini-challenge asks students to write a program that makes the robot navigate the course shown in Fig. 12.7. The robot must start from location "S," follow the path marked by the solid arrows, and then stops at location "D." The robot must display "Junction" at location "C" and "Dead End" at location "D." We provide a sequence of hints, as shown in Figs. 12.8, 12.9, 12.10, and 12.11, to help students solve the challenge.

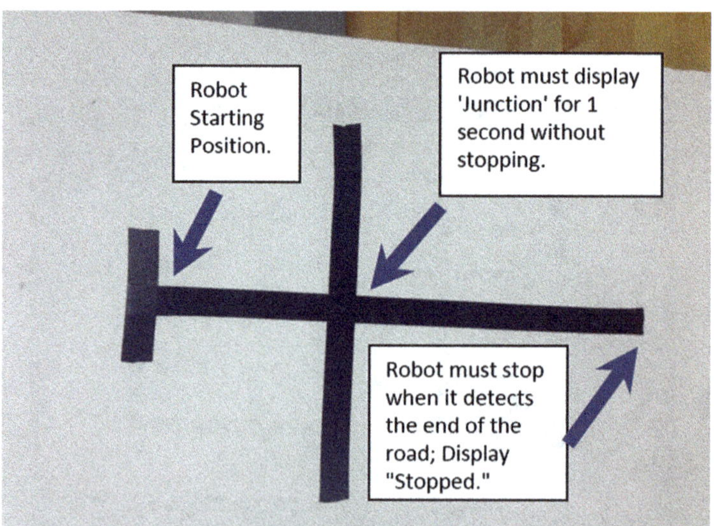

Fig. 12.6 Track of mini-challenge 1

Table 12.1 Explanation of the main statements of the application code for mini-challenge 1

Program functions and constructs	Explanation and comments
`initialize();`	This function initializes the 3pi robot, displays the welcome message on the LCD, and calibrates the reflectance sensors
`follow_segment();`	This function drives the 3pi robot straight along the black line
`set_motors(40,40);`	This function sets the speeds of the left and right motors of the 3pi robot
`delay_ms(50);`	This function allows the 3pi to keep on moving for a certain amount of time. In this example, the delay duration is set for 50ms
`unsigned int sensors[5];`	This statement declares the name of the array, "sensors," that will hold the readings of the 5 IR sensors
`read_line(sensors, IR_EMITTERS_ON);`	This function reads/updates the values received from the IR sensors and saves them in the array variable sensors

`if(sensors[0] > 100 ||`
` sensors[4] > 100)`
`{`
`clear();`
`print("Junction");`
`set_motors(40,40);`
`delay_ms(1000);`
`clear();`
`}`
	This If-construct checks if either the leftmost sensor (Sensor 0) of the rightmost sensor (Sensor 4) detects a black line. If they do, then this indicates that the 3pi robot has reached a junction. In this situation, (1) the LCD is cleared and the message "Junction" is displayed, and (2) the robot is allowed to keep on moving straight past the junction

`If(sensors[1]<100 &&`
` sensors[2]<100 &&`
` sensors[3]<100)`
`{`
`set_motors(0,0);`
`clear();`
`print("Stop");`
`break;`
`}`
	This If-constructs checks if the three middle IR sensors (sensors 1, 2, and 3) detect white surface ahead. If they do, then (1) the 3pi is stopped by setting the speed of both motors to 0, (2) the message "Stop" is displayed on the LCD, and (3) a break from the program loop is executed

Mini-Challenge 3

This mini-challenge asks students to extend the solution to mini-challenge 2 to make the robot take a 180-degree turn once it arrives at location "D." The robot must then continue to navigate the track back to location S following the path marked by dashed arrows as shown in Fig. 12.12. Once it arrives at location S, the robot must stop and display "Returned Home." In this mini-challenge, we provide minimal support to students and no extra hints. All student teams could complete this mini-challenge, demonstrating the effectiveness of the experiential learning model.

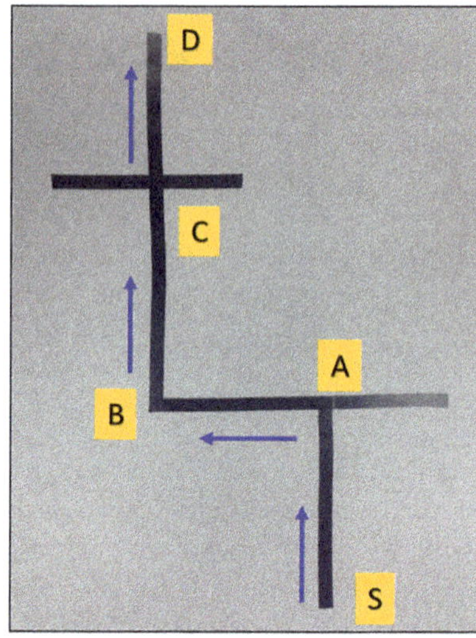

Fig. 12.7 Track for mini-challenge 2

Code

```
/* Code that detects T intersection (Location A) and makes the robot
turn "Left" at that intersection  */
   set_motors(50,50);
   read_line(sensors,IR_EMITTERS_ON);
   if(sensors[0] > 100)  /* Test if left most sensor detects black */
      found_left = 1;
   if(sensors[4] > 100)  /* Test if right most sensor detects black */
      found_right = 1;
/* Aligns wheels at an intersection if we wish to turn. */
   delay_ms(180);
/* Reads inside sensor data to determine if the path continues straight
ahead. */
   read_line(sensors,IR_EMITTERS_ON);
/* Test if any of the three inner sensors are detecting black */
   if(sensors[1] > 100 || sensors[2] > 100 || sensors[3] > 100)
      found_straight = 1;
/* If we arrived at our first intersection, we make a left turn. */
   if(found_left == 1 && found_straight == 0 && found_right == 1)
      turn('L');
```

T Intersection

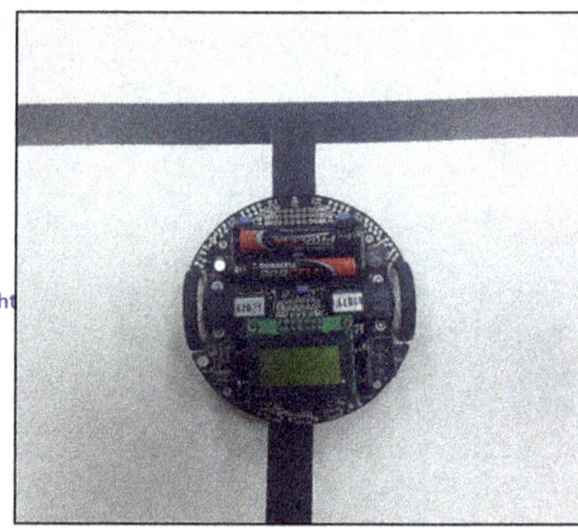

Fig. 12.8 Hint 1 for mini-challenge 2: detecting a T junction and taking a left turn

Code

```
/*  Code that detects L intersection (Location B) and makes the robot
turn "Right" at that intersection. */
   set_motors(50,50);
   read_line(sensors,IR_EMITTERS_ON);
   if(sensors[0] > 100)  /* Test if left most sensor detects black */
      found_left = 1;
   if(sensors[4] > 100)  /* Test if right most sensor detects black */
      found_right = 1;
/* Aligns wheels at an intersection if we wish to turn. */
   delay_ms(180);
/* Reads inside sensor data to determine if the path continues straight
ahead. */
   read_line(sensors,IR_EMITTERS_ON);
/* Test if any of the three inner sensors are detecting black */
   if(sensors[1] > 100 || sensors[2] > 100 || sensors[3] > 100)
      found_straight = 1;
/* If we arrived at the second intersection, we make a right turn. */
   if(found_left == 0 && found_straight == 0 && found_right == 1)
      turn('R');
```

L Intersection

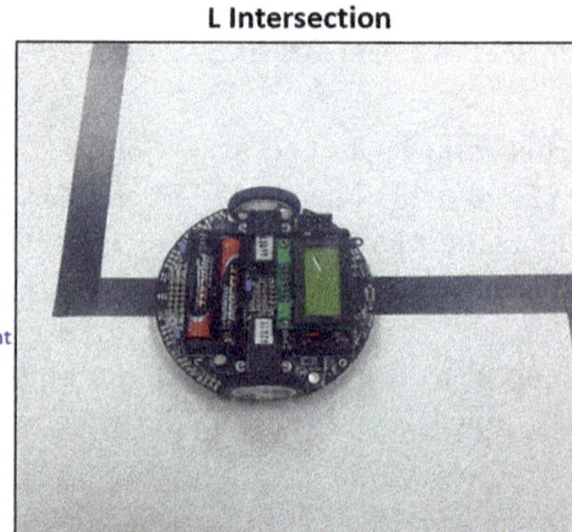

Fig. 12.9 Hint 2 for mini-challenge 2: detecting an L intersection and making a right turn

Code

```
/* This code detects + intersection (Location C) and makes the robot
move "Straight" past that intersection. */
   set_motors(50,50);
   read_line(sensors,IR_EMITTERS_ON);
   if(sensors[0] > 100)  /* Test if left most sensor detects black */
      found_left = 1;
   if(sensors[4] > 100)  /* Test if right most sensor detects black */
      found_right = 1;
/* Aligns wheels at an intersection if we wish to turn. */
   delay_ms(180);
/* Reads inside sensor data to determine if the path continues
straight ahead. */
   read_line(sensors,IR_EMITTERS_ON);
/* Test if any of the three inner sensors are detecting black */
   if(sensors[1] > 100 || sensors[2] > 100 || sensors[3] > 100)
      found_straight = 1;
/* If we arrived at the third intersection, we make a continue straight
ahead and
   display "JUNCTION". */
   if(found_left == 1 && found_straight == 1 && found_right == 1)
   {
      turn('S');
      print("JUNCTION");
   }
```

+ Intersection

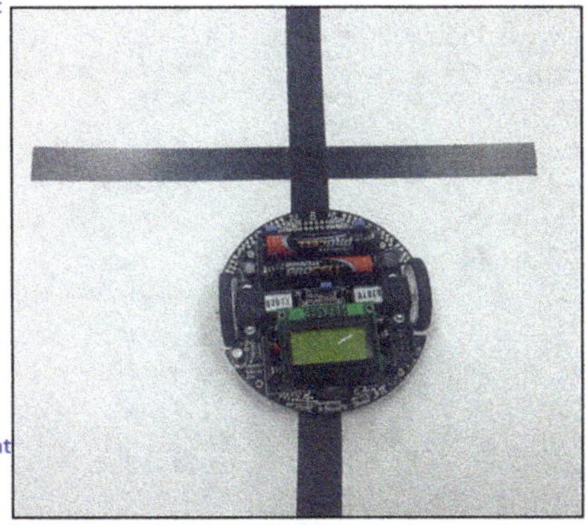

Fig. 12.10 Hint 3 for mini-challenge 2: detecting a + intersection and moving straight

Code

```
/* This code detects the Dead End (Location D) and makes the robot
stop at that location.*/
   set_motors(50,50);
   read_line(sensors,IR_EMITTERS_ON);
   if(sensors[0] > 100)  /* Test if left most sensor detects black */
      found_left = 1;
   if(sensors[4] > 100)  /* Test if right most sensor detects black */
      found_right = 1;
/* Aligns wheels at an intersection if we wish to turn. */
   delay_ms(180);
/* Reads inside sensor data to determine if the path continues straight
ahead. */
   read_line(sensors,IR_EMITTERS_ON);
/* Test if any of the three inner sensors are detecting black */
   if(sensors[1] > 100 || sensors[2] > 100 || sensors[3] > 100)
      found_straight = 1;
/* If we arrived at the last intersection, we make a hard stop and
display "DEAD END". */
   if(found_left == 0 && found_right == 0 && found_straight == 0)
   {
      set_motors(0,0);
      clear();
      print("DEAD END");
   }
```

Dead End

Fig. 12.11 Hint 4 for mini-challenge 2: detecting a dead end and stopping

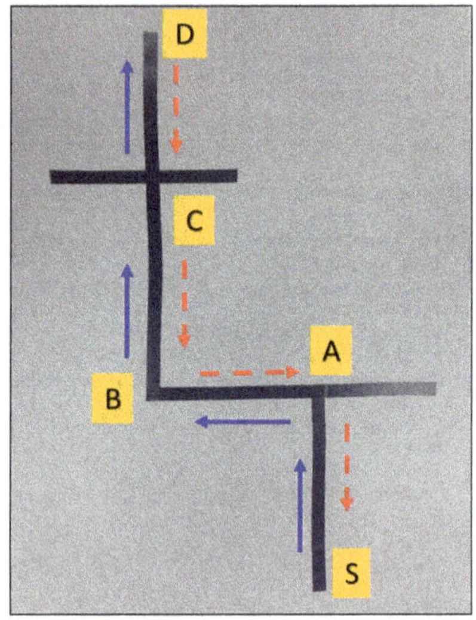

Fig. 12.12 Track for mini-challenge 3: dashed arrows indicate the return path of the robot

Main Robotics Project

Problem Statement

The culminating challenge was presented to students as follows: "Smart Highway Corporation (SHC) hires you to join a team of engineers who are working on developing prototypes of the next generation cars with auto-navigation powers. Your assignment is to develop a proof of concept where you will develop a software application for the robot to autonomously navigate a mock-urban course with traffic obstacles (Fig. 12.13)."

The detailed instructions given to students were as follows:

1. The robot must be placed at the START position before it starts moving through traffic following the path shown with arrows. The case where all reflective sensors are outside the START black position is an acceptable starting position.
2. The robot must stop at the STOP sign for 3 s and display STOP on its LCD.
3. The robot must stop at the traffic light for pedestrians for 3 s and display RED on its LCD. The message must then change to GREEN right before the robot starts moving again.
4. The robot must display RAILWAY CROSSING at the railway crossing sign, but without stopping.
5. The robot must stop inside the FINISH location and display FINISH on its LCD.

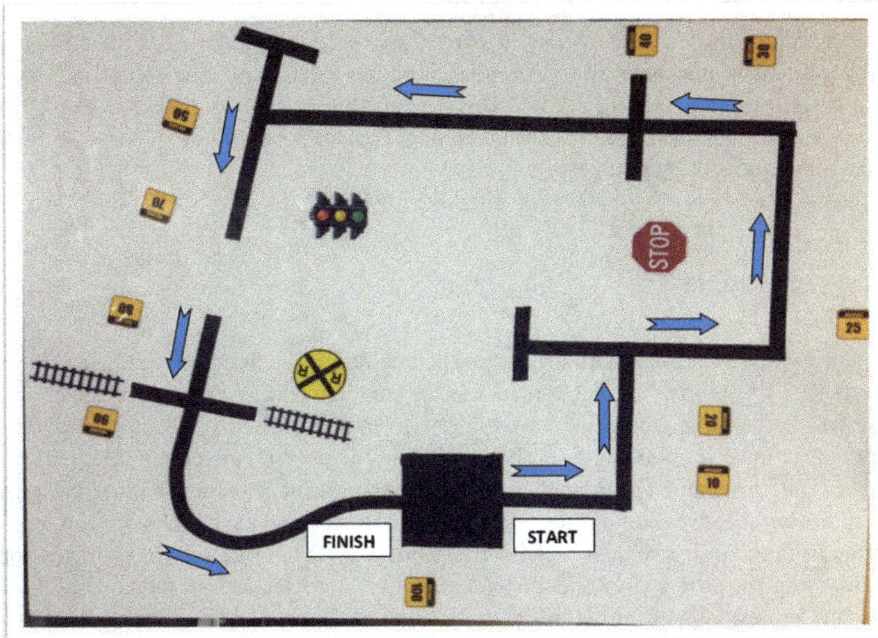

Fig. 12.13 Track for the main robotics challenge

Project Requirements and Scoring

Besides the two weeks reserved for the demo examples and mini-challenges, we assigned two laboratory sessions for about 5 h to allow students to design, develop, implement, and test their programs. During the third lab period, students were given the chance to make any final program adjustments and to conduct further testing before the official runs. Each team was allowed two possible runs, and the highest score achieved in the two runs is counted. The challenge scoring was based on the miles traveled by the robot as shown in Fig. 12.13. For instance, if the robot reaches the junction with the STOP sign, then the score obtained is 40 points (matching to 40 miles traveled). Therefore, a robot that reaches the FINISH line following the marked path will receive 100 points. If the robot fails to stop or display the correct message, then a 5-point deduction is applied for each instance.

We also ask each team to prepare a report that describes their step-by-step solution to the challenge. While there are of the track that students have seen in previous examples and mini-challenges, new difficulties were introduced in the main project. These difficulties consist of:

- The junction at the traffic light is not a perfect "T" junction.
- There is a white gap in the robot path before the railway crossing.
- The robot must stop at the destination location inside the black square.

We purposely introduced the new difficulties to test students' abilities to generalize the ideas they have learned following the experiential learning. About 80% of student teams managed to solve the complete challenge, demonstrating their improved problem-solving skills. We consider this to be remarkable because the first-year students had not yet taken formal computer programming courses or fundamental engineering courses.

Project Impact

We introduced the Pololu 3pi project in the course for the first time in Spring 2013. Figure 12.14 reports the total percentages of students who registered for the course but who either failed, dropped, or withdrew from the university. The historical data show that since introducing the robot project in the course curriculum, the percentage of fail, drop, and withdrawal has decreased for the next six academic semesters. While several causes could affect student performance and the need to drop or withdraw from the university, we believe the robot project has had a positive impact on motivating the first-year students. We offer only one section of the course every Spring semester, with an average enrollment of 20 students. The Fall semester enrollment is higher. In fact, we have been offering two sections of the course every Fall semester since 2013, with an average total enrollment of 50 students. In Spring 2013, we had only 13 students enrolled in the course and that could explain the 0% of fail, drop, and withdrawal. We believe the smaller class size combined with the exciting new robot project had a positive impact on student performance.

We have increased the complexity of the robot project since introducing it in the course curriculum. We developed extra mini-challenge materials to guide the stu-

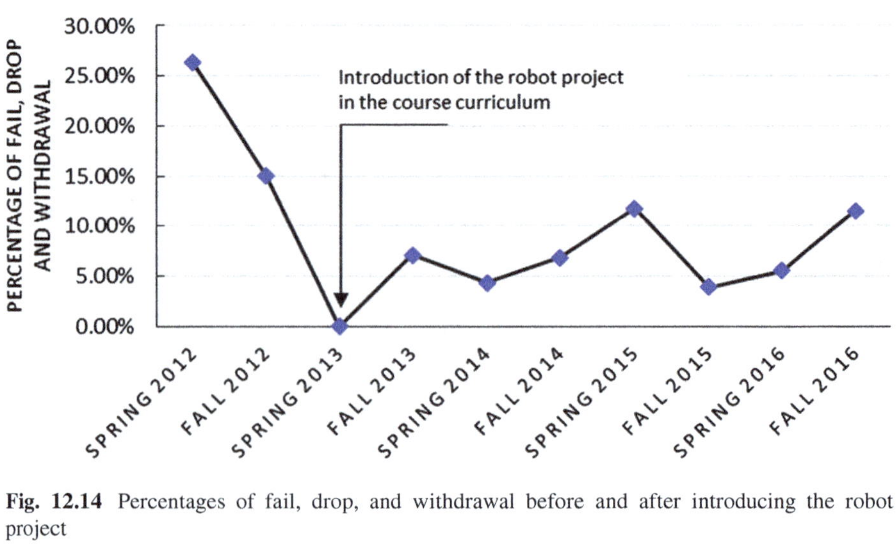

Fig. 12.14 Percentages of fail, drop, and withdrawal before and after introducing the robot project

dents to solve the main project successfully as discussed in sections "Preparing Students for the Robotics Project with Experimental Learning" and "Main Robotics Project". Even though we increased the complexity of the project challenge, the percentage of drop, fail, and withdrawal has remained below 15% as shown in Fig. 12.14.

Conclusion

This chapter described the curriculum of an introductory course for first-year students in the electrical engineering program at the University of Texas Rio Grande Valley. Besides basic laboratory experiments that introduce students to basic concepts of electrical circuits and digital systems, the course robot project gives students a glimpse into the different technical subjects within the electrical engineering field. Although most of our first-year students did not have a background in robot programming, we could guide them to write C++ programs that solved complex robot navigation challenges. The limited time reserved for the project motivated us to adopt the experiential learning model. We use this pedagogy to introduce students to robot programming through a series of demos and mini-challenges. The course main project allowed the students to demonstrate their understanding of robot programming and their abilities to solve problems. Students were enthusiastic and seemed motivated while working with the robots. We changed the time assigned for the project so students had a chance to complete their work. In the latest course offerings, we assigned four to five weeks to allow students to experience the demo examples, the mini-challenges, and the main project.

Often, students declare an engineering major without having an understanding of what that major entails or what their career goals are. This lack of understanding combined with the absence of proper mentoring results in a large percentage of students who either switch to other majors or drop out of college without receiving a degree. We presented data to show that an exciting robot project could contribute to keeping first-year students in the program. Robots provide a unique experience for first-year students to learn what engineering is about. Using robots in the classroom helps to support student learning and motivation – two important factors that affect student retention.

Acknowledgments The author would like to thank the following graduate students for their help in the laboratory: Jose Escobar, David Flores, Edgar Gil, Eric Gonzalez, Hugo Herrera, Everardo Ibarra, Christopher Martin, and Hector Trevino.

References

Aroca, R. V., Watanabe, F. Y., Ávila, M. T. D., & Hernandes, A. C. (2016). Mobile robotics integration in introductory undergraduate engineering courses. *2016 XIII Latin American robotics symposium and IV Brazilian robotics symposium (LARS/SBR)* (pp. 139–144). 8–12 Oct. 2016.

Behrens, A., Atorf, L., Schwann, R., Neumann, B., Schnitzler, R., Balle, J., et al. (2010). MATLAB meets LEGO mindstorms: A freshman introduction course into practical engineering. *IEEE Transactions on Education, 53*(2), 306–317. https://doi.org/10.1109/TE.2009.2017272.

Bernik, M., & Žnidaršič, J. (2012). Solving complex problems with help of experiential learning.

Blanding, W., & Meah, K. (2014). Laboratory-based project-oriented introductory course for electrical engineering. *8th international conference on electrical and computer engineering*. 20–22 Dec. 2014. pp. 832–835.

Carley, L. R., Khosla, P., & Unetich, R. (2000). Teaching "introduction to electrical and computer engineering" in context. *Proceedings of the IEEE, 88*(1), 8–22. https://doi.org/10.1109/5.811595.

Conger, A. J., Gilchrist, B., Holloway, J. P., Huang-Saad, A., Sick, V., & Zurbuchen, T. H. (2010). Experiential learning programs for the future of engineering education. *2010 IEEE transforming engineering education: Creating interdisciplinary skills for complex global environments* (pp. 1–14). 6–9 April 2010.

Hajshirmohammadi, A. (2017). Incorporating experiential learning in engineering courses. *IEEE Communications Magazine, 55*(11), 166–169. https://doi.org/10.1109/MCOM.2017.1700373.

Huettel, L. G., Brown, A. S., Coonley, K. D., Gustafson, M. R., Kim, J., Ybarra, G. A., et al. (2007). Fundamentals of ECE: A rigorous, integrated introduction to electrical and computer engineering. *IEEE Transactions on Education, 50*(3), 174–181. https://doi.org/10.1109/TE.2007.900020.

Karam, L. J., & Mounsef, N. (2011). Increasing retention through introduction to engineering design. *2011 digital signal processing and signal processing education meeting (DSP/SPE)* (pp. 186–191). 4–7 Jan. 2011.

Kolb, D. A. (1984). *Experiential learning: Experiences as the source of learning and development.* New Jersey: Prentice-Hall.

Mercede, F. J. (2008). Hands-on projects to introduce electrical and computer engineering. *2008 38th annual frontiers in education conference* (pp. F1F-9-F1F-14). 22–25 Oct. 2008.

Mix, D. F., & Balda, J. C. (1997). ELEG 1003-introduction to electrical engineering: An approach to motivate and teach EE Freshmen. *Proceedings frontiers in education 1997 27th annual conference. Teaching and learning in an era of change* (pp. 1215–1218, vol. 3). 5–8 Nov 1997.

Pololu Corporation. (2014). Polulu 3pi robot user's guide. Available at: https://www.pololu.com/docs/pdf/0J21/3pi.pdf.

Regev, G., Gause, D. C., & Wegmann, A. (2008). Requirements engineering education in the 21st century, an experiential learning approach. *2008 16th IEEE international requirements engineering conference* pp. 85–94). 8–12 Sept. 2008.

Roth, Z. S. (2002). The role of robotics in freshmen engineering curricula. *Proceedings of the 5th biannual world automation congress* (pp. 389–394). 9–13 June 2002.

Williams, D. B. (2009) A Freshman 'Introduction to ECE Design' course based on the LEGO mindstorms NXT. *Digital signal processing workshop and 5th IEEE signal processing education workshop,* 2009. *DSP/SPE* 2009. *IEEE 13th* (pp. 435–439). 4–7 Jan. 2009.

Chapter 13
Designing a Competition Robot as a Capstone Project for Electrical and Computer Engineering Students

Samuel Roberts, Joshua Acosta, Salvador Garza, Mounir Ben Ghalia, and Heinrich Foltz

Abstract Robotics-based capstone design projects provide unique educational opportunities for engineering students and help prepare them as future professional engineers. Robotics projects allow students to experience all steps of the engineering design cycle and to extend their knowledge in a wide range of subjects across multiple disciplines that include electrical, mechanical, and computer engineering. Designing and building robots for capstone design projects support a number of student learning outcomes. These include (i) the ability to apply engineering design to produce solutions that meet specified needs; (ii) the ability to function effectively on a team whose members together provide leadership, create a collaborative and inclusive environment, establish goals, plan tasks, and meet objectives; (iii) the ability to communicate effectively with a range of audiences; and (iv) the ability to create and use software both as an analysis and design tool and as part of systems containing hardware and software. Over the last 15 years, several engineering student teams from our robotics lab built mobile robots and competed in the annual robotics event organized by Region 5 (Institute of Electrical and Electronics Engineers [IEEE] is the world's largest technical professional organization dedicated to advancing technology for the benefit of humanity (https://www.ieee.org). IEEE is organized into ten regions worldwide. Region 5 is comprised of states within the southwestern region of the United States that include Arkansas, Colorado, Kansas, Illinois (southern), Louisiana, Missouri, Nebraska (western), New Mexico (southern), Oklahoma, South Dakota (western), Texas, and Wyoming (eastern) (http://ieeer5.org)) of the Institute of Electrical and Electronics Engineers, one of the world's largest technical professional organizations. Every year, the competition theme is different and the event hosts more than 30 student teams from different universities from Region 5. Robots designed for the competition are expected to autonomously complete a specified number of tasks on a playing field. This chapter

S. Roberts · J. Acosta · S. Garza · M. Ben Ghalia (✉) · H. Foltz
Department of Electrical Engineering, University of Texas Rio Grande Valley, Edinburg, TX, USA
e-mail: mounir.benghalia@utrgv.edu

L. Daniela (ed.), *Smart Learning with Educational Robotics*,
https://doi.org/10.1007/978-3-030-19913-5_13

presents an overview of our robotics lab educational program and chronicles the experience of a team of engineering students who built a mobile robot as part of their capstone design project for the 2018 robotics competition. The educational values gained from working on a robotics-themed capstone project and the lessons learned from the participation in the robotics competition are discussed.

Keywords Educational robotics · Robotics competition · Autonomous robots · Capstone design · Engineering design process · Engineering education

Introduction

Robots have become an indispensable part of making several industries, worldwide, competitive (Bekey and Yuh 2008). The word *robot* originated from *robota* which in Czech means *forced labor* or *servitude* and was the main theme in Karel Capek's 1921 play titled R.U.R, or *Rossum's Universal Robots* (Niku 2011). The play introduced the concept of manufacturing robots to do work. In the industrial sector, General Motors pioneered the efforts in manufacturing robots and installed the first industrial robot, *Unimate*, in one of its assembly lines over five decades ago (Engelberger 1980). Rapid advances in technology have made it possible for robots to cross over from automotive manufacturing to other applications such as healthcare, space exploration, and research and rescue. Technological progress made in computing, data communication, and computer vision has led to the development of surgical robots such as the *da Vinci* surgical robotic system that assists surgeons in performing minimally invasive surgeries, resulting in less trauma to patients and faster recovery times (Lanfranco et al. 2004). While industrial manipulators and surgical robots often work in a fixed environment and a constrained workspace, free-roaming mobile robots have led to new specialized applications. The National Aeronautics and Space Administration (NASA) developed several autonomous mobile robots such as the *Sojourner* rover and the twin Mars exploration rovers, *Spirit* and *Opportunity*, that successfully landed on Mars and contributed to the exploration of the planet (Bajracharya et al. 2008). Mobile robots have found important applications in search and rescue missions (Murphy 2012).

The market for robotics has been growing rapidly. Industrial robot sales is expected to increase worldwide at an annual rate of 14% for the period 2019–2021, with an estimated 2.1 million new industrial robots to be installed in factories in Europe, North and South America, and Asia (International Federation of Robotics 2018a). Professional service robots are seeing continuous growth that is estimated at 19% annually for the period 2019–2021 (International Federation of Robotics 2018b). The projected 5-year sales of service robots intended for domestic use and those marketed for entertainment are expected to increase at the annual rate of 31% and 12%, respectively (International Federation of Robotics 2018b).

The fast growing market for robots worldwide and the new robotics industries that are being created have led to an increased supply of jobs (World Economic Forum 2018). This opportunity, however, presents a challenge for these new specialized robotics industries that need to fill these positions with new types of skilled workforce in order to sustain their growth and remain competitive. In the United States, the Science and Engineering (S&E) proportion of all university undergraduate degrees has been decreasing (President's Council of Advisors on Science and Technology (PCAST) 2012). In order to meet the projected economic needs for S&E professionals, it is estimated that the number of students graduating with S&E degrees in the United States will need to increase by about 34% annually (President's Council of Advisors on Science and Technology (PCAST) 2012). Hence, to remain the global leader in technology innovations, the United States must produce and retain a higher number of S&E talent (National Science Board 2016). Robotics is an interdisciplinary field of S&E that includes electrical and electronics engineering, mechanical engineering, computer engineering, and computer science. Hence, it is important for strategies to be developed to transform S&E education to train future engineers and scientists in robotics in order to guarantee a continuing future supply of a skilled workforce to meet the increasing number of job opportunities in the robotics sector. Universities and education agencies play an important role in preparing the needed talent to meet the demand of the technology industries. A comprehensive and cohesive educational program that engages students in S&E must become the top mission of educators. Robotics presents a unique opportunity to motivate students in kindergarten (Di Lieto et al. 2017), primary schools (Scaradozzi et al. 2015), secondary schools (Cesaretti et al. 2017), and college (Yilmaz et al. 2013) to pursue S&E careers. Several robotics competitions are being held worldwide for students at different learning stages with the underlying goal to inspire and motivate students to pursue S&E studies and careers (Eguchi 2016; FIRST Robotics 2015).

This chapter presents initiatives taken at the University of Texas Rio Grande Valley (located in South Texas, USA) robotics laboratory to educate and train engineering students in the area of robotics. It chronicles the experience of a team of engineering students who built a mobile robot as part of their capstone design project for a regional robotics competition held in Spring 2018.

The main questions addressed in this chapter are:

- What key benefits and educational values do robotics competitions have on the professional preparation of engineering students?
- How does the integration of robotics competition into a capstone design course help motivate students to develop advanced solutions for their design problems?

The chapter is organized as follows. In section "Robotics in Education and Competitions," an overview of robotics in undergraduate education and the importance of competitions is presented. The section also highlights the robotics laboratory activities and participations in competitions. Section "The Chronicle of Designing a Competition Robot" chronicles the experience and participation of a team of our engineering students in the annual robotics competition organized by IEEE Region 5, one of the world's largest technical professional organizations. Section "Lessons Learned from Participation in Robotics Competitions" summarizes the educational values gained from working on a robotics-themed capstone project and the lessons learned from the participation in the robotics competitions. Finally, some concluding remarks are provided in section "Conclusions."

Robotics in Education and Competitions

Robotics in Undergraduate Education

Robotics education has been increasingly offered at various engineering and computer science programs around the world. To educate future robotics engineers and researchers, several robotics departments and programs have been established since the late 1970s. The first robotics department in any US university was the Robotics Institute at Carnegie Mellon University which was established in 1979 (Robotics Institute 2018). Its mission was to conduct robotics research in a wide variety of applications. Training graduate students as future roboticists has been an integral part of the institute's mission. In Japan, one of the first Department of Robotics was established at Ritsumeikan University in 1996 (Nagai 2001). Their robotics educational program offered several courses in the areas of robot hardware and software and human-machine interfacing. Robotics could be taught in a traditional method where lectures are augmented with laboratory experiments (Berry 2017). However, because robotics is an interdisciplinary field that includes electrical and electronics engineering, mechanical engineering, computer engineering, and computer science, several programs have adopted design courses for their robotics program (Piepmeier et al. 2003; Tur and Pfeiffer 2006; Bruder and Wedeward 2003; Jung 2013). In these design courses, students are given open-ended problems and are required to design their own robots. This practice exposes students to real-world problems and helps them develop advanced skills in robotics.

In our engineering program, the course ELEE 4325 *Introduction to Robotics* has been offered as an elective for senior students. The initial offering of the course focused primarily on industrial robot manipulators. In this course, students learn the kinematics and dynamics of a 6 degree of freedom (DOF) industrial robot arm, conduct computer simulations, and carry out experiments on an open-architecture 6 DOF robot arm located in our laboratory. However, there has been increasing interest by our students to learn about mobile robots. Designing and building autonomous robots requires funds and time. One way to make this possible in our undergraduate curriculum is to give students the opportunity to experience all the steps of designing and building a mobile robot as part of their capstone design project.

Participation of our Robotics Laboratory in Competitions

The Accreditation Board of Engineering and Technology[1] (ABET) that oversees the accreditation of engineering programs requires that an engineering curriculum include a culminating major engineering design experience that (1) incorporates

[1]ABET accredits postsecondary education programs in computing, engineering, and engineering technology in 32 countries. The accreditation of these programs occurs mainly in the United States (https://www.abet.org).

appropriate engineering standards and multiple constraints and (2) is based on the knowledge and skills acquired in earlier coursework (ABET 2018). Designing and building robots for capstone design projects support a number of student learning outcomes specified by ABET. These include[2]:

- The ability to apply engineering design to produce solutions that meet specified needs with consideration of public health, safety, and welfare, as well as global, cultural, social, environmental, and economic factors
- The ability to function effectively on a team whose members together provide leadership, create a collaborative and inclusive environment, establish goals, plan tasks, and meet objectives
- The ability to communicate effectively with a range of audiences
- The ability to create and use software both as an analysis and design tool and as part of systems containing hardware and software

Fifteen years ago, we decided to include the annual challenge proposed by the IEEE Region 5 Robotics Competition as one of the capstone design projects for our senior students. Students who are interested in robotics are given the opportunity not only to design and build their own robots but also participate in the regional robotics competition and compare their designs to those completed by students from other universities. Because the theme and requirements of the annual robotics competition change every year, this has allowed new student teams to come up with their own new designs. Several studies have emphasized the importance of design competitions as a tool for student learning (Bazylev et al. 2014; Michieletto and Pagello 2018; Murphy 2001; Chew et al. 2000; Kaiser and Troxell 2005). Competitions motivate students to excel in their designs and expose them to real-world scenarios where engineering companies strive to develop innovative new products and compete to outperform their rivals.

Our initial participations did not result in wins. However, our participations have allowed us to revise our strategy of mentoring students and to learn how to prepare for a competition. This learning curve has paid off. The record of our robotics laboratory in its participations in IEEE Region 5 robotics competitions includes:

- First place at the 2007 IEEE Region 5 Robotics Competition held in Fayetteville, Arkansas.
- Second place at the 2007 IEEE Region 5 Robotics Competition held in Fayetteville, Arkansas.
- First place at the 2008 IEEE Region 5 Robotics Competition held in Kansas City, Missouri.
- Fourth place at the 2018 IEEE Region 5 Robotics Competition held in Austin, Texas. This team also ranked first among the teams participating from Texas universities.

In the next section, we chronicle the robot design and participation in the robotics competition by our recent student robotics team.

[2]Based on the ABET revised student learning outcomes that become effective in the 2019–2020 cycle.

The Chronicle of Designing a Competition Robot

The Challenge

The 2018 robotics competition was held on April 7, 2018, in Austin, Texas, and the rules were made available in late August 2017 (IEEE R5 2018). The competition challenged student teams to build and program an autonomous robot to complete a set of tasks that involve picking up and placing a number of tokens on a playing field shown in Fig. 13.1. The field is made from two 4 ft × 8 ft sheets of plywood. The sheets are assembled together to create the 8 ft × 8 ft playing field. The black lines are 0.5 in wide and are intended to be used by the robot for navigation. The four-line boxes surrounding the center gray square are squares with sides that measure 2ft, 3 ft, 4 ft, and 5 ft.

The 24 gray disks represent shallow depressions that are cut into the playing surface. The diameter of each depression measures 1 in and its depth measures 1/16 in. Six colored squares (red (R), green (G), blue (B), cyan (C), magenta (M), and yellow (Y)) and two white squares (W) are located along the perimeter of the field. A gray square (GY) is located at the center of the field. Each square measures 1 ft × 1 ft.

Tokens to be picked up by the robot have a cylindrical shape (Fig. 13.2). The diameter of the base and the thickness of each token measure 1 in and 1/16 in, respectively. Tokens are made of magnetic steel so that they may be picked up using an electromagnet. Tokens are placed in the shallow circular depressions so that they are flush with the surface of the playing field. A total of 24 tokens are used in the competition. The top and the curved faces of all tokens are painted gray. The bottom face of 18 tokens are painted in red, green blue, cyan, magenta, and yellow (three each). The remaining 6 tokens are painted all gray.

The competition has three rounds with incremental difficulty and a possible tiebreaker.

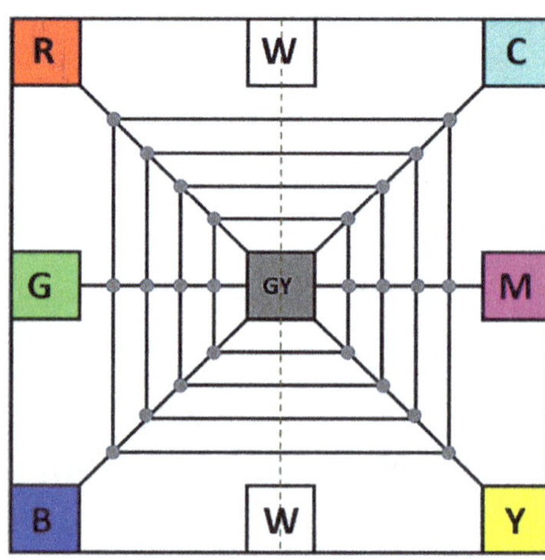

Fig. 13.1 Robot playing field layout. (IEEE R5 2018)

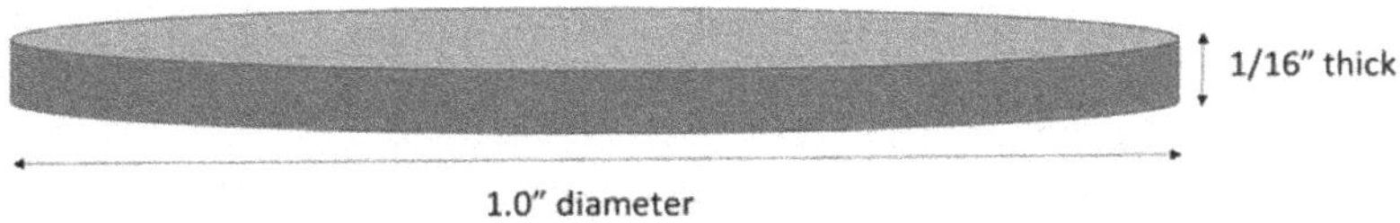

Fig. 13.2 Token. (IEEE R5 2018)

Round 1 Challenge and Scoring

Round 1 has a time limit of 5 minutes. In Round 1, a total of 12 tokens – 2 each of red, green, blue, cyan, magenta, and yellow – are placed on the field in the depressions with the colored face (front) down so that the color of the token cannot be seen without lifting the token. The 12 tokens are distributed in random order along the 2 ft line square and 4 ft line square, except that no token will be placed along the line from the center to its respective color square. The robot must start and end in one of the 1 ft × 1 ft white squares located on the edge of the playing field.

The challenge is for the robot to autonomously navigate the playing field, locate each of the 12 tokens, identify their color, and move the tokens and deposit them in the corresponding colored squares (i.e., the red token must be moved and placed in the red square). Each token correctly placed in its corresponding colored square scores 3 points. A token placed in the wrong colored square or in the center gray square scores only 1 point each. A bonus of 10 points is awarded if the robot successfully places all 12 tokens correctly in their corresponding squares. Another bonus score of 10 points is awarded if the robot successfully returns to one of the white squares. Hence, for Round 1 the maximum score awarded to a robot is 56 points.

Round 2 Challenge and Scoring

Round 2 has a time limit of 6 minutes. In Round 2, a total of 18 tokens – 3 each of red, green, blue, cyan, magenta, and yellow – are placed on the field in the depressions with the colored face (front) down so that the color of the token cannot be seen without lifting the token. The 18 tokens are distributed in random order along the 2 ft line, 3 ft line square, and 4 ft line square, except that no token will be placed along the line from the center to its respective color square. The robot must start and end in one of the 1 ft × 1 ft white squares located on the edge of the playing field.

The challenge is the same as in Round 1; the robot has to navigate the playing field, pick up and identify the colors of the tokens, and deposit them in their corresponding colored squares. The scoring system is the same as that used in Round 1, except for the bonus points and the introduction of a penalty. In this round, the bonus for successfully placing all 18 tokens in their corresponding squares is worth 20 points and the bonus given for a robot that successfully returns to one of the white squares located on the edge of the field is worth only 5 points. The robot receives a penalty of -1 point if it places a token outside any of the colored squares or the center gray square. The maximum score for Round 2 is 79 points.

Round 3 Challenge and Scoring

Round 3 has a time limit of 8 minutes and the robot has to deal with a total of 24 tokens. In addition to the 18 colored tokens used in Round 2, 6 all gray tokens are introduced in Round 3. The 24 tokens are distributed in random order in the 24 depressions. The gray tokens have to be picked up, identified, and placed in the center gray square. While the scoring system is almost similar to Round 2, there are a few variations: (i) a token placed in the wrong square earns 0 points, (ii) the bonus for successfully placing all 24 tokens in their corresponding squares is increased to 30 points, and (iii) a new penalty of –3 points is applied if the robot does not return to one of the white squares. The maximum score for Round 3 is 102 points.

After completing the three rounds, the robots are ranked based on their cumulative scores. The robot with the highest cumulative score is declared the winner. Two additional awards are given to the second and third highest-scoring robots.

Tiebreaker Round

The tiebreaker round involves a total of 14 tokens: 2 each of red, green, blue, cyan, magenta, yellow, and gray tokens. The 14 tokens are placed randomly in 14 of the 24 depressions. The time limit for this tiebreaker round is 4 minutes. The scoring system for this round is simple. A robot scores 1 point for each token successfully placed in its corresponding colored square. A bonus of 1 points is awarded if the robot returns to one of the two white squares. Hence, the maximum points that can be awarded for the tiebreaker round is 15 points.

Designing and Building the Robot

The Engineering Design Cycle

In addition to the description of the competition rounds, there are specific constraints regarding the robot to be designed by the student teams. The robot must be completely autonomous, its dimensions must not exceed 11 in × 11 in × 11 in, and its weight must not exceed 40 pounds. In previous competitions, student teams were given the option to compete with a set of cooperative robots. However, in this competition a system of swarm or multi-robotic system was not allowed.

When designing a solution to a complex problem, engineering students are trained to follow the engineering design cycle which is an iterative process that consists of multiple steps, providing a systematic guidance to solving the problem. While variations of the engineering design cycle may contain different number of steps, they always start with the identification of the problem and end with a solution to the problem. The engineering design cycle used in the robotics design is shown in Fig. 13.3. It consists of the following steps: (1) study the competition rules

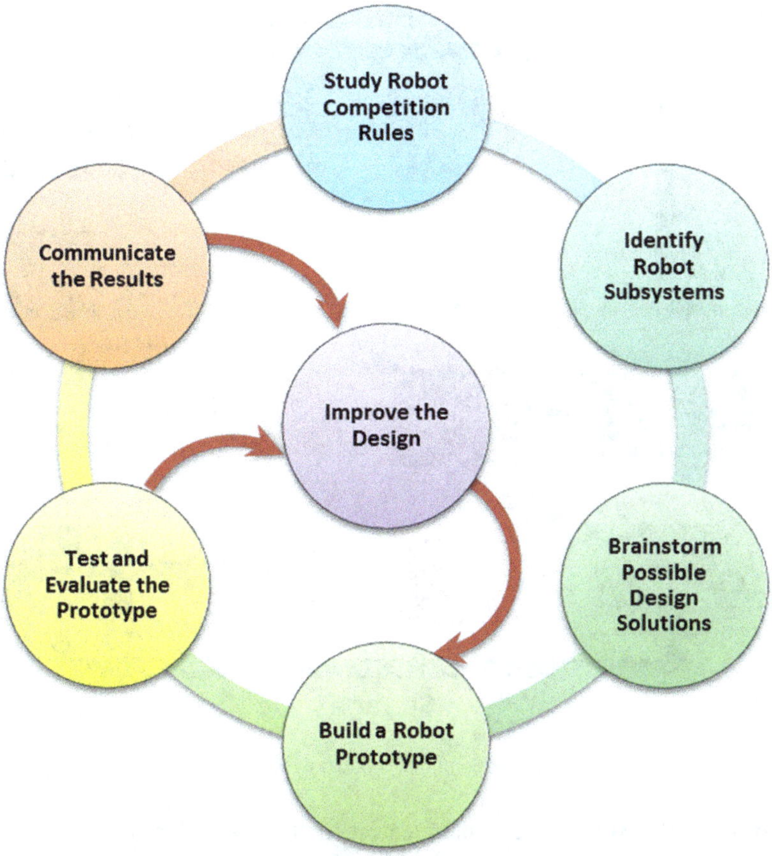

Fig. 13.3 Engineering design cycle for designing the robot

and the delineated requirements and rules; (2) identify all the subsystems needed to build the robot; (3) brainstorm possible design solutions; (4) build a robot prototype; (5) test and evaluate the prototype; (6) communicate the results; and (7) improve the design based on the test results and the feedback from the advisors.

Robot Subsystems

The design of the robot started in early September 2017. The following four main subsystems for the robot were identified: (1) navigation subsystem, (2) token retrieval and deposit subsystem, (3) token storage onboard of the robot chassis, and (4) color sensing subsystem. It was earlier on decided to use the Arduino microcontroller board because of its hardware and software capabilities. The overall block diagram of the mobile robot system is shown in Fig. 13.4.

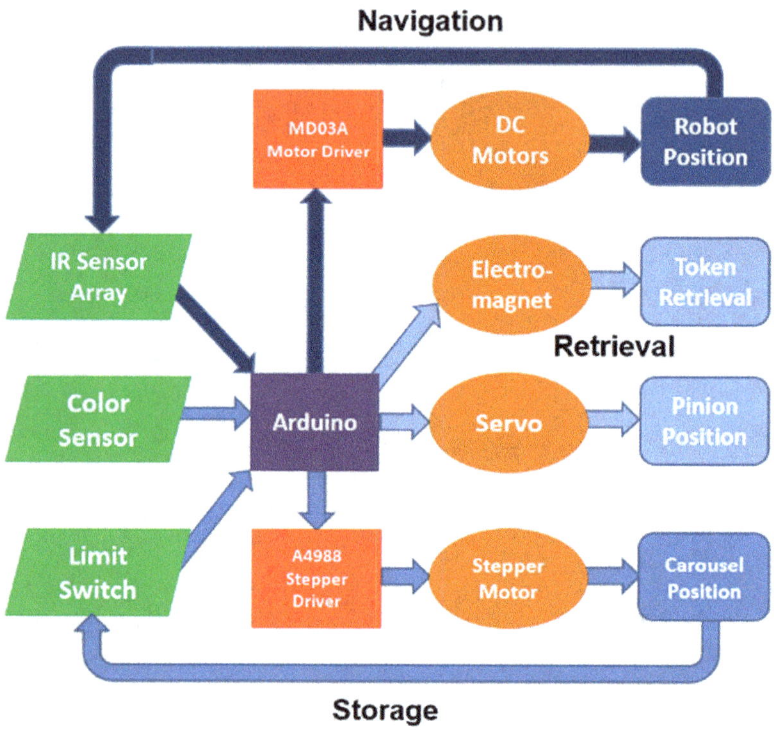

Fig. 13.4 Overall block diagram of the mobile robot system

Navigation Subsystem One of the greatest challenges of constructing an autonomous robot is the navigation. Since the user is not permitted to input real-time commands, the robot must be able to have quick decision-making capabilities that rely on input from various sensors. For this competition, we deemed it prudent to use the black lines forming a grid-like pattern on the playing field as a navigation tool. These black lines link the tokens and designated drop-off points. An array of reflectance sensors was used to follow a black line on a white background. The navigation subsystem was further decomposed into two modules: (1) straight line following and (2) detecting and maneuvering at junctions. The second navigation module is also used to detect the presence of a possible token in a depression.

The navigation of the mobile robot is achieved by implementing a proportional-integral-derivative (PID) control strategy. Readings received by the Arduino from the reflectance sensors are used by the PID controller to adjust the speed and the navigation of the mobile robot. In order to test the designed navigation subsystem, it was important to design a first prototype for the robot shown in Fig. 13.5.

The line following and maneuvering at junctions were thoroughly tested on a preliminary course built for the evaluation of the designed navigation subsystem (Fig. 13.6). The testing results of the navigation subsystem were evaluated and used to improve the navigation design.

Fig. 13.5 Robot prototype used to test the navigation subsystem

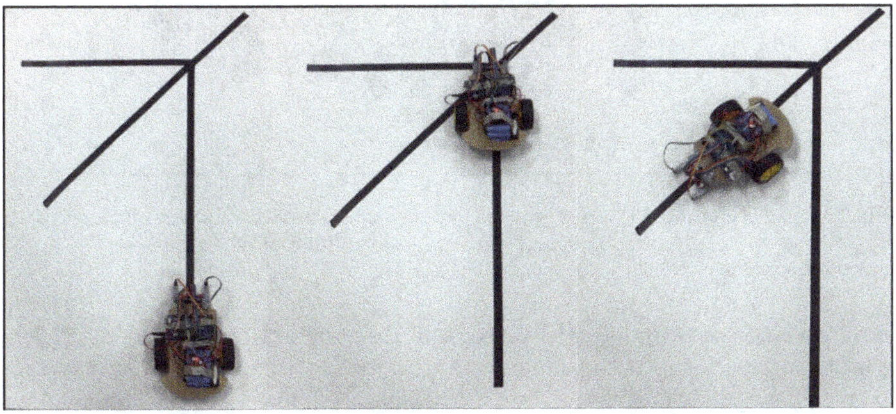

Fig. 13.6 Testing of the navigation subsystem using a first robot prototype

Token Retrieval and Deposit Subsystem This subsystem is responsible for picking up tokens and depositing them in the corresponding colored squares on the playing field. Several design solutions were brainstormed and evaluated for their advantages and disadvantages. The design that was selected consisted of a gear and pinion system actuated by a servo motor. The subsystem, when actuated, generates a vertical movement to lower or lift up an electromagnet that can pick up or drop off the metal tokens. A number of fine tunings were necessary to make the subsystem effective and reliable (Fig. 13.7).

Token Storage Subsystem To meet the competition challenge, the team's strategy was to collect all tokens from the playing field and store them onboard the robot before depositing them in their corresponding colored squares. This strategy was thought to minimize the travel time of the robot and complete the rounds within the time limits. Several design solutions were brainstormed. It was decided to use a rotating carousel mounted on the robot chassis and actuated by a stepper motor. The carousel has 9 positions: 7 are cylindrical containers for housing tokens of the 7

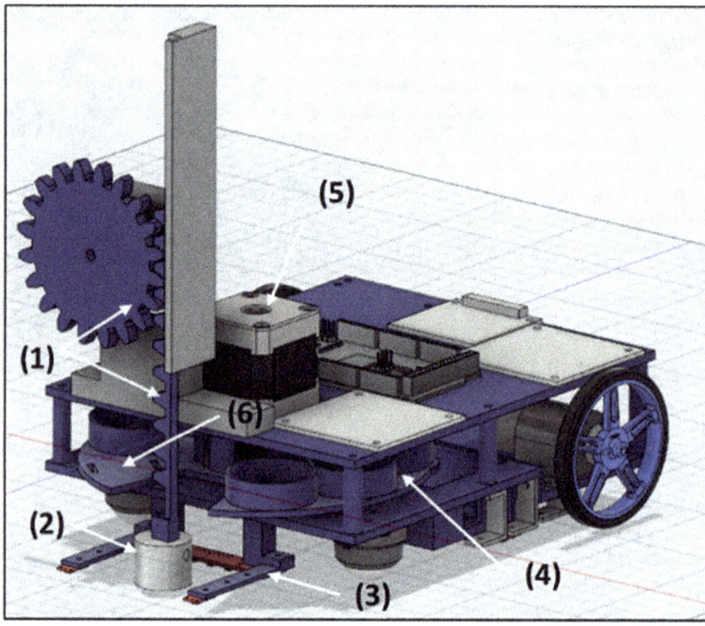

Fig. 13.7 A 3D model of the robot: gear and pinon system for retrieval and deposit of tokens *(1)*, electromagnet *(2)*, reflectance sensor array for navigation *(3)*, token storage carousel *(4)*, stepper motor to rotate the carousel *(5)*, location of the color sensor *(6)*

different colors, 1 is for the color sensor, and 1 is a cutout from the base (Fig. 13.7). The cutout position is the default or initial position of the carousel. When a junction on the playing field is reached, the electromagnet will lower through this cutout to the ground level, retrieve the token, and lift it above the height of the containers on the carousel. Then the carousel will rotate so that the color sensor is beneath the token. After the token's color is ascertained, the carousel will rotate again until the container assigned to that color is beneath the token. The electromagnet will then release the token, depositing it into the container. Once all the tokens have been collected, the robot will navigate to each colored square. When it reaches a colored square, the carousel will rotate to the container that matches the color of the square, and the electromagnet is lowered to reach the tokens in the container, pick them up all at once, and lift them above the container. Lastly, the carousel will rotate to the cutout, and the electromagnet will be lowered through the cutout and deposit the tokens at ground level.

Color Sensing Subsystem To detect the bottom colors of the tokens, an RGB color sensor was used. The sensor has an array of photodiodes, which are subdivided into four groups of 16 elements. Each group has a filter which only permits light of a certain color to pass. There is a red, blue, green, and clear filter group allowing for each base color to be measured. An Arduino code was developed to calibrate each of the photodiode group and to identify the different colored tokens used in the competition. The color sensor module is placed on the carousel as shown in Fig. 13.7.

Final Robot Design

The design and construction of the robot was completed in late March 2018. A 3D model of the designed robot is shown in Fig. 13.7 and the completely built robot is shown in Fig. 13.8. Modular testing protocols were developed to extensively teach each subsystem of the robot. It was important to check that all subsystems are working properly after being integrated together. The competition play field was constructed following instructions provided in the Robotics Competition Rules (IEEE R5 2018). This allowed to test the performance of the robot in all three rounds. Several runs of each round were performed to test the repeatability of the mobile robot. Overall, the robot performed well in all three rounds. Several runs of each round were successfully completed within the time limits. Hence, the robotics team felt confident about their prospects of doing well at the competition.

Experience at the Competition

The robotics team transported the robot, equipment, and spare parts to Austin, Texas, a day before the competition event. The plan was to have enough time to conduct some practice runs on the playing fields provided by the competition organizers. Thirty-three teams from twelve universities participated in the competition. Each team presents a different design solution for the competition. This is expected for open-ended engineering designs and the real world of engineering. Figure 13.9 shows a selection of the competing robots. During the practice runs, our robotics team encountered some issues with the robot that required some additional

Fig. 13.8 The final design of the mobile robot

Fig. 13.9 Display of the robots competing at the 2018 R5 IEEE robotics event

adjustments to various subsystems. However, overall our team performed well in all the three rounds during the practice runs. During the official runs as shown in Fig. 13.10, our robotics team encountered a few difficulties with the depressions holding the tokens on the playing field. In some instances the robot missed to retrieve a token, identified an incorrect color of a token, or did not unload all tokens from the storage carousel. After completing the three official rounds, our team was ranked fourth, but first among all teams who participated from universities in Texas. This was a great accomplishment given the degree of difficulty of the engineering design challenge.

The competition provided a unique platform to compare design solutions and to interact with students and faculty from other universities.

Fig. 13.10 Official round of our competing robotics student team in Austin, Texas

K-12 Outreach Activities

It has been a tradition in our robotics laboratory for our engineering students to showcase their robots to their peers and to K-12 students. Our university is located in Rio Grande Valley (RGV) of South Texas. Over 86% of the population in RGV is Hispanic. Most of the students who study at our university come from the local region and the majority of our student population are first-generation college students (who are in the first generation of their families to pursue a postsecondary education). Several outreach programs have been conducted to engage and motivate K-12 students to pursue S&E education and careers. We have had several visits by K-12 students and their teachers to our robotics laboratory. Over the years, our student robotics teams visited several K-12 schools to demonstrate their robots and talk about what it is like to be an engineer. Figure 13.11 shows our recent robotics student team demonstrating their robot and interacting with fourth- and fifth-grade level students in an elementary school located in Mission, Texas, during Career Day, an event aimed at exposing K-12 students to different professions.

Fig. 13.11 Robotics team participation in an outreach event in an elementary school in Mission, Texas

Lessons Learned from Participation in Robotics Competitions

Several lessons have been learned from our participation in the annual robotics competition. In this section, we present the lessons learned from faculty advisors and student teams. These lessons have helped our robotics laboratory achieve higher performance at a number of competitions.

- Over several competition cycles, the faculty advisors observed that many project teams did not initially plan sufficient time for reliability testing. In addition, the test plans initially developed often underestimated the importance of repeated trials with a statistically significant number of samples, or used nominal conditions rather than worst-case conditions. While the advisors did guide groups to more rigorous test plans, it was still apparent that the prior curriculum preparing students for senior design did not sufficiently emphasize these concepts.

- As a response, a project was modified in our third-year electrical engineering design laboratory course which is a prerequisite for the senior capstone design course. The original version of the project involved measurements on a simple integrated circuit logic gate to observe quantities such as threshold voltage, maximum available output current, input bias current, rise and fall times, and propagation delays. Measurements were conducted under nominal load and voltage levels. The modified version now requires students to determine, for each test, the most stringent conditions permitted by the datasheet and conduct the tests under those conditions. This requires students to analyze (for example) whether a particular test should be conducted under the minimum or the maximum allowed power supply voltage, or both. This practice has been extended and applied to various testing phases of the robotics design.
- When designing a robot for a competition, it is important to consider hardware and software redundancy during the design phase. During an official round at a robotics competition, a robot has only one chance to complete the challenge. If a critical hardware component fails, then another subsystem can be activated to carry over. Designing a fault-tolerant robotics system for a competition is considered essential.
- Forming an interdisciplinary student team to work on a robotics project has resulted in better results at competitions. For instance, student teams that included students from different engineering majors such as electrical, mechanical, and computer engineering came up with higher-performing robotics designs. Hence, we have made all possible efforts to form interdisciplinary student teams for the robotics design projects.
- The environment in the laboratory is completely different than the one experienced by students at the competition. Hence, we have learned that taking second year and third year students enrolled in the university engineering program and who have expressed interest in designing robots for their capstone design projects to the competition can help prepare them for what to expect when it's their turn to compete.
- Collaboration and time management between team members is important. Team members may propose different design ideas for a subsystem. Hence, it is important to promote communication and discussion of those ideas and select the best idea to implement.
- Participation in robotics competitions costs money. Hence, it is beneficial to seek industrial sponsors to help fund the cost of building the robot and the travel to the competition.

Conclusions

This chapter provided an overview of the robotics laboratory educational activities at the University of Texas Rio Grande Valley. The robotics lab has been providing robotics-based educational experiences for electrical and computer engineering senior students who select the topic of robotics for their capstone design projects. Robotics provide unique educational opportunity for senior students to experience all steps of the engineering design process and to extend their knowledge in a wide range of subjects across multiple disciplines that include electrical, mechanical, and computer engineering. Our educational efforts have aimed at training future engineers in robotics in order to guarantee a continuing future supply of a skilled workforce to meet the increasing number of job opportunities in the robotics sector.

This chapter also highlighted the achievements of our robotics laboratory in its participations in the annual robotics competition organized by Region 5 of IEEE, one of the world's largest technical professional organizations. To provide an example of our students' achievements, the chapter chronicled the robotics design by our recent student team that participated in the 2018 robotics competition. The chapter also presented the valuable lessons learned by faculty advisors and students from participation in the robotics competition.

The chapter emphasized the fact that robotics competitions motivate students to excel in their designs and expose them to real-world scenarios where engineering companies strive to develop innovative new products and compete to outperform their rivals. The fact that the design and performance of the robots built by the competing student teams from other universities are unknown prior to the competition day has provided a self-motivation for our students to strive to design and build the best possible robot that can have a chance to win the robotics competition. Our competing students realize that building a working robot prototype would satisfy the capstone senior design requirements of our engineering program. However, they understand that merely having a working robot prototype would not be enough to outperform the unpredictable competition. Over the years, the faculty advisors have noticed that students who selected competition-based capstone design projects tend to work harder and longer hours in the laboratories than noncompeting capstone design student teams. While the faculty evaluate all capstone senior design projects and assign course grades based on the completeness of the designed prototype, the demonstration of the working prototype, the end of the academic year oral presentation, and the technical report documenting the design and the testing of the prototype, the robotics competition teams go through an additional evaluation that is external, much stricter, and conducted in a high-pressure competition environment. Our experience has shown that when a capstone design project is integrated with an external competition, it pushes students to strive in developing advanced solutions for their design problems.

References

ABET. (2018). *Criteria for accrediting engineering programs*. Available at: https://www.abet.org. Accessed 2018.

Bajracharya, M., Maimone, M. W., & Helmick, D. (2008). Autonomy for Mars Rovers: Past, present, and future. *Computer, 41*(12), 44–50. https://doi.org/10.1109/MC.2008.479.

Bazylev, D., Margun, A., Zimenko, K., Kremlev, A., & Rukujzha, E. (2014). Participation in robotics competition as motivation for learning1. *Procedia - Social and Behavioral Sciences, 152*, 835–840. https://doi.org/10.1016/j.sbspro.2014.09.330.

Bekey, G., & Yuh, J. (2008). The status of robotics. *IEEE Robotics & Automation Magazine, 15*(1), 80–86. https://doi.org/10.1109/M-RA.2007.907356.

Berry, C. A. (2017). Robotics education online flipping a traditional mobile robotics classroom. *2017 IEEE Frontiers in Education Conference (FIE)*. 18–21 Oct. 2017. pp. 1–6.

Bruder, S., & Wedeward, K. (2003). Robotics in the classroom. *IEEE Robotics & Automation Magazine, 10*(3), 25–29. https://doi.org/10.1109/MRA.2003.1233554.

Cesaretti, L., Storti, M., Mazzieri, E., Screpanti, L., & Paesani, A. (2017). An innovative approach to School-Work turnover programme with Educational Robotics. *Mondo Digitale*, (72), 16.

Chew, M. T., Demidenko, S., Messom, C., & Gupta, G. S. (2000). Robotics competitions in engineering eduction. *2009 4th International Conference on Autonomous Robots and Agents*. 10–12 Feb. 2009. pp. 624–627.

Di Lieto, M. C., Inguaggiato, E., Castro, E., Cecchi, F., Cioni, G., Dell'Omo, M., et al. (2017). Educational Robotics intervention on Executive Functions in preschool children: A pilot study. *Computers in Human Behavior, 71*, 16–23. https://doi.org/10.1016/j.chb.2017.01.018.

Eguchi, A. (2016). RoboCupJunior for promoting STEM education, 21st century skills, and technological advancement through robotics competition. *Robotics and Autonomous Systems, 75*, 692–699. https://doi.org/10.1016/j.robot.2015.05.013.

Engelberger, J. F. (1980). *Robot use in manufacturing*. Boston, MA: Springer.

FIRST Robotics. (2015). *FIRST alumni study: brief summary of findings*. Available at: https://www.firstinspires.org/sites/default/files/uploads/resource_library/impact/FIRST-Alumni-Study-Summary-of-Finding.pdf

IEEE R5. (2018). *IEEE Region 5 Conference: 2018 Competition Description and Rules*. Available at: http://r5conferences.org/documents/2018/04/robotics-rules-april2018.pdf

International Federation of Robotics. (2018a). *Executive Summary World Robotics 2018 Industrial Robots*. Available at: https://www.ifr.org/downloads/press2018/Executive_Summary_WR_2018_Industrial_Robots.pdf

International Federation of Robotics. (2018b). *Executive Summary World Robotics 2018 Service Robots*. Available at: https://www.ifr.org/downloads/press2018/Executive_Summary_WR_Service_Robots_2018.pdf

Jung, S. (2013). Experiences in developing an experimental robotics course program for undergraduate education. *IEEE Transactions on Education, 56*(1), 129–136. https://doi.org/10.1109/TE.2012.2213601.

Kaiser, C., & Troxell, W. (2005). Student design competitions in undergraduate engineering education. *Proceedings Frontiers in Education 35th Annual Conference*. 19–22 Oct. 2005. pp. S3J–S3J.

Lanfranco, A. R., Castellanos, A. E., Desai, J. P., & Meyers, W. C. (2004). Robotic surgery: a current perspective. *Annals of Surgery, 239*(1), 14–21. https://doi.org/10.1097/01.sla.0000103020.19595.7d.

Michieletto, S., & Pagello, E. (2018). Competitions and industrial tasks as a way to learn basic concepts in robotics. *2018 IEEE International Conference on Autonomous Robot Systems and Competitions (ICARSC)*. 25–27 April 2018. pp. 173–178.

Murphy, R. R. (2001). "Competing" for a robotics education. *IEEE Robotics & Automation Magazine, 8*(2), 44–55. https://doi.org/10.1109/100.932757.

Murphy, R. R. (2012). A decade of rescue robots. *2012 IEEE/RSJ International Conference on Intelligent Robots and Systems*. 7–12 Oct. 2012. pp. 5448–5449.
Nagai, K. (2001). Learning while doing: practical robotics education. *IEEE Robotics & Automation Magazine, 8*(2), 39–43. https://doi.org/10.1109/100.932756.
National Science Board. (2016). *Science and Engineering Indicators 2016*. Available at: https://www.nsf.gov/statistics/2016/nsb20161/#/report
Niku, S. B. (2011). *Introduction to robotics: Analysis, control, applications*. Hoboken, N.J: Wiley.
Piepmeier, J. A., Bishop, B. E., & Knowles, K. A. (2003). Modern robotics engineering instruction. *IEEE Robotics & Automation Magazine, 10*(2), 33–37. https://doi.org/10.1109/MRA.2003.1213614.
President's Council of Advisors on Science and Technology (PCAST). (2012). *Engage to excel: Producing one million additional college graduates with degrees in science, technology, engineering, and mathematics*. Available at: https://obamawhitehouse.archives.gov/sites/default/files/microsites/ostp/pcast-engage-to-excel-final_2-25-12.pdf
Robotics Institute. (2018). *Robotics Institute at Carnegie Mellon University*. https://www.ri.cmu.edu/. Accessed 2018.
Scaradozzi, D., Sorbi, L., Pedale, A., Valzano, M., & Vergine, C. (2015). Teaching robotics at the primary school: An innovative approach. *Procedia - Social and Behavioral Sciences, 174*, 3838–3846. https://doi.org/10.1016/j.sbspro.2015.01.1122.
Tur, J. M. M., & Pfeiffer, C. F. (2006). Mobile robot design in education. *IEEE Robotics & Automation Magazine, 13*(1), 69–75. https://doi.org/10.1109/MRA.2006.1598055.
World Economic Forum. (2018). *The future of jobs report*. Cologny/Geneva, Switzerland. Available at: http://www3.weforum.org/docs/WEF_Future_of_Jobs_2018.pdf
Yilmaz, M., Ozcelik, S., Yilmazer, N., & Nekovei, R. (2013). Design-oriented enhanced robotics curriculum. *IEEE Transactions on Education, 56*(1), 137–144. https://doi.org/10.1109/TE.2012.2220775.

Chapter 14
Future Class Teachers and Educational Robotics: Current State and Possible Future Use

Ivana Đurđević Babić

Abstract The teaching profession demands constant education and training, so it is not surprising that study programmes that provide certain sets of skills to future teachers are under constant change. In this chapter, the focus is on students of class teacher studies who are trained for teaching educational programming languages, but did not have a lot of or any opportunity to work with educational robots. Since educational robotics is gaining more attention at all levels of education, it is almost certain that they will not only have the desire but also the need to use educational robotics in some segment of their future work.

Students' overall attitudes regarding the integration of educational robotics into higher education and primary school curriculum, including applicability of educational robots concerning different subjects in primary education and different university classes, were examined. In addition, their general familiarity with educational robotics, their motivation and interest for its use, subjective impression about their own educational background in a sense of preparedness for future use of educational robots in classroom, their suggestions for improvement of their knowledge and their estimation of potential positive impacts of educational robots use are presented. The aim of this chapter is to contribute to greater understanding of these students' needs and the challenges they face regarding this issue to help the students to overcome them.

Keywords Higher education · Class teacher students · Educational robots · Attitudes

---

I. Đurđević Babić (✉)
University of Osijek, Osijek, Croatia
e-mail: idjurdjevic@foozos.hr

L. Daniela (ed.), *Smart Learning with Educational Robotics*,
https://doi.org/10.1007/978-3-030-19913-5_14

Introduction

Construction and programming of robots in education is generally referred to as *educational robotics (ER),* and this term will be used in that context in this chapter. In the past decade, researchers have reported about a globally increased interest for ER in diverse countries (e.g. see Park et al. 2015; Alimisis 2013) and the growth of activities which serve to promote robotics and engage students in ER (e.g. see Petrovič and Balogh 2008). It is safe to say that the popularity of educational robotics in Croatia is at its peak. Since the information technology (IT) sector had an important role in the economic recovery from the recent global financial crisis and still lacks a full workforce, it is easy to understand why science, technology, engineering and mathematics (STEM) education has an important part in Croatia's current educational reform. This reform evidently tries to direct elementary school pupils from early days towards STEM occupations, paying attention to the use of information and communication technology (ICT), particularly programming. For instance, in an experimental programme called "School for Life", whose holder is the Ministry of Science and Education of the Republic of Croatia, Informatics is a course suggested for experimental implementation in formal education as early as the 1st grade of elementary school (MZO 2018b). Later, upon completion of secondary education, with different scholarships, future university students are encouraged and motivated to get a degree in STEM fields of study (e.g. MZO 2018a).

It was revealed that teachers have an impact on the students' eagerness towards learning, and the teachers' approach, along with their interest in a particular subject, have been shown to be important in a classroom environment (Omolara and Adebukola 2015). Besides, one could argue that implementation of educational robotics in elementary education is closely related to implementation of educational robotics in class teacher studies programmes, since teachers will hardly implement something in their classrooms if they do not feel competent enough to use it on their own. In Croatia, class teachers teach from 1st to 4th grade of primary school, and this type of teaching is usually carried out by one teacher who teaches primary students several subjects prescribed by the national curriculum (Eurydice - European Commission 2018) in one classroom. The national law and regulations prescribe what kind of professional qualifications class teachers must have to conduct class teaching. Currently, the situation at the Faculty of Education University of Osijek is such that students have only one compulsory course that enables them to teach educational programming languages (EPL) and to attract elementary school children to computer programming. Even though educational robotics can also be used for teaching programming, it is still not a part of their study programme. Students are expected to have non-formal education from this area, and they have little or no knowledge about it; therefore, the need for educational projects in this area still exists. While presenting the "Teacher Education on Robotics-Enhanced Constructivist Pedagogical Methods" (TERECoP) project, which recommends a constructivist model of teacher education concerning the implementation of educational robotics, Alimisis et al. (2007) draw attention to a relatively small number of projects that are engaged in the education of future and present school teachers regarding educational robotics.

The rest of this chapter is organized as follows: a concise overview of the conducted research from this area is given in the section "Literature Review", methodology of the performed research is stated in the section "Research Methodology and Sample Description", after which comes the section "Results and Discussion", followed by the "Conclusion" of this chapter.

Literature Review

A lot of research deals with the problem of the implementation of robotics in primary and secondary school education. Robotics is then perceived as a *learning object* in robotics classes where robotics is taught as a subject or as a *learning tool* in other subjects where it is used (Alimisis and Kynigos 2009). As pointed out by Barak and Assal (2018), educational robotics is linked with the constructivism and constructionism learning theories, which support an active environment where the students can learn while working with robots. Eguchi (2017) indicated that there are a variety of reasons preventing teachers from using robotics in classrooms, and in that context, he underlined the fact that robotics is not a part of the ongoing formal curriculum in average schools; therefore, the potential of educational robotics in classes is not acknowledged. While considering the use of robotics in primary and secondary education, researchers developed new educational robots (e.g. see Park et al. 2015; Naya et al. 2017), used a robot and described its use through several lessons (e.g. see Bellas et al. 2017), explored the impact of educational robotics on early school leaving (e.g. Daniela et al. 2017; Daniela and Strods 2018), students' attitudes or motivation towards STEM learning (e.g. see Holmquist 2014; Barak and Assal 2018) and found evidence that the integration of ER has a beneficial influence on the students' motivation towards learning (Chin et al. 2014) and the development of different skills. For instance, it has been established that ER improves the students' fluency and originality (Park et al. 2015) and has a positive impact on the development of spatial abilities (Julià and Antolí 2016). Polishuk and Verner (2017) conducted a research study in which 346 elementary school students participated. Students were divided into different groups in order to examine various influences of educational robotics on students. As part of this research, researchers analysed the development of systems thinking skills of 86 elementary school students who participated in a robotics workshops "Animal-like robots". They observed the systems thinking skills through structural, dynamic, generic, operational, scientific, closed-loop and continuum thinking as described by Richmond (1993), and at the end of the research study they detected an improvement of the students' system thinking skills.

Additionally, educational robotics was also explored in a higher education environment. Working on robotics projects, university students gained problem-oriented skills, and a positive impact on their theoretical knowledge was confirmed (Damaševicius et al. 2017). College students, using robotics as a learning object, expressed a positive learning experience during the course that altered their impression of collaboration with fellow students, increased their interest in robotics and technology and affected their problem-solving and creative thinking skills (Eguchi 2014). Furthermore, the usefulness of educational robotics in fostering pre-service elementary school teachers' enjoyment and interest in STEM is reported by Kim et al. (2015).

Researchers also tackled the perception of different stakeholders in education towards educational robotics. Liu (2010) put focus on elementary school students' perception, namely early adolescents, while Lin et al. (2012) explored the perception of 39 parents whose children were enrolled in secondary school. Their research revealed the parents' positive attitudes towards the usefulness of educational robots and their willingness to encourage their children towards activities concerning educational robotics. Khanlari (2013) investigated the robotic class teachers' impressions about educational robotics when it is used for learning STEM subjects. After interviewing six teachers with experience in teaching robotics classes and analysing their responses, he concluded that his participants believe that robotics is useful for students when they learn STEM subjects and that it causes or strengthens their interest in STEM subjects. Moreover, all participants in his study agreed that robotics should be implemented in all grades of elementary school education. In a later conducted research, in which 11 elementary school teachers participated, Khanlari (2016) pointed out that the teachers perceived the positive effects of educational robotics on the students' problem-solving skills and interpersonal skills. As a major barrier stopping them from using educational robotics, participants in his research indicated the insufficient number of educational robots and that suitable programmes were needed, as well as other physical components. Kaya et al. (2017) marked that in their research, 11 pre-service elementary teachers changed their views towards the nature of engineering (NOE) after working with robots. Smyrnova-Trybulska et al. (2016) discussed some technical and legal points related to educational robotics implementation in education in Poland. They presented the results of a research study in which primary school teachers and those who are studying to become primary school teachers from Poland and Ukraine participated. In their research, a small percent of participants from Poland (15.6%) and from Ukraine (7.8%) considered that robotics can be taught in school within the mandatory curricula.

Although teachers' and future teachers' viewpoints were covered in the existing literature, owing to the relatively small number of conducted researches there is still the need to address and explore these issues, which will result in better understanding of the current situation concerning their education, assertiveness and enthusiasm about educational robotics.

Research Methodology and Sample Description

The research study was conducted in the summer semester of the academic year 2017/2018 in which students from class teacher studies from the Faculty of Education University of Osijek participated. From this targeted population, a total

Table 14.1 The students' distribution according to enrolled study year

Study year	N	%
1.	22	18.49
2.	21	17.65
3.	33	27.74
4.	23	19.33
5.	20	16.81

of 119 students from all study years took part. The distribution of these students by year of study is presented in Table 14.1.

The vast majority of respondents (92.44%) were female and 7.56% were male, which is considered a reasonably acceptable gender representativeness of students enrolled at the Faculty of Education where, in general, more female students are enrolled.

Since there are three study modules (developmental module, computer science, foreign language) currently being carried out at the Faculty of Education, it was important that the sample included representatives from all three modules as they differ in some courses. Overall, 41.18% of respondents were enrolled in developmental, 34.45% in computer science and 24.37% in the foreign language module.

An online questionnaire was used to gather data about the students' perception concerning educational robotics. This questionnaire was composed of questions that could be classified into three blocks according to their objectives. The first block was constructed of questions which examined demographic characteristics of respondents (gender, study year, study module). The second block was used to obtain the students' general attitudes towards educational robotics as well as to gather information about their previous knowledge and education from this field. Questions and statements in the third block were used to acquire the students' attitudes and opinions regarding the use of educational robotics in their future professions. Before students started filling out the questionnaire, the term *educational robotics* was shortly explained to all of them.

Results and Discussion

Related to the students' familiarity with the term *educational robotics*, the results showed that more than half of respondents (59.66%) had heard this term before participating in the research, but 40.34% had not. The participants were asked to assess their level of knowledge of educational robotics and educational programming languages. Little more than 60% of the respondents self-assessed their knowledge of educational robotics as insufficient (60.50%). This percentage changes notably when it comes to the self-assessment of their knowledge of educational programming languages. Little more than one fifth of respondents (21.85%)

Table 14.2 The frequency table of participants' self-assessments about their knowledge from EPL and ER

	ER		EPL	
Assessed level of knowledge	n	%	n	%
Insufficient	72	60.50	26	21.85
Sufficient	26	21.85	35	29.41
Good	14	11.76	26	21.85
Very good	5	4.20	19	15.97
Excellent	2	1.68	13	10.92

assessed that they have insufficient knowledge from this area. The distribution of all students' responses regarding this question can be seen in Table 14.2.

More than 60% of students (61.34%) reported a generally positive attitude towards the integration of educational robotics into the education of future teachers, 35.29% reported neutral and only 3.36% reported a negative attitude. On the basis of the majority of the respondents' opinion (63.87%), future teachers should know how to work with educational robotic sets. However, more than one fourth of the respondents (27.73%) do not know if future teachers should be provided with this knowledge (see Fig. 14.1).

When it comes to their opinion within which group of subjects should educational robotics be taught in class teacher studies, more than half of the respondents (57.14%) believe that it should be done within the computer science (informatics) subject area.

Concerning their preparedness for efficient knowledge transfer, 47.90% of them think that they did not adopt enough knowledge during their education (formal, non-formal and informal) to effectively teach younger school-age children how to use educational programming languages. In addition, the majority of them (68.91%) see the knowledge of how to use educational programming languages as a precondition for the efficient use of educational robots, while 9.24% do not and 21.84% remained neutral concerning this issue.

More students expressed themselves positively (27.73% are somewhat satisfied and 14.29% are very satisfied with the level of their knowledge) towards their level of knowledge of educational programming languages than negatively (11.76% are very dissatisfied and 18.49% are somewhat dissatisfied with their level of knowledge), while more than one fourth of respondents (27.73%) remained neutral (neither satisfied nor dissatisfied).

Although slightly more than half of participants believe that prior education prepared them for the integration of educational robotics in primary school classrooms, at least to some extent (24.37% satisfactory, 15.13% good, 8.40% very good and 2.52% excellent), almost half of participants (49.58%) believe that their prior education was inadequate. The vast majority of participants (92.44%) never had the opportunity to work with an educational robot kit, and only one participant worked with a humanoid robot. Thus, it is not surprising that the majority of them (81.51%) feel that they should get additional education from the field of educational robotics

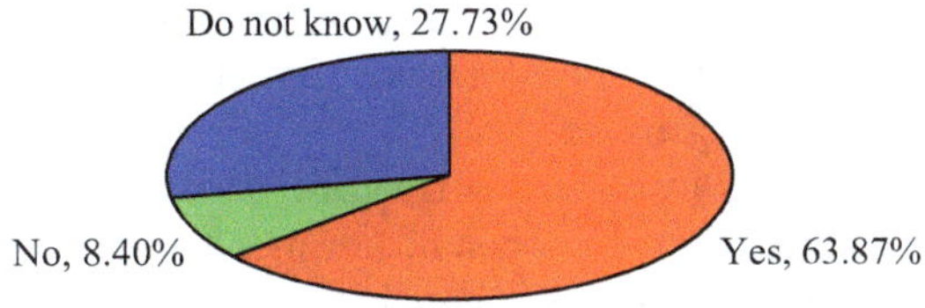

Fig. 14.1 Pie chart of students' responses on the question "In your opinion, do future teachers need to know how to work with educational robotic sets?"

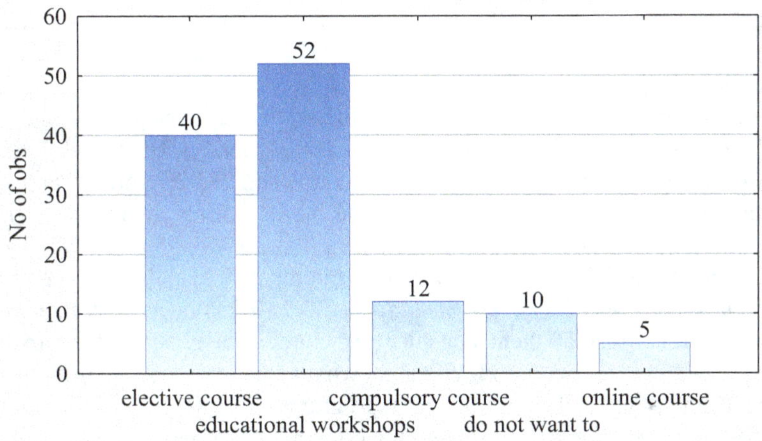

Fig. 14.2 Graphical representation of students' responses to the question "Indicate in which way you would like to deepen your knowledge from educational robotics"

in order to improve and increase their knowledge. However, not the same percentage of students wants to get additional education; 68.91% of participants indicated that they want to get additional education, 17.65% revealed that they do not want to get additional education at the moment and 13.45% completely discarded the idea of any form of additional education. Yet, when they were asked to indicate the way they would like to deepen their knowledge of educational robotics, some of them changed their mind (see Fig. 14.2) and almost the same number of students expressed that they would like to do so during their studies as part of their study programme (33.61% in an elective course and 10.08% in a compulsory course) and by attending educational workshops (43.70%).

More than half of participants were not acquainted with the activities in educational robotics workshops for students (58.83%), and only a very low percentage of participants indicated that they were partially (7.56%) and fully acquainted (2.52%). Even a higher percentage of participants pointed out their unfamiliarity with educational robotic workshops that are specially intended for teachers (68.07% fully unacquainted). The majority of participants were fully unfamiliar with the existence and work of robotic camps (73.11%), 59.66% of participants were fully unacquainted with the existence of national and 62.18% with the existence of world robotic competitions. The participants also indicated low familiarity with the work of civil society organizations that encourage programming and robotics (63.03% fully unacquainted).

When participants were asked to indicate the importance of factors that would be essential for them when acquiring an educational robotic kit to use in their future primary classrooms, the majority of them indicated that price is an important factor (38.66% somewhat important and 28.57% absolutely essential) as well as ease of use (35.29% somewhat important and 38.66% absolutely essential) and availability of instructional materials for teaching (41.18% somewhat important and 27.73% absolutely essential). For 39.49% of participants, the appearance of the educational kit is neither an important nor unimportant factor when acquiring an educational robotic kit.

Furthermore, 47.90% of participants neither agreed nor disagreed with the statement *It is hard to choose an adequate educational robotics kit*. More participants expressed disagreement (18.49% strongly disagree and 23.52% somewhat disagree) than agreement (14.29% somewhat agree and 5.88% strongly agree) with the idea that *Educational robotics should only be taught in some form of informal education*. The majority of participants agreed with the statement *Working with an educational robot kit has many positive effects on students* (30.25% somewhat agree and 31.09% strongly agree). Most of them consider and agree that it encourages individual learning as well as teamwork (35.29% somewhat agree and 37.82% strongly agree), enables young school children to learn programming easily (35.29% somewhat agree and 36.97% strongly agree), empowers young students for successful problem solving and decision making (31.09% somewhat agree and 25.21% strongly agree), encourages creativity in children (27.73% somewhat agree and 38.66% strongly agree) and develops their competitive spirit (37.82% somewhat agree and 23.53% strongly agree). More results can be seen in Tables 14.3 and 14.4. Most of the participants agreed that children have to think critically while working with the educational robotics kit (36.97% somewhat agree and 29.41% strongly agree).

Their agreements with the individual Likert-type statements (1 – strongly disagree to 5 – strongly agree) suggest that participants believe that children are motivated to work with educational robots (33.61% somewhat agree and 31.93% strongly agree), that educational robotics helps teachers to achieve educational goals more easily (38.66% somewhat agree and 21.85% strongly agree), helps to get young school children interested in the STEM area (28.57% somewhat agree and 38.66% strongly agree) and stimulates inventive thinking (23.53% somewhat agree and 40.34% strongly agree). Nearly half of participants (48.74%) neither agree nor disagree with the statement *Managing educational robots is quite simple,* and more than a third of participants neither agree nor disagree with the statement *Working with educational robots helps in the socialization of children* (35.29%). Slightly less than one third of participants are convinced that they can successfully implement working with the educational robotics kit (20.17% somewhat agree and 12.61% strongly agree) in primary school teaching. Participants believe that it is the teachers' obligation to encourage the interest of younger school children in educational robotics (32.77% somewhat agree and 23.53% strongly agree) (see Fig. 14.3).

Table 14.3 Agreement and disagreement of students on individual statements that explore students' perception about potential benefits of working with educational robot kits

	Disagree			Agree	
Working with an educational robot kit…	Strongly (%)	Somewhat (%)	Neutral (%)	Somewhat (%)	Strongly (%)
… has many positive effects on students.	2.52	6.72	29.41	30.25	31.09
… encourages individual learning as well as teamwork.	1.68	7.56	17.65	35.29	37.82
… enables young school children to learn programming easily.	1.68	5.88	20.17	35.29	36.97
… empowers young students for successful problem solving and decision making.	5.04	12.61	26.05	31.09	25.21
…develops students' competitive spirit.	5.88	5.88	26.89	37.82	23.53
… encourages creativity in children.	4.20	5.88	23.53	27.73	38.66

Table 14.4 Students' perception about additional usefulness of ER in education

	Disagree			Agree	
Statement	Strongly (%)	Somewhat (%)	Neutral (%)	Somewhat (%)	Strongly (%)
Children have to think critically while working with the educational robotics kit.	0.84	7.56	25.21	36.97	29.41
Educational robotics helps teachers achieve educational goals more easily.	1.68	10.92	26.89	38.66	21.85
Educational robotics helps to get young school children interested in the STEM area.	2.52	9.24	21.00	28.57	38.66
Educational robotics stimulates inventive thinking.	1.66	6.72	27.73	23.53	40.34
Working with educational robots helps in the socialization of children.	6.72	13.45	35.29	31.09	13.45

However, a vast majority of them have the opinion that educational robotics is not well known in the teaching profession (27.73% somewhat agree and 50.42% strongly agree). Almost half of participants consider that it is necessary to introduce educational robotics in the elementary school curriculum as soon as possible (31.09% somewhat agree and 16.81% strongly agree). A little less than one third of participants (10.92% strongly disagree and 21.85% somewhat disagree) disagreed with the statement *I am convinced that I can successfully implement the work with*

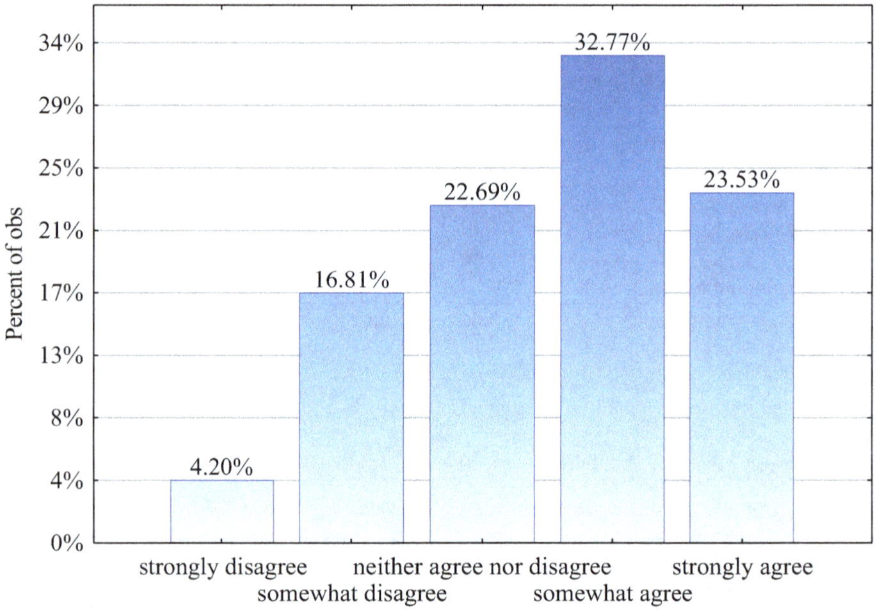

Fig. 14.3 Graphical representation of students' agreement with the statement "It is the teachers' obligation to encourage the interest of younger school children in educational robotics"

an educational robotic kit in my future teaching. A little more than 40% disagreed (15.13% strongly disagreed and 25.21% somewhat disagreed) with the thought that everybody can learn how to assemble and program an educational robot. More than half of participants (33.61% somewhat agree and 24.37% strongly agree) trust that it takes a lot of time and patience to learn how to do so. The majority of them are convinced that teachers need to put in an extra effort if they want to use educational robots in the classroom (38.66% somewhat agree and 39.50% strongly agree). However, they mostly agree (25.21% somewhat agree and 45.38% strongly agree) with statements *By programming a robot, students develop creativity* and *Students acquire important skills while working with the robotics kit* (25.21% somewhat agree and 44.54% strongly agree). Although, 42.01% of participants acknowledge that they do not know how to implement educational robotics in classroom teaching (25.21% somewhat agree and 16.81% strongly agree) and 46.22% (24.37% somewhat agree and 21.85% strongly agree) agree that it is fun to work with educational robotic sets. Slightly more than half of participants agree (28.57% somewhat agree and 21.85% strongly agree) that educational robotics helps in the realization of inclusive education (see Table 14.5).

More than a third of participants think that educational robotics could currently be most easily implemented in the fourth-grade curriculum (36.13%) and more than half of participants consider that it could be done within the content of an elective

Table 14.5 Students' opinion about the use of ER in their future profession

	Disagree			Agree	
Statement	Strongly (%)	Somewhat (%)	Neutral (%)	Somewhat (%)	Strongly (%)
Educational robotics is not well known in the teaching profession.	3.36	5.88	12.61	27.73	50.42
It is necessary to introduce educational robotics in the elementary school curriculum as soon as possible.	5.88	13.44	32.77	31.09	16.81
Everybody can learn how to assemble and program an educational robot.	15.13	25.21	31.93	21.01	6.72
It takes a lot of time and patience to learn how to assemble and program an educational robot.	2.52	8.40	31.09	33.61	24.37
Teachers need to put in an extra effort if they want to use educational robots in the classroom.	1.68	4.20	15.97	38.66	39.50
By programming a robot, students develop creativity.	1.68	7.56	20.17	25.21	45.38
Students acquire important skills while working with the educational robotics kits.	2.52	6.72	21.00	25.21	44.54
I do not know how to implement educational robotics in classroom teaching.	5.04	15.97	36.97	25.21	16.81
It is fun to work with educational robotic sets.	4.20	8.40	41.18	24.37	21.85
Educational robotics helps in the realization of inclusive education.	3.36	9.24	36.97	28.57	21.85

course (57.98%) in lower grades of elementary school (see Fig. 14.4). The majority of participants (64.71%) stated that they would like to use educational robotics in educating young school children, and 66.39% of participants recognize their inadequate education about the use of educational robotic kits as the biggest obstacle in doing so.

Conclusion

This chapter gives insights into attitudes and viewpoints of students who are training to become class teachers regarding educational robotics in general, their knowledge of it, as well as their previous education regarding it and their way of looking at the use of educational robotics in the profession they are being educated for. In other words, the integration of ER into elementary school education.

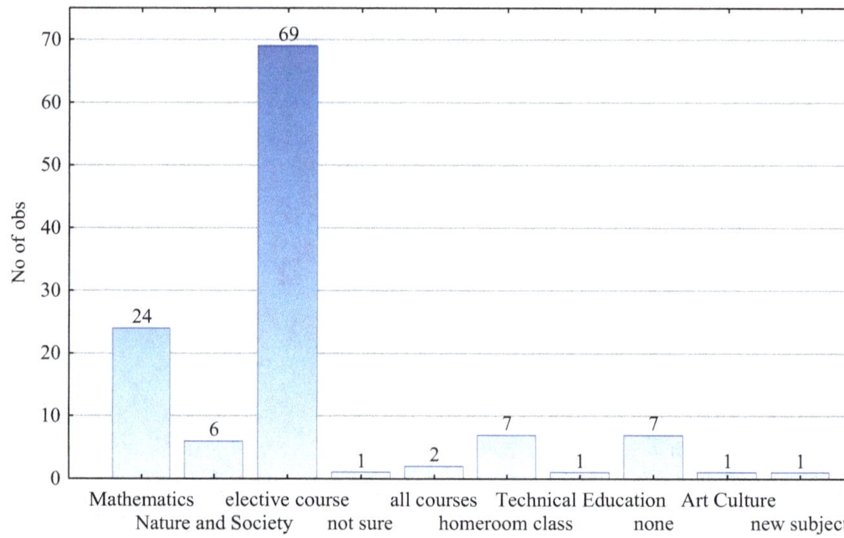

Fig. 14.4 Students' opinions regarding the subject in which educational robotics could be most easily implemented in primary school class teaching

It can be said that the obtained results regarding the opinions and viewpoints of students from class teacher studies are consistent with the conclusions and results from previous research that have examined the views of different stakeholders in education. For instance, with Khanlari (2016), regarding the perceived usefulness of educational robotics in gaining valuable skills.

The results derived from the conducted research lead to the conclusion that future teachers all in all have a positive opinion about educational robotics and its integration in elementary school education, acknowledge possible benefits of using educational robotics, believe that future teachers need to know how to work with educational robotic kits and, although they are aware of their lack of knowledge concerning educational robotics, they are not sufficiently familiar with the possibilities of non-formal and informal education in this area. Therefore, it is necessary to additionally familiarize students with the currently available informal and non-formal learning opportunities, as well as to include educational robotics in some aspect of their formal education. Also, the results suggest that special attention should be paid to increasing the students' motivation for any form of additional education to help them overcome their uncertainties and prejudices towards educational robotics. Research results confirm that students, regardless of their education, still perceive the price of educational robotic kits as one of the biggest barriers that would inhibit them from integrating educational robotics, although some low cost educational robotics kits are available for class use.

The obtained results should be viewed in the context of the current situation in Croatia's primary education and the current class teacher studies programme that is being conducted at the Faculty of Education, University of Osijek. These students did not have educational robotics or computer programming as part of their lower grades primary school curriculum while they were attending. The situation in Croatia is changing with new educational reform and the experimental programme "School for life" with the subject of Informatics where computer programming is integrated in experimental programme curriculum from the first grade of primary school (MZO 2018b). This experimental curriculum, which is now carried out in some primary schools in Croatia, will become effective in the school year 2020/2021 for all lower grades (1st–4th) of primary schools (MZO 2018c). Also, in the recommendations for achieving educational objectives in this national curriculum, even for the first grade of primary school, it is suggested to use robots for visualization of programming (MZO 2017).

As mentioned by Bredenfeld et al. (2010), some schools, e.g. in Germany, integrated robotics into their curricula, but Alimisis (2013) noticed the lack of systematic introduction of robotics in European schools. Balanskat and Engelhardt (2014) reported that programming has been part of England's national curriculum since 2014 where computer programming is compulsory for students entering primary school. In 2014, the situation was such that only two more counties in the European Union (Estonia and Greece) participated in research integrated computer programming and coding in their primary school educational system (Balanskat and Engelhardt 2014). Balanskat and Engelhardt (2014) also reported that nine of the countries which had integrated computer programming in the curriculum of primary or secondary education supported teacher training in some manner (e.g. Cyprus and Ireland have compulsory education as part of general initial teacher training or in-service training). At the Faculty of Education University of Osijek, computer programming has been a part of the class teacher studies programme since 2005 as a compulsory subject for all students when this faculty was still called the Faculty of Teacher Education in Osijek and students who enrol in Computer Science module at this faculty have one more mandatory computer programming course (Faculty of Teacher Education in Osijek 2005). However, educational robotics is not integrated in computer science courses currently being held at the Faculty of Education.

In their final report of the project "Creativity and Innovation in Education and Training in the EU27 (ICEAC)", Cachia et al. (2010) emphasized that teachers who had training in ICT will have a more positive view of its benefits for learning and suggested revision of teacher training programmes to encourage the use of innovative teaching methods and development of digital competences among other things. The results of their study also indicated that there is the need for enhancing teachers' ICT skills to exploit the possibilities of new technologies for creative learning and innovative teaching. Numerous teacher education programmes are failing to properly prepare their participants for technology integration (Fishman and Davis 2006; Moursund and Bielefeldt 1999; Willis and Mehlinger 1996; Zhao et al. 2002 as cited in So and Kim 2009). As shown by Zhu et al. (2013), teachers' technological competencies are positively related with their innovative teaching. Owing to its educational benefits, robotics is used in various schools for innovative learning

(Blanchard et al. 2010). Kvesko et al. (2018) note that robotics as a technology can contribute to a better comprehension of theoretical and practical knowledge underlying that "... the special course on robotics will allow to implement the ideas of the cross-disciplinary complex, and to promote practical use of theoretical knowledge..." (p. 3).

The limitations of this research are mostly related to sample size and profile; therefore, future research should include a larger sample size and a more diverse profile of students from various faculties that educate future teachers. Also, different types of research could be conducted, including the comparison between the students' beliefs before working with ER and after its use.

References

Alimisis, D. (2013). Educational robotics: Open questions and new challenges. *Themes in Science and Technology Education, 6*(1), 63–71.

Alimisis, D., & Kynigos, C. (2009). Constructionism and robotics in education. In D. Alimisis (Ed.), *Teacher education on robotic-enhanced constructivist pedagogical methods* (pp. 11–26). Athens: School of Pedagogical and Technological Education (ASPETE).

Alimisis, D., Moro, M., Arlegui, J., Pina, A., Frangou, S., & Papanikolaou, K. (2007, August). Robotics & constructivism in education: The TERECoP project. In I. Kalas (Ed.), *Proceedings of the 11th European Logo Conference, Bratislava, Slovakia* (Vol. 40, pp. 19–24). Bratislava: Faculty of Methematics, Physics and Informatics Comenius University.

Balanskat, A., & Engelhardt, K. (2014). Computing our future: Computer programming and coding-Priorities, school curricula and initiatives across Europe. European Schoolnet. Brussels. Belgium. http://www.allyouneediscode.eu/documents/12411/14644/Coding+initiative+report-European+Schoolnet-October2014.pdf/66475be8-cc31-429c-a5cc-32767366c8c2

Barak, M., & Assal, M. (2018). Robotics and STEM learning: Students' achievements in assignments according to the P3 Task Taxonomy—Practice, problem solving, and projects. *International Journal of Technology and Design Education, 28*(1), 121–144.

Bellas, F., Naya, M., Varela, G., Llamas, L., Prieto, A., Becerra, J. C., et al. (2017, April). The Robobo project: Bringing educational robotics closer to real-world applications. In W. Lepuschitz, M. Merdan, G. Koppensteiner, R. Balogh, & D. Obdržálek (Eds.), *Robotics in education. RiE 2017. Advances in intelligent systems and computing* (Vol. 630). Cham: Springer. https://doi.org/10.1007/978-3-319-62875-2_20.

Blanchard, S., Freiman, V., & Lirrete-Pitre, N. (2010). Strategies used by elementary schoolchildren solving robotics-based complex tasks: Innovative potential of technology. *Procedia-Social and Behavioral Sciences, 2*(2), 2851–2857.

Bredenfeld, A., Hofmann, A., & Steinbauer, G. (2010). Robotics in education initiatives in europe-status, shortcomings and open questions. *Proceedings of SIMPAR* 2010 *Workshops. International conference on simulation, modeling and programming for autonomous robots (SIMPAR 2010)* (pp. 568–574). Darmstadt, Germany, November 15–16, 2010. https://www.sim.informatik.tu-darmstadt.de/simpar/ws/sites/TR-TWR-2010/22-TeachingRobotics.pdf

Cachia, R., Ferrari, A., Ala-Mutka, K. and Punie, Y. (2010). *Creative learning and innovative teaching: Final report on the study on creativity and innovation in education in the EU Member States*. Luxembourg.

Chin, K. Y., Hong, Z. W., & Chen, Y. L. (2014). Impact of using an educational robot-based learning system on students' motivation in elementary education. *IEEE Transactions on Learning Technologies, 7*(4), 333–345.

Damaševicius, R., Narbutaite, L., Plauska, I., & Blažauskas, T. (2017). Advances in the use of educational robots in project-based teaching. *TEM Journal, 6*(2), 342.

Daniela, L., & Strods, R. (2018). Robot as agent in reducing risks of early school leaving. In L. Daniela (Ed.), *Innovations, technologies and research in education* (pp. 138–156). Newcastle upon Tyne: Cambridge Scholars Publishing.

Daniela, L., Strods, R., & Alimisis, D. (2017). *Analysis of robotics-based learning interventions for preventing school failure and early school leaving in gender context.* 9th International Conference on Education and New Learning Technologies (Edulearn17), Proceedings, 9th International Conference on Education and New Learning Technologies, 3–5 July, 2017, Barcelona, Spain, pp. 810–818.

Eguchi, A. (2014). Educational robotics for promoting 21st century skills. *Journal of Automation Mobile Robotics and Intelligent Systems, 8*(1), 5–11.

Eguchi, A. (2017). Bringing robotics in classrooms. In M. Khine (Ed.), *Robotics in STEM education* (pp. 3–31). Cham: Springer.

Eurydice - European Commission. (2018). *Initial education for teachers working in early childhood and school education.* Retrieved from https://eacea.ec.europa.eu/national-policies/eurydice/content/initial-education-teachers-working-early-childhood-and-school-education-14_en. January 3, 2019.

Faculty of Teacher Education in Osijek. (2005). *Integrated undergraduate and graduate five-year university Class teacher studies.* Osijek.

Holmquist, S. (2014). *A multi-case study of student interactions with educational robots and impact on Science, Technology, Engineering, and Math (STEM) learning and attitudes.* (Graduate Theses and Dissertations). https://scholarcommons.usf.edu/etd/5043

Julià, C., & Antolí, J. Ò. (2016). Spatial ability learning through educational robotics. *International Journal of Technology and Design Education, 26*(2), 185–203.

Kaya, E., Newley, A., Deniz, H., Yesilyurt, E., & Newley, P. (2017). Introducing engineering design to a science teaching methods course through educational robotics and exploring changes in views of preservice elementary teachers. *Journal of College Science Teaching, 47*(2), 66–75.

Khanlari, A. (2013, December). Effects of educational robots on learning STEM and on students' attitude toward STEM. *2013 IEEE 5th Conference* on *Engineering Education (ICEED)* (pp. 62–66). IEEE. doi: https://doi.org/10.1109/ICEED.2013.6908304.

Khanlari, A. (2016). Teachers' perceptions of the benefits and the challenges of integrating educational robots into primary/elementary curricula. *European Journal of Engineering Education, 41*(3), 320–330. https://doi.org/10.1080/03043797.2015.1056106.

Kim, C., Kim, D., Yuan, J., Hill, R. B., Doshi, P., & Thai, C. N. (2015). Robotics to promote elementary education pre-service teachers' STEM engagement, learning, and teaching. *Computers & Education, 91*, 14–31.

Kvesko, S. B., Kvesko, N. G., Korniyenko, A. A., & Kabanova, N. N. (2018, May). Educational robotics as an Innovative teaching practice using technology: minimization of risks. In *IOP Conference Series: Materials Science and Engineering* (Vol. 363, No. 1, p. 012004). IOP Publishing.

Lin, C. H., Liu, E. Z. F., & Huang, Y. Y. (2012). Exploring parents' perceptions towards educational robots: Gender and socio-economic differences. *British Journal of Educational Technology, 43*(1), E31–E34.

Liu, E. Z. F. (2010). Early adolescents' perceptions of educational robots and learning of robotics. *British Journal of Educational Technology, 41*(3), E44–E47.

[MZO] Ministarstvo znanosti i obrazovanja Republike Hrvatske. (2017). Nacionalni kurikulum nastavnog predmeta Informatika [National curriculum of school subject Informatics]. Retrieved from: https://mzo.hr/sites/default/files/dokumenti/2017/OBRAZOVANJE/NACION-KURIK/PREDMETNI-KURIK/Informatika/informatika2.pdf. January 3, 2019.

[MZO] Ministarstvo znanosti i obrazovanja Republike Hrvatske. (2018a). STEM stipendije [STEM scholarships]. Retrived from https://stemstipendije.mzo.hr/. September 13, 2018.

[MZO] Ministarstvo znanosti i obrazovanja Republike Hrvatske. (2018b). Eksperimentalni program "Škola za život". [Experimental program "School for Life"]. Retrived from https://mzo.hr/sites/default/files/dokumenti/2018/OBRAZOVANJE/Nacionalni-kurikulumi/Skola-za-zivot/eksperimentalni_program-skola_za_zivot.pdf. September 13, 2018.
[MZO] Ministarstvo znanosti i obrazovanja Republike Hrvatske. (2018c). Odluka o donošenju kurikuluma za nastavni predmet Informatike za osnovne škole i gimnazije u Republici Hrvatskoj [Decision on curriculum for subject Informatics for elementary schools and high schools in the Republic of Croatia]. Retrived from https://narodne-novine.nn.hr/clanci/sluzbeni/2018_03_22_436.html. January 3, 2019.
Naya, M., Varela, G., Llamas, L., Bautista, M., Becerra, J. C., Bellas, F., et al. (2017, September). A versatile robotic platform for educational interaction. In *Intelligent Data Acquisition and Advanced Computing Systems: Technology and Applications (IDAACS), 2017 9th IEEE International Conference on* (Vol. 1, pp. 138–144). IEEE. doi: https://doi.org/10.1109/IDAACS.2017.8095065
Omolara, S. R., & Adebukola, O. R. (2015). Teachers' attitudes: A great influence on teaching and learning of social studies. *Journal of Law, Policy and Globalization, 42*, 131–137.
Park, I., Kim, D., Oh, J., Jang, Y., & Lim, K. (2015). Learning effects of pedagogical robots with programming in elementary school environments in Korea. *Indian Journal of Science and Technology, 8*(26).
Petrovič, P. & Balogh, R. (2008). *Educational robotics initiatives in Slovakia.* Workshop Proceedings of SIMPAR 2008 International Conference on Simulation, Modeling and Programming for Autonomous Robots, Venice, Italy 2008 November, 3–4, pp. 122–131.
Polishuk, A., & Verner, I. (2017). Student-robot interactions in museum workshops: Learning activities and outcomes. In M. Merdan, W. Lepuschitz, G. Koppensteiner, & R. Balogh (Eds.), *Robotics in education*.. Advances in Intelligent Systems and Computing (Vol. 457, pp. 233–244). Cham: Springer. https://doi.org/10.1007/978-3-319-42975-5_21.
Richmond, B. (1993). Systems thinking: Critical thinking skills for the 1990s and beyond. *System Dynamics Review, 9*(2), 113–133.
Smyrnova-Trybulska, E., Morze, N., Zuziak, W., & Gladun, M. (2016). Robots in elementary school: Some educational, legal and technical aspects. In E. Smyrnova-Trybulska (Ed.), *E-learning methodology–implementation and evaluation*. Poland: Studio NOA for University of Silesia in Katowice.
So, H. J., & Kim, B. (2009). Learning about problem based learning: Student teachers integrating technology, pedagogy and content knowledge. *Australasian Journal of Educational Technology, 25*(1), 101–116.
Zhu, C., Wang, D., Cai, Y., & Engels, N. (2013). What core competencies are related to teachers' innovative teaching? *Asia-Pacific Journal of Teacher Education, 41*(1), 9–27.

Chapter 15
EduRobot Taxonomy

Dave Catlin, Martin Kandlhofer, John-John Cabibihan, Julian Angel-Fernandez, Stephanie Holmquist, and Andrew Paul Csizmadia

Abstract For many years you could count the number of companies selling education robots on two hands. Since 2010, we've seen this number grow at an ever-increasing rate. What do these robots have in common and how do they differ? Reading the literature focused on robots in education, we see authors using different words for the same idea. We need a standard grammar. Education robots involve multiple strands of intellectual effort, for example, from psychology to teaching practice and classroom assessment to high-stakes testing. Trying to fit all these ideas into one theory isn't the way to solve the problem. We need to isolate and explore different features. EduRobot is one strand: it's concerned with the nature of robots – not the way we use them or their benefits – other papers do that. This chapter briefly explains EduRobot, its ideas and arguments before using it to classify some of the robots cited in this book.

Keywords Education robots · EduRobot taxonomy · Robot classification

---

D. Catlin (✉)
Valiant Technology, London, UK
e-mail: dave@robots-for-education.com

M. Kandlhofer
Graz University of Technology, Graz, Styria, Austria
e-mail: mkandlho@ist.tugraz.at

J.-J. Cabibihan
Qatar University, Doha, Qatar
e-mail: john.cabibihan@qu.edu.qa

J. Angel-Fernandez
Technische Universität Wien, Vienna, Austria
e-mail: Angel-Fernandez@acin.tuwien.ac.at

S. Holmquist
University of South Florida, Tampa, FL, USA
e-mail: sh2@usf.edu

A. P. Csizmadia
Newman University, Birmingham, UK
e-mail: A.P.Csizmadia@staff.newman.ac.uk

L. Daniela (ed.), *Smart Learning with Educational Robotics*,
https://doi.org/10.1007/978-3-030-19913-5_15

Introduction

Developing EduRobot needs an international effort. Its current authors represent the USA, Europe, the Middle East and Asia.[1] The first efforts showed a paper to the Robotics in Education Conference held in Malta in 2018 (Catlin et al. 2018a). This prompted the expert delegates to debate the issues and suggest improvements to EduRobot. Later that year, the expanded team presented EduRobot version 2.01 to the Constructionism Conference in Vilnius, Lithuania (Catlin et al. 2018b).

Both these conferences urged the authors to create EduRobot online which they've done and you can access at www.robots-for-education.com. This site:

1. Defines the terms used in EduRobot
2. Explains decisions about its nature and organisation
3. Presents an encyclopaedic list of education robots
4. Most importantly provides community forums to debate and develop EduRobot and education robots

In this chapter, we present a brief description of EduRobot and then use it to classify the robots mentioned in the rest of the book. We finish with a short discussion prompted by this classification effort.

About EduRobot

Life scientists worked thousands of years sorting out the living world. Yet, they're still improving their ideas – this against the slow backdrop of evolution. Animals mutate from a common ancestor to different species. Technology diverges because of innovation, but it also converges as people copy the best ideas from one design to another. The breakneck pace of these technological changes sees exciting ideas fleeting shine before heading off to the museum. Marketers have a habit of exaggerating and inventing new words for old ideas, just to make their product offerings more attractive. None of this helps the serious scientific study of our topic.

EduRobot concerns itself with all education robots, including the extinct ones. For our understanding of these robots to improve, we need stability. What does innovation change? If a developer adds a new feature, or more radically revamps the product, is it (in taxonomic terms) a different robot? Can we find a universally accepted way of talking about education robots? A debate in the EduRobot forums will lead to clarity by sorting out issues like this. As the community wrestles with such matters we expect EduRobot to change.

Right now, our thinking behind EduRobot aims to preserve simplicity as shown in Fig. 15.1.

[1] We're actively seeking contributions from other parts of the world; contact info@robots-for-education.com if you're interested.

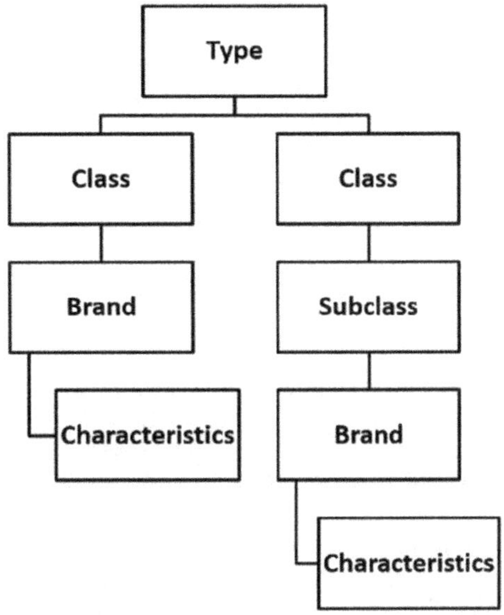

Fig. 15.1 Basic EduRobot taxonomy. (Courtesy of Robots for Education)

Classifying robots has three levels: type, class and subclass. Development doesn't change these: they describe the brand – the name of the robot – for example, Lego Mindstorms. Characteristics add substance to our understanding of a robot, but they don't change its nature. By analogy, think of a man – he can be fat or thin – and the fat man can become thin or the thin man become fat. None of these changes stops him from being a man. So a developer adding, for example, Bluetooth to a robot won't alter the robot's classification, but it does help you understand it.

Figure 15.2 shows EduRobot's latest taxonomy.

One way to test and improve a taxonomy is to use it: what problems did it raise? Table 15.1 classifies most of the robots mentioned in this book. Our efforts to classify them raised some issues which we highlight in the discussion.

Figure 15.3 shows an example of a full classification.

Discussion

This chapter presents the results of classification in action. It revealed a number of insights that triggered a few issues in need of further debate.

Mark Pesce wrote about how technology would infiltrate a child's world of play (Pesce 2000). The EduRobot analysis reveals a significant number of toy companies now producing products which can, at last, support their marketing department's claim that their product is educational. However, we do need more clarity about what merits the label toy.

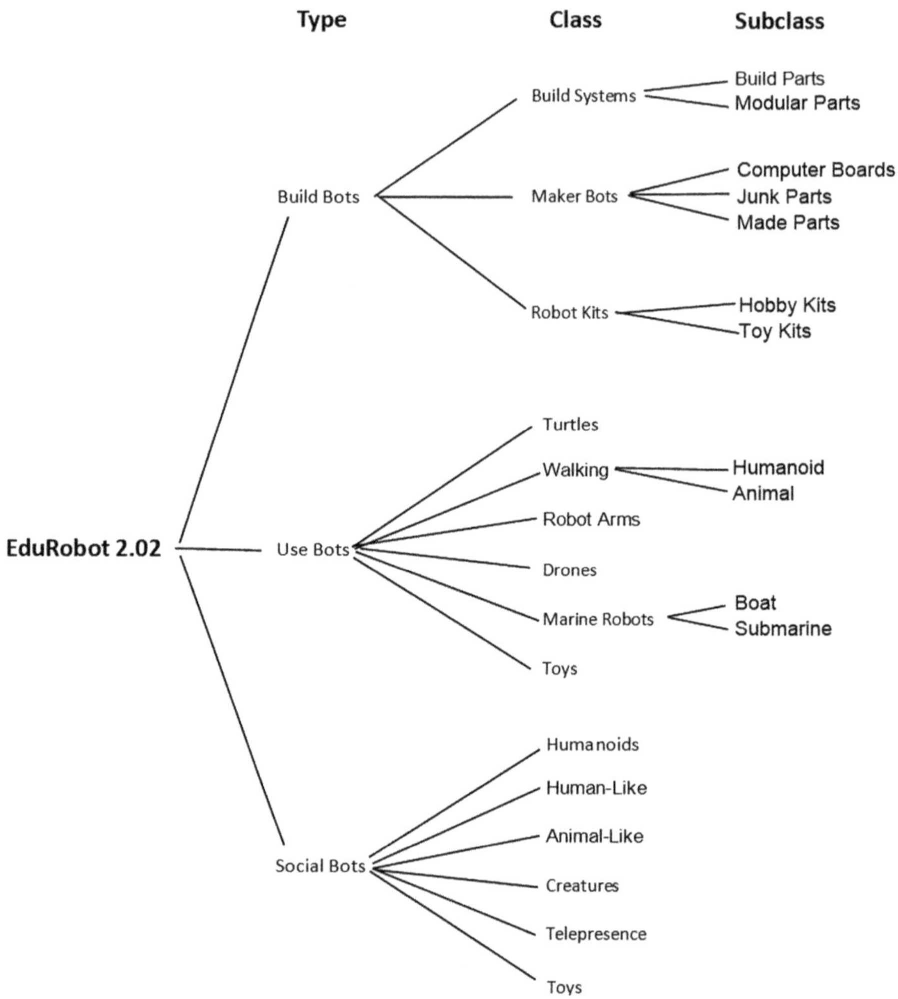

Fig. 15.2 The type, class and subclass for EduRobot version 2.02. (Courtesy of Robots for Education)

A rule states you can't classify a creature a mammal and a reptile – which is why it took 70 years to sort out the duckbilled platypus. Our original classification of Sphero called it a Turtle, but we've revised that decision and included it in the User Bot Toy class. We can only consider our listing as provisional subject to a review process which the EduRobot team need to develop.

Many of the walking robots hover between Use Bots and Social Bots. Forum discussions have started with the aim of clarifying the difference.

The guidelines for the 300-character description of the robot are too vague. Our aim is to keep present a factual description devoid of marketing excess.

Table 15.1 Classification of robots cited in this book according to EduRobot 2.02

Type	Class	Subclass	Robots
Build Bot			
	Build systems	Build parts	Little Bit Droid, Lego WeDo, Lego Mindstorms EV3, Pi2Go, ROBO TX Explorer, Sam Lab Kits, BYOR
		Modular parts	Cubelets, Moss Robots, AirBlock
	Maker Bots	Computer boards	
		Junk parts	
		Made parts	Miniskybot
	Robot kits	Hobby kits	Boe Bot, Pololu 3pi Robot, mBot Kits, Openrov 2.8
		Toy kits	Jimu, Angel Fish, Puffer Fish, Tinkerbots, RoboMaker Pro
User Bots	Turtles		Blue Bot, Bee Bot, Turtle[a], Jessup Turtle, BBC Buggy, Tasman Turtle, Turtle Tot, Valiant Turtle, Roamer, Thymio, Matatalab, Dash and Dot
	Walking robots	Humanoid	
		Animal	
	Robot arms		Kuka Youbot, Dobot Magician
	Drones		Parrot Mambo
	Marine robots	Boats	
		Submarines	
	Toys		Doc, Cubetto, Ozobot, Sphero, Codey Rocky, Mind Designer
Social Bots	Humanoid		Nao, RoboThespian, Alpha 1 Pro
	Humanlike		Kaspar
	Animallike		
	Creatures		
	Telepresence		Pebbles
	Toys		

[a]The world's first education robot invented by Seymour Papert in 1969

Continuing Research

EduRobot is an ongoing project and anyone wishing to contribute is welcome. They can do this through the web site www.robots-for-education.com by:

1. Classifying a robot and adding it to the library
2. Starting a debate in the forums on any relevant topic

Valiant Turtle

February 12, 2019 by Dave Catlin

Description: First sold in 1983 the Valiant Turtle became the most popular and enduring of the classic turtle robots. In 1989, when Valiant launched the Classic Roamer, they stopped the Turtle's production line. But they did make them to order until 2014 when the supply of key parts ran out.

Type: Use Bot

Version:

Other Versions:

Class: Turtles

Subclass:

Characteristics

Locomotion: Wheels
Power: Batteries
Command and Control: Computer, Tangible Computing
Communication: Infrared
Sensors: None
Outputs: Drawing
Programming: Text , Tangible
Architecture: Direct
Modularity: Fixed
Morphology: Fixed

Manufacturer: Valiant Technology Ltd

Website: http://www.roamer-educational-robot.com/the-valiant-turtle

Availability: Historical

Fig. 15.3 An example of a completed EduRobot classification. (Courtesy of Robots for Education)

References

Catlin, D., Kandlhofer, M., & Holmquist, S. (2018a). EduRobot taxonomy a provisional schema for classifying educational robots. *9th International Conference Robotics in Education*. Qwara, Malta. Retrieved November 12, 2018, from https://www.researchgate.net/publication/324786505_EduRobot_Taxonomy_A_Provisional_Schema_for_Classifying_Educational_Robots

Catlin, D., Kandlhofer, M., Holmquist, S., Csizmadia, A. P., Angel-Fernandez, J., & Cabibihan, J. (2018b). Edurobot Taxonomy and Papert's Paradigm. In V. Dagienė, & E. Jasutė (Ed.), *Constructionism 2018: Constructionism, computational thinking and innovation* (pp. 151–159). Vilnius, Lithuania. Retrieved November 12, 2018, from http://www.constructionism2018.fsf.vu.lt/file/repository/Proceeding_2018_Constructionism.pdf

Pesce, M. (2000). *The playful world*. New York\Toronto: Ballantine Publishing Group.

Index

L. Daniela (ed.), *Smart Learning with Educational Robotics*,
https://doi.org/10.1007/978-3-030-19913-5

CPSIA information can be obtained
at www.ICGtesting.com
Printed in the USA
LVHW080351050719
623228LV00003B/41/P

9 783030 199128